Applied remote sensing

C. P. Lo
University of Georgia

Longman
Scientific &
Technical

Longman Scientific & Technical,
Longman Group UK Limited,
Longman House, Burnt Mill, Harlow,
Essex CM20 2JE, England
and Associated Companies throughout the world.

Published in the United States of America
by Longman Inc., New York

© Longman Group UK Limited 1986

First published 1986

British Library Cataloguing in Publication Data

Lo, C. P.
   Applied remote sensing,
   1. Remote sensing
   I. Title
   621.36′78   G70.4
   ISBN 0-582-30132-7

Library of Congress Cataloging-in-Publication Data

Lo C. P. (Chor Pang), 1939–
   Applied remote sensing.

   Bibliography; p.
   Includes index.
   1. Remote sensing.   I. Title.
G70.4.L6   1987   621.36′78   85–16611
ISBN 0–582–30132–7

Set in 9½/11 pt Linotron 202 Century Schoolbook
Produced by Longman Group (FE) Limited
Printed in Hong Kong

For my mom,
Christine
and Kit

Contents

References to colour illustrations are indicated thus – 1.
Colour illustrations are located between pages 36 and 37, 212 and 213,
364 and 365, and 380 and 381.

Preface

Interest in remote sensing has grown so rapidly in recent years that the literature on this subject has become staggering. Application using remotely sensed data from aerospace platforms are increasingly diverse especially because scientists from different disciplines have found these data useful. The interdisciplinary nature of remote sensing has made it a valuable integrating tool in the study of the environment. In view of these, this book has been written stressing the practical applications of remote sensing to mankind in different parts of the terrestrial environment. The applications have been selected carefully from among those that have proven to be significant. These are employed as case studies to demonstrate the theoretical underpinnings of the technique. These applications have been classified by subjects and then organized according to the portion of the electromagnetic spectrum within which the data are acquired. It is hoped that through this approach a better understanding of the operational aspect of the technique and an appreciation of the interdisplinary nature of the subject can be achieved. This book has been aimed at all scientists interested in the environment, who also wish to know how others in different disciplines utilize the remotely sensed data.

Acknowledgements

This book could not have been written without the cooperation of the many investigators in different disciplines who readily responded to my request for materials. The help rendered by these individuals are acknowledged at appropriate places in the text. This book was written at a most difficult time of my career – a move across the Pacific from the University of Hong Kong to the University of Georgia. Technical supports from the departments of geography of both universities are gratefully acknowledged. I also wish to thank Professor James Fisher, Head, Department of Geography, University of Georgia, for his support of this project. Typing assistance from Ms Betty Chun (University of Hong Kong), Ms Audrey Hawkins (University of Georgia) and Ms Peggy Peters (University of Georgia) are gratefully acknowledged. To my wife Christine I am, as always, deeply indebted for her help and forbearance during this difficult period.

Last but not least, I wish to acknowledge the permission given by the following societies, journals, firms and publishers for using their copyright figures and photographic plates in the book.

The author, R E W Adams and the American Association for the Advancement of Science for fig 2.3 from fig 1 (Adams et al 1981) Copyright 1981 American Association for the Advancement for Science; American Association of Petroleum Geologists for figs 4.47, 4.48, 4.51 (Ford 1980); American Congress on Surveying and Mapping for figs 6.23–6.25 from figs 1, 2, 4 pp. 123, 124, 127 (Jensen 1978); American Geophysical Union for figs 4.21, 4.22 from figs 1, 6, 7 pp 3045, 3048, 3049 Copyright by the American Geophysical Union (Idso et al 1975); American Meteorological Society for figs 3.5 & table 3.1 from fig 2, table 1 (Jacobwitz et al 1979), 3.8, 3.9 from figs 2, 17 (Gube 1982), 3.13 from figs 8, 9, 10 (Rockwood & Cox 1978), 3.14–3.16 from figs 1, 3, 5, 7, 8 (Carlson et al 1981), 3.17 from fig 5 (Harris & Barrett 1978), 3.19, 3.20 from figs 7, 9 (Park et al 1974), 3.27 from fig 9 (Rodgers et al 1979), 3.29, 3.31 from figs 3, 4 (Steranka et al 1973), 3.32 from figs 2, 3, 4, (Cadet & Desbois 1980); the American Society of Photogrammetry for figs 1.17 from fig 2 p 49 (Thompson 1979), 2.11 from fig 5 p 206 (Welch & Zupko 1980), 4.5 from fig 1 p 298 (Smith 3967), 4.25–4.27 from figs 1, 2, 4 pp 190, 193 (Lewis & Waite 1973), 4.53, 4.54 from figs 5, 6 p 1906 (Heilman and Moore 1982), 5.7 from fig 1 p 1149 (Murtha 1978), 5.13 from fig 2 p 103 (Ulaby et al 1980), 5.18 from fig 3 p 319 (Pearson et al 1976), 6.4, 6.5 from figs A-1, 2, 3, 4 pp 1455, 1456, 1462 (Adeniyi 1980), 6.6 from fig 3 p 1306 (Brown and Holz 1976), 6.13, 6.14 & tables 6.8, 6.9 from figs 1, 2 pp 302, 305 tables 1, 2 p 306 (Lins 1976), 7.26 from fig 10 p 933 (Klooster & Scherz 1974), 7.27 from fig 2 p 1541 (Ritchie et al

1976), 7.37 from figs 1, 6 pp 534, 538 (Klemas & Philpot 1981), 8.11 from fig 8 p 848 (Masry & Gibbons 1973), 8.13–8.15 from figs 3, 4, 12 pp 1085, 1091 (Leberl 1979), 8.16 from fig 1 p 1658 (Friedmann et al 1983), 8.17 from fig 14 p 1169 (Chevrel et al 1981), tables 5.2, 5.3 & plate 7 from tables 1, 3 pp 274, 280, plate 1 p 278 (Carter, Malone & Burbank 1979) Copyright by the American Society of Photogrammetry: Association for Computing Machinery for fig 9.2 (Nagy & Wagle 1979); Association of American Geographers for fig 2.12 (Lo and Welch 1977), plate 11 (Welch, Pannell & Lo 1975); the author, H C Blume for figs 7.33, 7.34 (Blume et al 1978); the author N A Bryant for figs 9.10–9.15 & plate 17 (Bryant & Zobrist 1976); Cambridge University Collection for fig. 2.1; Canadian Forestry Service for figs 5.1, 5.2 (Sayn-Wittgenstein, 1978), 5.3, 5.4 (Aldred & Lowe 1978), 5.5 (Sayn-Wittgenstein et al 1978); the author, W G Collins and Elsevier Science Publishers BV for figs 6.1, 6.2, tables 6.3, 6.4 from figs 1.2, tables I, II pp 72, 72, 75, 76 (Collins & El-Beik 1971); the author, P Curran and The Royal Society for figs 5.14–5.17, table 5.6 from figs 1, 2, 3, 7, table 1 (Curran 1983); the author, M. Deutsch and American Water Resources Association for plate 6 (Deutsch and Ruggles 1978); the author, M. Dunbar and International Glaciological Society for fig 7.5 (Dunbar 1975). Elsevier Science Publishing Co Inc for figs 2.9, 2.10 from figs 6, 7 pp 109 (Welch 1980, 2.16, 2.17 from figs 6.10 (Iisaka & Hegedus 1982), 4.28 from fig 4 (Dean et al 1982), 6.10, 6.11 from figs 2a, 3 (Bryan 1975), 7.1, 7.2 from figs 1, 2, 16 (Zeng & Kelmas 1982), 7.4 from fig 1 (Ketchum & Lohanick 1980), 7.7–7.10 from figs 1, 2, 3, 6 (Hunter & Hill 1980), 7.11, 7.12 from figs 8a, 12a (Finley & Baumgardner 1980), 7.14, 7.15 from figs 1, 2 (Gower (1979), 7.35 from fig 2 (Khorram 1982), table 5.7 from table 4 (Walsh 1980) Copyright 1980 by Elsevier Science Publishing Co. Inc; The Geological Society of America for figs 4.2, 4.3, 4.4 (Howard 1965); Geological Survey of Canada for fig 4.50 (Slaney 1981); the author, B Holt for figs 7.6, 7.13, 7.18, 7.19, 7.32, 7.38, 7.39 (Fu & Holt 1982); Mark Hurd Aerial Surveys for figs 4.6–4.8; the author, S B Idso and the American Association for the Advancement of Science for fig 5.9 from fig 2 (Idso et al 1977); The Institute of Electrical and Electronics Engineers Inc for figs 4.18, 4.19 from figs 1, 6 pp 130, 134 (c) 1975 IEEE (Watson 1975); the author, R W Johnson and the American Society of Photogrammetry for fig 7.24 from figs 5, 6 pp 622–3 (Johnson 1978); the author, V Klemas for figs 7.29, 7.30 (Klemas et al 1974), 7.36 (Klemas and Philpot 1981); Laboratory for Computer Graphics and Spatial Analysis, Harvard University and the American Society of Civil Engineers for figs 9.3, 9.4 from figs 12–17 pp. 131–133, 143 (Teichoz 1980); National Council for Geographic Education for table 6.5 from fig 1 p 219 (Snyder 1981); Optical Society of America for figs 3.6, 3.7 redrawn from figs 7.14 (Smith et al 1977); the author, D E Parry and the Royal Geographical Society for fig 5.11 from fig 3 (Parry & Trevett 1979); the author J T Parry and Elsevier Science Publishing Co Inc for fig 4.31 from fig 2 p 186 (Parry et al 1980); The Photogrammetric Society for figs 8.9 from fig 1 p 759 (Leatherdale & Keir 1979), 9.16 from fig 3 p 313 (Lo 1981); The Remote Sensing Society for figs 4.10–4.12 from figs 3, 4, 5 pp 25, 26 (Lewin & Weir 1977); the Royal Geographical Society for fig 6.16, tables 6.6, 6.7 from fig 2, tables I, II pp 416, 411, 417 (Deane 1980); Royal Meteorological Society for fig 3.4, 3.11 from figs 3, 10 (Houghton 1979); Taylor and Francis Ltd for fig 3.36 from fig 7 p 78 (Hung & Smith 1982), 4.42, 4.43 from figs 3,

4 pp 152–3 (Schanada et al 1983), 6.26 from figs 2, 3 pp 280–1 (Howarth & Wickware 1981); the author, D Q Wark and the American Association for the Advancement of Science for fig 3.10 from fig 1A Copyright 1969 American Association for the Advancement of Science (Wark & Hilleary 1969); the author, J Welsted and Elsevier Science Publishers B.V. for figs 4.13–4.15 from figs 3A, 3B, 9A, 9B (Welsted 1979).

C. P. Lo
Athens, Georgia, USA

Chapter 1 Nature of remote sensing data

Introduction

Remote sensing refers to the group of techniques of collecting information about an object and its surroundings from a distance without physically contacting them. Normally, this gives rise to some form of imagery which is further processed and interpreted to produce useful data for application in agriculture, archaeology, forestry, geography, geology, planning and other fields. The prime objective of remote sensing is to extract environmental and natural resources data related to our earth. Information about the object concerned is conveyed to the observer through electromagnetic energy which is the information carrier and thus provides the communication link. We can therefore regard remote sensing data as basically wavelength-intensity information which needs to be decoded before the message can be fully understood. This decoding process is analogous to the interpretation of the remotely sensed imagery which impinges heavily on our knowledge of the properties of electromagnetic radiation.

Principle of electromagnetic remote sensing

Physical characteristics of electromagnetic remote sensing

Electromagnetic radiation is a form of energy transfer in the free space, which exhibits both wave and particle properties (Hunt 1980). According to its wave properties, the electromagnetic energy is seen to travel through space in a plane harmonic wave pattern at the velocity of light, i.e., about 3×10^{10} cm/s. The wave consists of one electric (vertical) and one magnetic (horizontal) field, which are orthogonal to each other and to the direction of wave propagation (Fig. 1.1). The wave can be described in terms of the *wavelength* (λ) which is the distance of separation between adjacent wave peaks or its *frequency* (f) which is the number of wave peaks passing a fixed point in a given time. They are related as:

$$\lambda f = c$$
$$\text{or } f = \frac{c}{\lambda} \qquad [1.1]$$

Fig. 1.1
Electromagnetic wave showing
the electric and magnetic fields

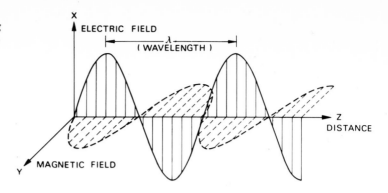

where c is the universal constant or the speed of light in the case of electromagnetic radiation. It is clear from equation [1.1] that wave frequency varies inversely as the wavelength and directly as the speed of wave propagation. These wave properties as exhibited by electromagnetic radiation represent interaction between energy and matter on a macroscopic scale.

On the other hand, electromagnetic energy is also observed to be transferred in discrete units called *quanta* or *photons*. This gives rise to the particle properties of electromagnetic radiation. The energy of radiation (E) is related to the frequency (f) via a constant value known as Planck's constant (h) which is 6.6256×10^{-27} erg sec such that

$$E = hf \tag{1.2}$$

By substitution into equation [1.1] one can relate the particle and wave properties of electromagnetic radiation as:

$$E = \frac{hc}{\lambda} \tag{1.3}$$

Thus, the energy of a photon varies directly with the frequency and inversely with the wavelengths of the radiation. In other words, the greater the energy, the higher its frequency and the shorter its wavelength. These properties represent interactions between energy and matter on an atomic and molecular scale.

It is observed that all bodies with a temperature of above absolute zero (0 K or -273 °C) will emit electromagnetic radiation over a broad range of wavelengths. As the absolute temperature (T) of a body changes, the wavelength of the maximum radiation (λ max), or the *dominant wavelength*, also shifts or is displaced according to the following relationship:

$$\lambda_{max} = \frac{\alpha}{T} \tag{1.4}$$

where α is a constant equal to 2,898 μm K. This is known as *Wien's displacement law*. The relationship is shown graphically in Fig. 1.2 which shows the shift of the dominant wavelengths to the shorter region as the *blackbody* temperatures increases from 300 K to 6,000 K. The black body is a hypothetical body which absorbs all the radiation

Fig. 1.2
Spectral radiant emittance
curves for 6,000 K, 3,000 K and
300 K blackbody sources (*Source*:
Slater 1980)

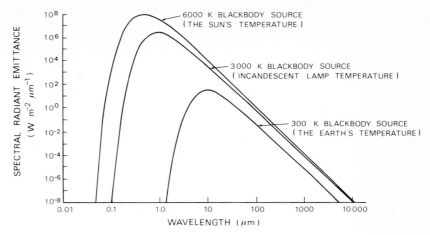

incident upon it and emits the maximum amount of radiation at all
wavelengths. It would reflect no light, hence its black colour. We
regard this simply as a perfect emitter. The total amount of radiation
emitted by a black body, i.e. the radiant emittance (W), is found to be
proportional to the fourth power of the absolute temperature (T in K)
of the body such that

$$W = \sigma T^4 \qquad [1.5]$$

where W is the energy radiated per second per unit area of the surface
of the body, and σ is the Stefan–Boltzmann constant (5.669×10^{-12} W/cm^2 K^4).

Therefore, as the temperature increases, the radiant energy emitted
by an object increases very quickly. This is known as the
Stefan–Boltzmann law. But for a *real* body, the equation becomes

$$W = \varepsilon\sigma T^4 \qquad [1.6]$$

where ε is the emissivity of the body which lies between 0 and 1.

The Wien's displacement law and the Stefan–Boltzman law are
combined together into a more general law defined from quantum
theory known as *Planck's law* which expresses the amount of energy
$E(\lambda)d\lambda$ present in the radiation from a perfect emitter and having a
wavelength in the range from λ to $\lambda + d\lambda$ by:

$$E(\lambda) = \frac{C_1}{\lambda^5[\exp(C_2/\lambda T)-1]} \qquad [1.7]$$

where C_1 and C_2 are constants. A more useful computational form of
the formula is:

$$E(\lambda) = \frac{3.74151 \times 10^8}{\lambda^5[\exp\{(1.43879 \times 10^4)/\lambda T\}-1]} \qquad [1.8]$$

where $E(\lambda)$ is in W m^{-2} μm^{-1} and λ in μm. Therefore, the computed
$E(\lambda)$ is the radiant energy given in watts, within a spectral interval
of 1 μm, emitted from a blackbody having an area of 1 m^2 into a

hemisphere (Slater 1980: 37). The form of this curve is shown in Fig. 1.2. The area under the curve can be evaluated as:

$$\int_0^\infty E(\lambda)\,d\lambda = \sigma T^4 \qquad\qquad [1.9]$$

The result corresponds to that given by Stefan–Boltzmann law.

By arranging the electromagnetic radiation according to wavelength, frequency or energy, a continuum of energy called *electromagnetic spectrum* can be defined (Fig. 1.3). The spectrum is usually divided into arbitrary regions named either after the sources producing them, such as the gamma ray and X-ray; or as extensions from the visible region (0.4–0.7 μm) such as the ultraviolet and the infrared regions; or according to how the wavelengths are being utilized, such as the radio wave region. In remote sensing, the more useful regions of the electromagnetic spectrum are the visible (0.4–0.7 μm), the reflected infrared (0.7–3 μm), the thermal infrared (3–5 μm and 8–14 μm) and the radar (0.3–300 cm) regions.

Atmospheric effects on electromagnetic radiation

For terrestrial remote sensing, the most important natural source of electromagnetic radiation is the sun which has a surface temperature of about 6000 K and radiates energy covering wavelength regions in the ultraviolet, visible and infrared. The dominant wavelength occurs at 0.5 μm (green light) which is visible to human eyes (Fig. 1.3). This represents the *reflected* solar energy which permits us to see earth features. On the other hand, the surface temperature of the earth is much lower, about 300 K and according to Wien's displacement law, its dominant wavelength is shifted to the longer 9.7 μm wavelength, (Fig. 1.2) which happens to fall in the thermal infrared region of the electromagnetic spectrum (Fig. 1.3). This represents the *emitted* energy from the earth and is more easily detected at night because of its much lower magnitude in comparison with the reflected solar energy. In remote sensing, the wavelength of 2.5 μm is used as the upper limit for reflected solar energy whilst the wavelength of 6 μm is the lower limit for self-emitted thermal energy. The wavelength region between 2.5 μm and 6 μm contains some of each.

Fig. 1.3
The electromagnetic spectrum: wavelength, frequencies and band designations

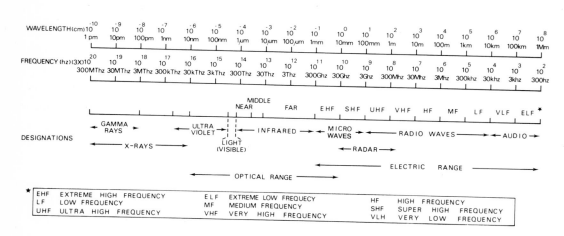

Electromagnetic energy, as it carries information from the earth to a remote sensor in the air or space, is seriously attenuated by its passage through the atmosphere. The atmosphere contains gas molecules and aerosol particles such as water droplets, dust and smoke, which give rise to *scattering* and *absorption* of the electromagnetic energy. Scattering causes changes in direction and intensity of radiation. It is wavelength dependent. Generally, it decreases with an increase in radiation wavelengths. If the molecules and small particles in the atmosphere have diameters much less than the wavelength of the radiation, *Rayleigh scatter* is present. Its intensity is inversely proportional to the fourth power of the wavelength. This effect explains the blue sky on a clear day when the ultraviolet radiation (blue) is scattered much more than radiation in longer visible wavelengths. When aerosol particles are present with diameters that are about equal in size to the wavelength of the radiation, *Mie scatter* occurs. This tends to affect radiation of longer wavelengths than in the case of Rayleigh scatter. When the aerosol particles in the atmosphere are very much larger than the radiation wavelengths, such as water droplets, the scatter is non-selective with respect to wavelengths. Hence, it is called *non-selective scatter*. The *white* cloud is the result of this kind of scatter which scatters all wavelengths of visible light (i.e. blue, green and red) in equal amounts. In order to obtain good quality imagery, proper filter systems have to be employed on the remote sensors to eliminate or minimize these scattering effects. The impact of different sized atmospheric particles and constituents on different portions of the electromagnetic spectrum is shown in Fig. 1.4,

Fig. 1.4
Attenuation to radiation of different wavelengths by common atmospheric constituents and other particles (*Source*: Tomlinson 1972)

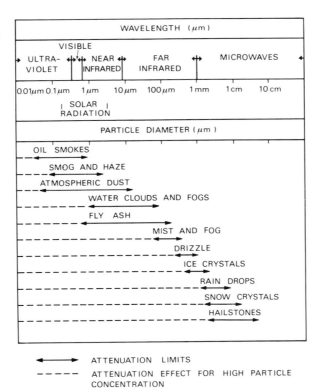

which provides a guide to the selection of proper remote sensing systems to avoid the undesirable scattering effects.

Absorption is the retention of radiant energy by a body or a substance. In the atmosphere, electromagnetic radiation is effectively absorbed by gas molecules, notably water vapour (H_2O), carbon dioxide (CO_2), oxygen (O_2) and ozone (O_3). The result is that the energy is attenuated or lost. Absorption, however, is limited to radiation in certain wavelength regions only (Fig. 1.5). For example, wavelengths shorter than 0.3 μm are completely absorbed by the ozone (O_3) layer in the upper atmosphere whereas water particles in clouds absorb and scatter electromagnetic radiation at wavelengths less than about 0.3 cm. Therefore, there are certain 'spectral windows' in the atmosphere through which the electromagnetic energy of certain wavelengths can be fully transmitted. The notable examples are the two spectral windows at 3 to 5 μm and 8 to 14 μm (Fig. 1.5) through which emitted thermal energy from the earth can be transmitted. Similarly, the window in the 1 mm to 1 m wavelength region also permits radar and microwave energy to be transmitted. The combined effects of scattering and absorption on electromagnetic radiation can be expressed in terms of an 'extinction coefficient' (or 'attenuation coefficient') which is the factor by which the radiation through a medium is reduced in travelling a distance of one wavelength in the medium.

Environmental impacts on electromagnetic radiation

When the electromagnetic energy eventually reaches the earth, it is further modified through interacting with features on the earth's surface. Some of the energy is *reflected* (if the features are solids) and the rest enters the features as a *refracted* wave front which may either be *absorbed* or retained by the features, or *transmitted* through the features, depending on the nature of the materials making up these features and the wavelengths of the energy. The reflection of the energy depends on the degree of surface roughness of these features in relation to the wavelength of the energy incident on them. When a surface is smaller in height variations than the wavelength of the incident energy, it acts as a smooth reflector and most of the energy is reflected in a forward direction called *specular* reflection. If the

Fig. 1.5
Radiation transmission through the atmosphere by radiant energy of different wavelengths (*Source*: Tomlinson 1972)

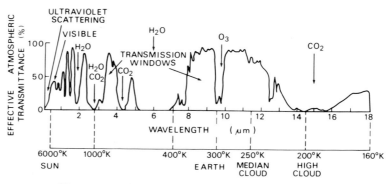

surface is greater in height variations than the wavelength of the incident energy, it is a rough surface and the energy is reflected more or less uniformly all around and is called *diffuse* reflection. Most of the earth's features lie between these two extremes and the directions of reflected energy are difficult to predict. The diffuse reflections of terrestrial objects give rise to colour information, and are most useful as a source of spectral data in remote sensing. A *reflection coefficient* or *albedo* is usually employed to denote the percentage of the radiant energy which is reflected back to space by a surface. Finally, the energy absorbed by the earth will eventually be released again in the form of heat as *emitted* energy. Added to the geothermal energy emitted from the interior of the earth, they form another important source of electromagnetic radiation for remote sensing.

In general, one can summarize the total interactions described above in the form of an equation as follows:

Total energy detected = source energy – absorbed energy – transmitted energy – reflected energy + emitted energy [1.10]

This equation provides the basis for *passive* remote sensing, i.e. employing radiation naturally reflected or emitted from the terrain. On the other hand, *active* remote sensing utilizes electromagnetic radiation provided by the sensor itself.

Characteristics of major imaging sensor systems

In order to detect the electromagnetic energy which exists in a wide range of wavelengths and frequencies, different varieties of remote sensing instruments are required. In this section, only those types of remote sensing instruments capable of producing images of the environment in a form suitable for visual interpretation will be discussed.

Photographic cameras

The most commonly employed sensor is the conventional photographic camera designed to detect energy in the visible (0.4–0.7 μm) and near infrared (0.7–0.9 μm) portions of the electromagnetic spectrum. This has been made possible with the use of suitable black-and-white or colour films sensitized to these spectral regions. In order to enhance the contrast between the object and background of interest in the resultant photograph a minus-blue (i.e. yellow) filter is normally employed at the time of photography to eliminate the Rayleigh scatter in the atmosphere. An important component of the photographic camera is the lens system which can cause images of points to be displaced on the photograph if *lens aberrations* are present. However, in aerial cameras designed for topographic mapping purposes (Fig. 1.6), the lens distortion is cut down to ±6 μm, which is

Fig. 1.6
Components of an aerial camera
(*Source*: Moffitt and Mikhail
1980)

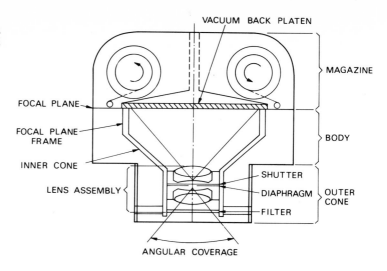

small enough to regard the lens as distortion-free. Most of the photography is obtained from aircraft with such a metrically accurate camera for photogrammetric purposes. This usually involves vertical aerial photography of the terrain with 60 per cent forward overlap between successive frames of photographs along a flight strip and 15 to 20 per cent lateral lap between two adjacent flight strips (Fig. 1.7). This flight configuration permits aerial photographs to be viewed *stereoscopically* to give an impression of the third dimension, thus facilitating the interpretation and mapping of the terrain features by the analogue approach. Geometrically, the vertical aerial photograph gives a central perspective projection with the lens idealized as a point fixed at a distance from the negative plane (Fig. 1.8). The electromagnetic energy is reflected from the terrain to pass through the lens and is regarded as a bundle of light rays travelling in straight-line paths. Thus, by similar

Fig. 1.7
Flight configuration in stereo-scopic aerial photography

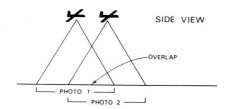

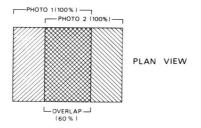

Fig. 1.8
The central perspective projection of a vertical photograph over the terrain

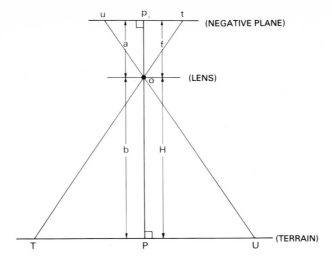

triangles, the nominal scale (S) of the vertical aerial photograph can be found by the following equation:

$$S = f/H \qquad [1.11]$$

where f is the focal length of the camera and H the flying height from the ground. However, if the relief of the terrain is taken into account (Fig. 8.2, p. 340), the scale of the photograph at point A will be

$$S = f/(H - h_A) \qquad [1.12]$$

where h_A is the height of the point A from the datum. On the whole, the average scale of the photograph (S_{AV}) should take into account the average terrain height (h_{AV}) so that

$$S_{AV} = f/(H - h_{AV}) \qquad [1.13]$$

The central perspective projection gives rise to relief displacements caused by the heights of the terrain from the datum plane, such as aa' and bb' which are seen to be radial from the centre of the photograph. It is important to note that in aerial photography the whole area covered by the photograph is acquired instantaneously.

Apart from the lens, another important component of photographic remote sensing systems is the photographic film which consists of a base and a thin layer of chemical *emulsion* on top. The emulsion is made up of silver halide crystals or grains of different sizes ranging from 0.1 to 1 μm^3 embedded in a solidified gelatine. The halide ions are commonly bromine or iodine. When the silver halide crystals are exposed to light, a *latent image* is formed. When a suitable alkaline reducing agent is added in the *development process*, those silver halide crystals that have been exposed to light are reduced to metallic silver which appears as black grains to form a visible image. This gives rise to a negative from which positive prints of the photograph on paper, film or glass can be made for interpretation or topographic mapping purposes.

In order to understand the relationship between the resultant photographic tone (density) and the duration of exposure of the chemical emulsion to light in a film, a *characteristic curve* as introduced by Hurter and Driffield in 1890 is normally used (Fig. 1.9). From this, one can see that the optical density (photographic tone) varies linearly with the log exposure over a wide range of exposures and that differences in density correspond well with differences in visual appearance of the objects. The slope of the characteristic curve differs from one type of film to another, and, by finding the tangent value of the angle of the slope (θ) made with the horizontal axis, a parameter called *gamma* (γ) indicates the *contrast* of a film. As the slope gets steeper, the gamma is higher, and the greater is the contrast, i.e. the exposure range being distributed over a larger density range. The gamma can also be varied by changing the developer, the development time and temperature.

Photographic emulsions have different degrees of sensitivity to light according to their *film speeds*. In Fig. 1.9, the film speed is graphically represented by point *i* or the point of intersection of the extension of the straight line of the characteristic curve with the horizontal line. This point is called the *inertia*. The reciprocal of the inertia is taken as proportional to the film speed. A fast film has only low exposure level and is characterised by much larger film grains of silver halide crystals than a slow film. The *graininess* (as it refers to a visual impression) or *granularity* (as it refers to the objectively measured density) of the film is closely related to the spatial resolution of the photographic recording.

The standard film type for aerial photography is the *panchromatic black-and-white film* which has a spectral sensitivity ranging from 0.35 to 0.70 μm. A minus-blue (yellow) filter is normally used to absorb the atmospheric haze found in the blue portion of the spectrum. Another film type readily available for use is the *infrared black-and-white film* which has a spectral sensitivity in 0.34–0.52 μm and 0.60–0.90 μm. A red filter is normally used to eliminate atmospheric haze so that the effective spectral range is from about 0.70–0.90 μm. Leaf mesophyll of vegetation is highly reflective and water bodies are highly absorptive of infrared radiation. These factors, coupled with its ability to penetrate haze make infrared black-and-white film ideal for rural land use and land cover mapping.

Nowadays, increasingly more remote sensing applications make use of *colour films* because the human eyes can distinguish a greater

Fig. 1.9
Characteristic curve. Note that gamma (γ) = tanθ = $\Delta D/\Delta \log E$

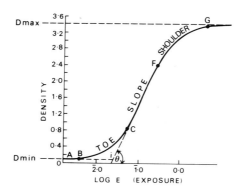

number of colour shades than grey tones. A *true* colour film consists of three layers of emulsion each sensitized to one of the three primary colours, blue, green and red (Fig. 1.10(a)). Because the green and red layers are also sensitive to blue light, a yellow filter is inserted between them to eliminate this undesirable blue light. After exposure, each layer is developed out in the three complementary colours of yellow, magenta and cyan corresponding to the colour sensitivity of each layer. When these are combined together, the original colour of the scene is visualized. A *false colour film (or colour infrared film)* can also be manufactured by shifting the colours. Instead of being sensitized to blue, green and red, the three layers of emulsion are made to sensitize to green, red and infrared respectively (Fig. 1.10(b)). A minus-blue (yellow filter) has to be used at the time of photography to cut out the blue light. Colour infrared film reacts most strongly with vegetation which is exhibited in varying shades of red, according to the amount of reflectance from the leaves. This kind of film has been most extensively used in environmental remote sensing applications because of its acclaimed all round capability.

Multiband photography is possible with the use of an array of synchronized cameras, such as the four-Hasselblad-camera array used in the S065 experiment on Apollo 9 in March 1969, or a multi-lens camera, such as the Itek S190A six-lens camera employed in various Skylab missions from 14 May 1973 to 8 February 1974 (Slater 1980). Different film-filter combinations were used. The multiband photographs are in smaller format such as 60 mm or 90 mm, than the usual 230 mm format of the mapping aerial camera. However, for most applications, four-band photography in blue, green, red and infrared is adequate. The film commonly employed is a black-and-white

Fig. 1.10
Sensitivity and cross-section of (a) a true colour film and (b) a colour infrared film

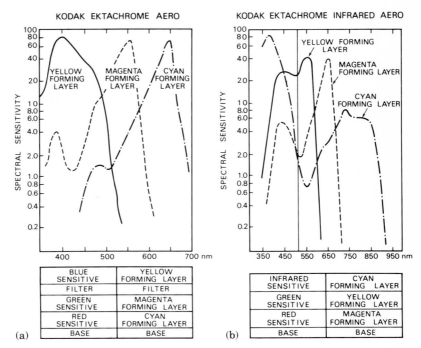

infrared film from which the desired spectral bands can be obtained with the use of suitable filters.

Although conventional aerial photography invariably maintains a single-point perspective projection, it is also possible to acquire photography by either moving the film inside the camera at a rate proportional to the aircraft ground speed or by swinging the camera lens from side to side (i.e. scanning) during exposure. Both systems are designed for military reconnaissance applications. The former, known as continuous strip photography, is designed for low-altitude, high-speed photography by passing the film over a narrow slit in the focal plane whilst the latter, known as *panoramic photography*, makes use of a rotating lens assembly to expose the film on a cylindrical surface (Lo 1976; Slater 1980). Both systems are capable of producing a strip of photograph of wide coverage (greater than 100 °) with long focal length cameras (from a few centimetres to more than a metre). For panoramic photography, the high resolving power is constant over the large angular field of the camera. It is noteworthy that both systems produce a highly distorted image of the real world and that the single-point perspective projection is destroyed.

Electro-optical detectors

Electro-optical detectors are transducers which transform electromagnetic radiation into electrons or electrical signals from a scene viewed. The detector elements behave like the grains of the photographic film and produce an analogue electrical record of each point of the scene scanned according to the nature of the incoming electromagnetic radiation. There are two basic classes of detectors – *quantum* and *thermal* – depending on the physical processes by which the electromagnetic radiation input is converted to electrical output. In a *quantum detector*, the input radiation in the form of photons interacts directly with the electronic energy levels within the detector material to produce *free-charge carriers* (electrons or holes). It is characterised by a high sensitivity and speed of response, but the detector materials have a limited spectral region of response. A *thermal detector*, on the other hand, makes use of heat-sensitive materials which absorb the incident radiation according to the temperature. An increase in temperature causes a change in resistance or produces a voltage, which can be monitored electrically. The advantage of thermal detectors is their extended range of spectral sensitivity. However, their sensitivity which is completely independent of the wavelength of the radiation is low. They are therefore less suitable than quantum detectors for use in remote sensing systems and so this section focuses only on quantum detectors.

Quantum detectors can be further subdivided into two groups: *photo-emissive detectors* and *solid-state detectors*.

In photo-emissive detectors, the photo-electrons are emitted into a vacuum or gas upon absorption of photons from incident radiation. This limits their operation to wavelengths less than about 1 μm where the energy is sufficient to release the electrons from the atoms of the photosensitive material. Therefore, they can only be employed for sensing in the ultraviolet and visible spectrum. Examples of photo-emissive detectors are photo tubes and photomultipliers.

The solid-state detectors comprise both *photoconductive* and *photovoltaic* devices which use similar materials but operate differently. In photoconductive detectors, the incident photons free the electrons to produce a change in resistance of the photosensitive material, which varies inversely with the number of incident photons. Much less energy is therefore required to free the electrons from the atoms as compared with photo-emissive detectors, hence their sensitivity to long wavelengths (up to several tens of micrometres). These detectors are typically made from silicon and germanium doped with various materials. In photovoltaic detectors, the incident photons generate a difference in voltage between the electrodes, producing a current which varies directly with the incident light intensity.

The electro-optical detectors are subject to unpredictable fluctuations in their electrical output known as *noise*, which is analogous to granularity in the photographic film. According to Slater (1980) five main sources of noise can be found in electro-optical detectors: (1) *thermal noise* caused by the random motion of charge carriers in a resistive element resulting in a random electrical voltage across the element; (2) *temperature noise* caused by the fluctuations in the temperature of the detector; (3) *current noise* caused by surface irregularities and contact noise; (4) *generation–recombination noise* caused by fluctuations in the rates of generation and recombination of charged particles in the sensitive element of all photodetectors. It is a result of the random arrival rate of photons from the background and hence is also known as *background noise*; and (5) *photon noise or shot noise* caused by current pulses in the electric current produced by individual electrons. The usual measure of performance of a detection system is the *signal-to-noise ratio* or S/N which is the ratio of the signal to the combined noises. The performance of the detectors for remote sensing applications should also be evaluated with respect to their spectral responsivity and the speed, linearity and dynamic range of response. The resolving power of the system is determined by the packing density of the detector elements.

In order to illustrate the working of electro-optical detectors as imaging sensors, two examples are described below:

Vidicon television camera

The vidicon camera is a form of image tube which can be regarded as the electronic counterpart of photographic cameras, because its principal application is sensing in the visible and near infrared portions of the electromagnetic spectrum (up to 0.85 μm). It makes use of an optical system (camera lens) to focus an optical image onto a photosensitive surface of an electron gun which then converts the image into electronic form, either as charge pattern on a storage device or in a non-storing mode, travelling electrons. The vidicon television camera is based on the photoconductive process (Fig. 1.11). A photoconductive target is coated on a transparent electrode which has a positive charge. Before exposure, the target is first scanned by an electron beam in the dark, which deposits a uniform layer of negatively charged electrons on its rear surface. As the mechanical or electronic shutter is opened for a short while, an image is focused on the photoconductive target. Light incident on the target increases its conduc-

Fig. 1.11
Diagrammatic representation of a vidicon camera (*Source*: Tomlinson 1972)

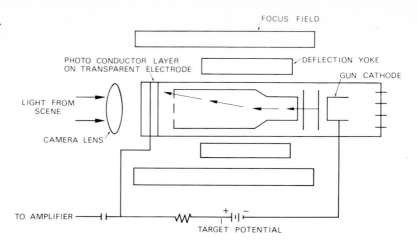

tivity locally and electrons escape to the electrode which also acts as a conductor. The cathode of the electron gun produces an electron beam which is attracted to the positively charged electrode. In the dark areas, the electron beam passes through in proportion to the brightness of the illumination. This creates a current at the electrode. By scanning the image, line by line with the aid of the deflection coils, the image is displayed in a cathode ray tube (CRT) and can be photographically recorded. After the completion of scanning, the photoconductive target is removed of residual electrons and is ready for another new picture. It is noteworthy that scanning at high velocity results in higher spatial resolution but decreased spectral sensitivity. Therefore, for remote sensing from space a low velocity scanning is normally used with the additional advantage that the video signals can be transmitted to the earth with small frequency bandwidth and low power requirements. The great advantage of a vidicon camera is its light weight and eraseable surface. As soon as one picture is taken and recorded on tape or transmitted to a receiving station, another can be taken. It permits real-time viewing because no time is lost for processing. It is also noteworthy that the image tube is about three times more sensitive than photographic film. However, its main disadvantages are the much poorer resolving power and a limited sensor surface.

Vidicon television cameras are used on meteorological satellites. The Advanced Vidicon Camera System (AVCS) is used on the ESSA, NIMBUS and ITOS/NASA meteorological satellite series. It produces an image of 800 lines in a 12.7 mm format giving an effective ground resolution from about 1 km to 16 km per line pair, according to the imaging scales ranging from 1 : 25 million to 1 : 250 million. A great improvement to this kind of camera is the Return Beam Vidicon (RBV) camera used in the Landsat series. The RBV cameras in Landsat 1 and 2 give a picture of about 4,200 lines per 25 mm frame which is equivalent to an effective ground resolution of about 180 m at an image scale of 1 : 7.3 million. In Landsat 3, the RBV cameras have produced pictures of even better ground resolution (about twice as good or 90 m) by doubling the camera focal length at a scale of about 1 : 3.6 million.

The vidicon television camera still produces pictures in the single perspective projection as in the case of the photographic camera.

Fig. 1.12
Typical electronic distortion pattern of a vidicon camera (*Source*: Tomlinson 1972)

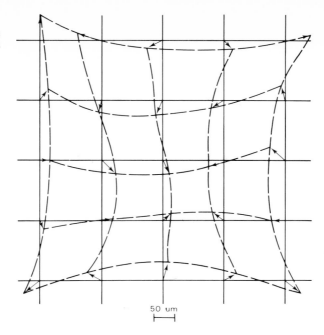

50 um

However, its electronic scanning by the electron beam across the photosensitive tube surface causes geometric errors such as size change, centering shift, deflection skew, barrel-shaped and S-shaped distortions (Fig. 1.12). The resultant pictures have tonal properties similar to those of the conventional photographs.

Thermal infrared scanner

The thermal infrared scanner is a good example of the optical–mechanical line scanners developed for generating imagery outside the spectral region of the photographic film. The characteristic of the line scanner is the multifaced rotating mirror inclined at 45° to the rotation axis, which scans the ground along lines perpendicular to the flight direction (also known as 'whiskbroom' scanning). The radiation emitted or reflected from the ground is projected by the rotating mirror onto the detector which senses the intensity of the radiation in a specific part of the electromagnetic spectrum and converts it to electrical signals. These signals are used to modulate the intensity of a single-line cathode ray tube (CRT) to expose an image line on the photographic film which is made to advance at a rate proportional to flight velocity of the sensor platform. Thus, as the sensor platform moves forward, successive new scan lines of the ground surface are swept. Together the continuous scan lines give a real-time image of the ground surface sensed (Fig. 1.13). It is also possible to record the scanner data on magnetic tape which is later employed to reconstruct the image.

In the case of the thermal infrared scanner, the detectors commonly used belong to the photoconductive type. For sensing in the 1–5.5 μm infrared window, indium antimonide (InSb) or indium arsenide (InAs)

Fig. 1.13
Operation of a line scanning system

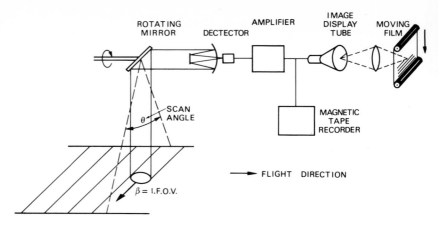

detectors are used, whilst for the 2–14 μm window, mercury-doped germanium (GeHg) or mercury-cadmium-telluride (HgCdTe) detectors are employed. It is worthy to note that the thermal detector has to be cooled to a required operating temperature (mostly about 77 K). It also has to be calibrated *radiometrically* with reference to known thermal sources both in the laboratory and in the field so that the image tones can be related to the appropriate temperatures. On the whole, within the recordable temperature range of the detector, warm objects appear in lighter grey tones than cool objects, whilst objects cooler and hotter than the recordable temperature will be imaged as black and white respectively.

The spatial resolution of thermal imagery is determined by the instantaneous field of view (IFOV), the platform height and the scan angle (θ) of the scanner (Fig. 1.13). The IFOV is the solid angle subtended by the detector in scanning, which is determined by the physical size of a detector element. For civilian applications, the IFOV ranges between 1.5 to 2.5 milliradians, and the best ground resolution is obtained at the point directly beneath the sensor (i.e. the nadir). The resolution cell size decreases laterally from the nadir as a function of $sec^2\ \theta$ (Y-resolution) and also decreases longitudinally as a function of $sec\ \theta$ (azimuthal resolution) where θ is the scan angle (Fig. 1.14). In order to improve the spatial resolution, a smaller IFOV needs to be used, but this would mean an increase in scanning speed, which is currently at 6,000 scans per minute.

Because of the non-linear relationship between the scanning mirror and the film recording device, the thermal imagery exhibits scale compression along the scan direction, which increases from the nadir towards the edges of the film; but in the flight direction, the image scale remains constant. As a result, a linear feature which makes an acute angle with the flight direction will appear as an elongated S-shaped feature (Fig. 1.15). Apart from this, incorrect film moving speed relative to the velocity and altitude of the sensor platform also causes scale distortion (either compression or elongation) in the flight direction. The variations in attitude (tilts and rotations) of the sensor platform can give rise to further errors by causing gaps in the scan coverage. In addition, one should not ignore the effect of relief

Fig. 1.14
Ground resolution of the Infrared
Line Scan System imagery
(*Source*: Leberl 1971)

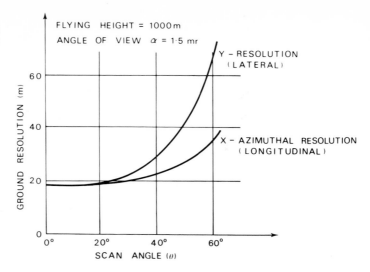

Fig. 1.15
Distortion characteristics of
thermal infrared scanner images
(*Source*: Sabins 1978)

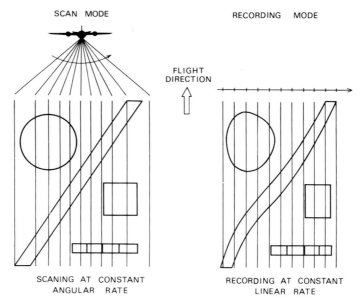

displacement which occurs only along the scan direction at right angles from the nadir (unlike the relief displacement radial in all directions from the nadir in a vertical aerial photograph) and increases with increasing range.

The thermal resolution or the ability of the system to detect small radiometric temperature differences is dependent on the sensitivity of the detector element employed, which is dependent on the detector area and frequency bandwidth of operation. It is possible to record temperature differences of 0.1 K or less, but this would require larger detector elements and a slower scan rate for longer exposure to the radiation, to the detriment of spatial resolution.

Multispectral scanners

The optical-mechanical or 'whiskbroom' scanner can operate in any wavelength region so that by using detectors sensitized to radiation in several spectral channels, multispectral data can be collected simultaneously. Early multispectral scanners made use of the method of beam splitting, but advances in detector fabrication technology makes possible the construction of multi-element detector arrays. One popular method is to use a dispersing spectrometer whose entrance slit serves as the scanner field stop. By placing a detector array in the image plane of the spectrometer, each detector sees the same slit image in its proper spectral region (Fig. 1.16). It is possible to manufacture multispectral scanners to sense in 24 separate spectral channels, but such a large amount of spectral data normally requires the use of a high-speed digital computer to help in the analysis. Clearly, much of the spectral data are redundant so that for practical application a smaller number of spectral bands is used. A good example is the multispectral scanner system (MSS) employed in the Landsats 1, 2, 3 and 4, which is designed to provide images of the earth simultaneously in four spectral bands (0.5–0.6 μm, 0.6–0.7 μm, 0.7–0.8 μm and 0.8–1.1 μm). Altogether 24 detectors are used, 6 in each band. The characteristics of the Landsat MSS will be discussed in a later section.

Apart from the optical-mechanical scanner using whiskbroom mode of scanning, it is also possible to make use of solid-state detectors operating in the *pushbroom* mode. In pushbroom scanning, only the forward motion of a sensor platform is used to sweep a linear array of detectors orientated perpendicularly to the flight direction across a scene being imaged (Thompson 1979; Fig. 1.17). One array is used for each spectral channel. Thus, there are no moving parts in this system. This will eliminate complex geometric distortions inherent in the

Fig. 1.16
Scheme of multispectral dispersive scanner (*Source*: Lowe 1980)

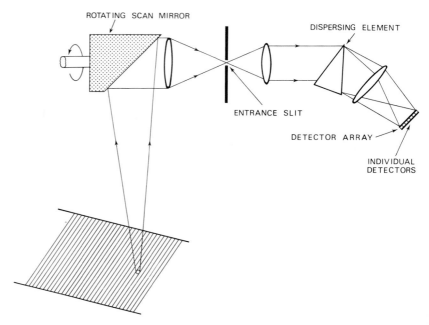

ROTATING SCAN MIRROR

DISPERSING ELEMENT

ENTRANCE SLIT

DETECTOR ARRAY

INDIVIDUAL DETECTORS

Fig. 1.17
Geometry of pushbroom scan technique (*Source*: Thompson 1979)

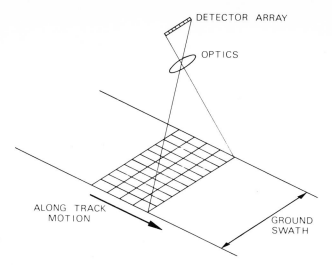

optical-mechanical scanner system. The signal-to-noise ratio is also significantly improved because of the longer integration time now possible. However, the pushbroom scanning system suffers from the need to have many more detectors for the array (e.g. 6,300 detectors per spectral band required to achieve a 30-m resolution for a similar Landsat type swath), and the lack of infrared capability at the present level of detector technology.

Microwave imagers

Microwave imagers make use of *antennae* to collect radiation information in the microwave wavelengths (i.e. 100 cm–0.1 cm or 0.3–300 GHz in frequencies) from the ground and are generally classified into passive and active types.

Passive microwave imagers

Passive microwave sensing collects thermal emission from the earth's surface in the microwave spectrum. The commonly employed spectral bands are separated into microwave regions, each of which is given a letter designation (Table 1.1). In the microwave spectrum the frequency of the wave is more commonly employed than its wavelength, probably because of the association of microwave techniques with radio technology. The distinct advantage of microwave sensors is their capability to penetrate clouds, although water vapour and oxygen can still hamper them, especially if microwaves at higher frequencies (above 206 Hz or wavelength below 1.5 cm) are used. Unlike optical sensors, the radiant energy being detected is low level. Passive microwave sensors detect emitted, reflected and transmitted radiation within the 1 mm to 300 mm wavelengths. The strength of the passive microwave radiation largely depends on the *temperature* and *dielectric* properties of the material rather than on the surface roughness. The dielectric property of a material is characterized by its

Table 1.1
Microwave and radar wave-
bands
(*Source*: Ulaby *et al.* 1981)

Designations	Frequency (GHz)	Wavelength (cm)
P	0.225–0.390	133–77
L	0.390–1.550	77–19
S	1.550–4.20	19–7.1
C	4.20–5.75	7.1–5.2
X	5.75–10.90	5.2–2.8
K	10.90–36.0	2.8–0.83
Ku	10.90–22.0	2.8–1.36
Ka	22.0–36.0	1.36–0.83
V	46.0–100.0	0.65–0.54
W	56.0–100.0	0.30–0.54

dielectric constant which measures the response of the material to the electromagnetic wave, i.e. the propagation and energy loss behaviour of the wave in the material. The large dielectric constant for water at 20 °C (about 80 at 1 GHz frequency) contrasts with other natural materials which have dielectric constants ranging from 3 to 8 only, thus making passive microwave remote sensing useful in monitoring water resources. It should, however, be noted that a high dielectric constant caused by a high moisture content results in a low thermal emissivity of the material. However, microwaves with lower frequencies (or longer wavelengths) are more sensitive to moisture content changes. Finally, one should also note that the radiation emitted in the microwave region from most surfaces is *polarized*. If the electric field vector in the electromagnetic wave is perpendicular to the earth's surface, it is called *vertical* (V) *polarization*. If the electric vector is parallel to the earth surface, it is called *horizontal* (H) *polarization*. The nadir angle of sensing and the degree of surface roughness can have an impact on the direction of polarization. In general, microwave remote sensing is employed to detect the H component of the emitted radiation only.

The passive microwave imager is a mechanical scanning device which consists of a parabolic reflector. At the centre of the reflector is a small antenna which collects the radiant energy and feeds it into the receiver-amplifier. Scanning is achieved by mechanically rotating the antenna to produce a *conical* scan at a constant angle of tilt with respect to the direction of motion of the sensor platform (Fig. 1.18(a)). A good example is the Electrically Scanning Microwave Radiometer (ESMR) operating in the 37 GHz frequency on board the Nimbus-5 meteorological satellite, which scans in azimuth ±35 ° about the forward direction at a constant tilt angle of 45 ° with the direction of motion. This is more preferable to conventional scanning, perpendicular to the direction of motion which results in a variation of the angle of incidence (Fig. 1.18 (b)).

The spatial resolution of the resultant imagery is dependent on the antenna size so that the ground resolution is determined by the *antenna beamwidth*, sensor platform height and scan angle. The beamwidth (*b*) is expressed as

$$b \doteq \lambda/d \qquad [1.14]$$

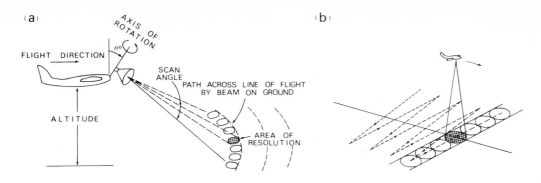

Fig. 1.18
Microwave radiometer scanning: (a) conical scanning and (b) cross-track scanning in the plane normal to the direction of flight (Tomlinson 1971 and Ulaby *et al.* 1981)

where λ is the wavelength and d is the antenna diameter. As in the case of the thermal infrared scanner, high spatial resolution requires a fast scan rate which is detrimental to temperature resolution.

Geometrically, the passive microwave imager shows similar characteristics of those in the optical-mechanical thermal infrared scanner, i.e. correct scale in the flight direction but incorrect in the scan direction (panoramic distortion), in addition to the errors introduced by variations in sensor platform height and attitude.

Active microwave imagers

Active microwave remote sensing involves the sending out of a pulse of microwave energy to a target from the sensor and then measuring the return or reflected signal. This method of sensing is more commonly known as *radar* (the acronym for *Radio Detection and Ranging*) which was rapidly developed during the Second World War for military applications. As its name implies, radar was designed for measuring distances and determining locations of objects. Current radar systems operate in single wavelength (monochromatic illumination) within the 35–9.1 GHz frequencies (or 0.86–3.3 cm wavelengths) in the microwave spectrum, which fall mainly in the X and K bands. A major advantage of radar is its all-weather, day and night operation capability which makes its application particularly valuable in areas with perennial cloud covers where conventional aerial photography cannot be applied. Unlike passive microwave sensing, the strength of the return signal in radar is determined by the surface roughness amd the dielectric properties of the terrain rather than the temperature of the material.

A commonly used radar imaging system is the *Side-Looking Airborne Radar* (SLAR) which produces imagery of the terrain on one or both sides of the aircraft flight line. An antenna is mounted on the belly of the aircraft parallel to its longitudinal axis and hence to the flight direction over the terrain. From the antenna, a short pulse of energy of a specific wavelength is transmitted in a fan which lies approximately in a plane perpendicular to the flight line (Fig. 1.19). When the pulse strikes a target, a signal is returned and detected by the receiver on the aircraft which converts it to a video signal to modulate the intensity of the electron-beam of a cathode ray tube to give a line trace of the terrain. This is transferred by a lens to a photographic film which is moved forward at a velocity proportional

Fig. 1.19
Operation characteristics of Side-Looking Airborne Radar (SLAR);
(*Source*: Cooke and Harris 1970)

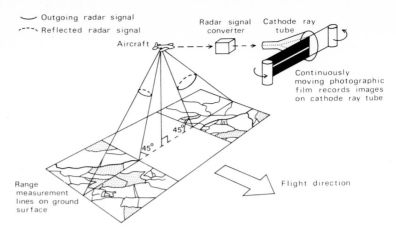

to the velocity of the aircraft. As the aircraft moves forward in its flight, scanning of the terrain in the flight direction (or azimuth) is accomplished and a series of lines is recorded onto the film, which builds up a two-dimensional image of the terrain.

In recording the line trace of the terrain, either a *slant range* or *ground range* presentation is possible (Fig. 1.20). The slant range sweeps refer to linear scan across the cathode ray tube so that spacing between features on the image is directly proportional to the time interval between the return signals received by the radar. In other words, the slope distances between the radar and the objects are recorded. As a result, objects equally spaced on a level terrain will not be equally spaced as they appear on the radar image. For ground range sweeps, the velocity of the recording sweep by the cathode ray tube is varied, using a hyperbolic sweep signal. This is so that the spacing of objects on the radar image is modified to the image scale that the objects would actually have on the terrain, if the terrain were both level and at a fixed height beneath the aircraft. The slant range can be corrected to ground range provided that the altitude of the aircraft above the datum is accurately known. From the range marks and

Fig. 1.20
Geometric characteristics of the SLAR system showing the pulse length *l*, beam width θ, angular width γ, slant range and ground range (*Source*: Leberl 1971)

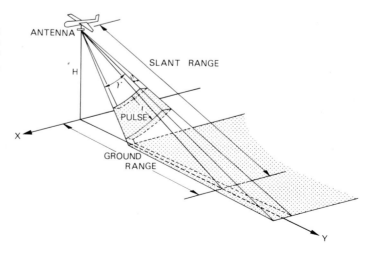

azimuth timing marks placed on the film at the time of exposure, the scale and sweep linearity of the radar image can be established.

Because radar determines the distances to various objects by measuring the time for the signals reflected from these objects to return, a near object is therefore imaged earlier than a far object. Thus in Fig. 1.21, the reflected radiation is received earlier from points B and C than from points D and E. Since the return signals from B and C are received simultaneously, the two points are recorded as one point on the image. However, since the point B which is the peak of the mountain is imaged earlier than the point E, the base of the mountain, the result is the leaning of the mountain towards the nadir of the aircraft. This is known as *layover* which is analogous to relief displacement in aerial photography, but in the opposite direction. Clearly, the appearance of the terrain on the radar image depends on the incidence angle of the radar beam rather than the absolute range to the aircraft. Therefore, in order to avoid ambiguity in imaging, the terrain directly beneath the aircraft is not imaged, and the SLAR receiving antenna will not accept return signals from angles of more than 45 ° from the horizontal (Fig. 1.19). This gives rise to a blank strip or blind area in the radar imagery directly beneath the aircraft.

SLAR images also exhibit shadows analogous to those long shadows found in low sun-angle aerial photography because of the oblique radar beam illumination to targets. In general, the shadow length of terrain features will increase *directly* with relative relief and incidence angle and *inversely* with depression angle. Radar shadow are useful in enhancing subtle terrain features. By using either the layover or the shadow length or both, according to whether the layover or shadow is visible in the image or not, the height of a terrain feature can be determined (La Prade and Leonardo 1968).

The spatial resolution of the SLAR system is determined by the antenna beamwidth in the along-track or *azimuth* direction and the *pulse length* in the across-track or *range* direction. According to equation [1.14] mentioned before, a small beamwidth is desirable for improving the azimuth resolution, which is best achieved with a longer antenna. A 15-m antenna is the longest that an aircraft can carry. The pulse length or the duration of the transmitted pulse and the depression angle combine to determine the range resolution. A shorter transmitted pulse gives better resolution because it travels a shorter two-way distance. The typical pulse lengths are between 0.05 and 0.1 μs. On the whole, at short ground ranges, the range resolution

Fig. 1.21
Diagrammatic explanation of radar layover (*Source*: La Prade and Leonardo 1968)

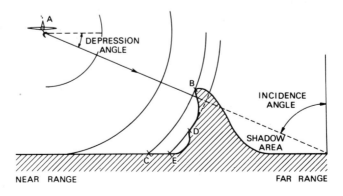

Fig. 1.22
Ground resolution of the SLAR
system (*Source*: Leberl 1971)

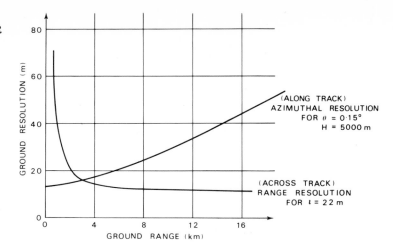

is poorer than the azimuth resolution, but as the range increases, the range resolution becomes far superior to the azimuth resolution (Fig. 1.22; Leberl 1971).

There are two types of radars designed for use with the SLAR systems: the *real aperture or brute force radar* and the *synthetic aperture radar* (SAR). The real aperture radar makes use of the physical antenna to transmit and receive the signal. However, as it was noted above, the antenna aperture size cannot be increased excessively to improve the azimuth resolution, and one has to fly low to obtain better azimuth resolution in detriment to the range resolution. The synthetic aperture radar overcomes the problem of antenna length in the real aperture radar. It makes use of only a small antenna (approximately 1 metre in length) which forms a part of a large antenna to record the return signals. This is made possible by storing the phases and amplitudes of the returning pulses, which are later put together to create a synthetic antenna of 60–90 m long. The synthetic antenna length is really determined by the length of time taken (i.e. distance travelled). The aircraft is receiving signals from a target (Fig. 1.23). Thus, finer azimuth resolution is obtained independent of the distance. This is a major advantage that makes it useful for high-altitude long-range and short–range operations. Hence, the synthetic aperture side-look radar is preferred for use in spacecraft, such as the Seasat SAR.

Fig. 1.23
Synthetic aperture configura-
tion (*Source*: MacDonald and
Lewis 1976)

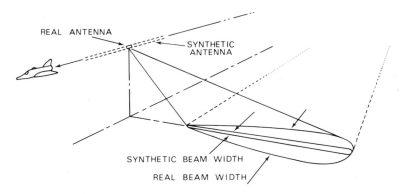

Geometrically, the radar imagery exhibits a scale distortion in the y or the across-track direction which causes distances at the near portion to appear to be compressed (i.e. at a smaller scale), relative to those at the far range portion on slant range display (Fig. 1.24). However, for the real aperture radar system, this distortion can be corrected during the recording phase by means of a non-linear sweep of the cathode ray tube to produce a ground range display as explained before. As for the synthetic radar system, the raw data may be processed either optically or digitally. The optical processing makes use of a laser beam to illuminate the return signal recorded on photographic film through a complicated optical system to produce images. The digital processing is to make use of the computer to carry out these optical manipulations. With more data control, the digital method produces better quality images but at a higher cost.

The microwave energy from the radar can be transmitted in either the vertical or horizontal plane known as *polarization*. The traditional SLAR system makes use of a single polarization in the horizontal field, i.e. a horizontally plane-polarized pulse is transmitted and upon striking the ground the return signal received also has the same horizontal polarization. This is known as HH or like-polarized return. It is possible for the radar to transmit horizontally plane-polarized energy by one antenna and receive vertically plane-polarized energy by another. Thus, HV or *cross-polarization* return is obtained. Both types of radar imagery are very different in information content and complement each other.

Finally, one should note that as in aerial photography, stereo radar imagery can be obtained by taking two strips of overlapping radar images from the same side of the terrain with some convergence angle and a 60 per cent overlap (de Loor 1969).

Space platforms and imaging systems

The different types of imaging systems possess characteristics which tend to complement each other. For example, the photographic

Fig. 1.24
Geometric distortion in radar imagery due to the non-linear relationship between time delay and ground range (*Source*: Tomlinson 1972)

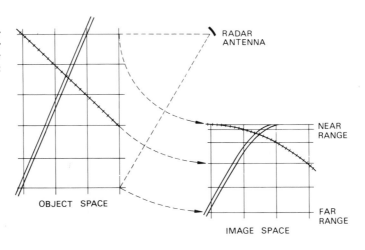

RADAR ANTENNA

NEAR RANGE

OBJECT SPACE

FAR RANGE

IMAGE SPACE

systems give photographs with a simple central perspective projection and a high spatial resolution, which make them suitable for topographic mapping and reconnaissance purposes. On the other hand, the non-photographic systems using the extraction of spectral information are specially suited to thematic mapping applications. For these different imaging systems to work properly the choice of the sensor platform is crucial. In conventional photographic systems, aircraft are normally employed although other less expensive aerial platforms such as tethered balloons (Whittlesey 1970; Vozikis 1983) and radio-controlled model helicopters have been used for various purposes (Wester-Ebbinghaus 1980). However, all these aerial platforms are limited by the heights they can reach in the sky and the short-duration capability of these flying missions. Only the NASA U-2 aircraft which can reach a height of 20 km is specially developed for high altitude, long duration and long range operation (NASA Ames Research Center 1978). In recent years, increasing attention has been paid to spacecraft as suitable sensor platforms because they have overcome difficulties of the ceiling limit and operation duration. The use of spacecraft orbiting regularly around the earth from a height of several hundreds of kilometres makes regular surveillance of the earth with suitable remote sensing devices possible. The term 'spacecraft' refers here to *rockets, artificial satellites* as well as *manned or unmanned space vehicles*.

Rockets

Rockets (which can be regarded as a combustion chamber to produce a thrust of hot gases) play an important part in space exploration. At the early stage of development in the 1950s and early 1960s, rockets were used as platforms for photographic systems, but in most cases the photography acquired was only used for recording the orientation of the rocket in flight. The cameras used were mostly non-metric in nature with focal lengths ranging from 38 mm–163 mm and formats in 16 mm, 35 mm and 102 × 127 mm. The films were basically panchromatic black-and-white, infrared and colour, and a number of different film-filter combinations were also tried.

In general, the photographs were fairly clear and cloud-free because the rockets were usually fired in a period of fine weather, exemplified in the photography from Vikings 11 and 12 in the USA. But the coverage of these rocket photographs was very restricted because the rocket had to be fired vertically upwards and it had to land not far from its launching site. Also, the rocket went up and down so fast that the camera had at the most, only 8 minutes to take photographs above the earth. Investigations of these early rocket photographs seemed to reveal that natural features could be more easily detected than cultural features (Bird and Morrison 1964). In 1972 the idea of using a rocket as a sensor platform was taken up by the Royal Aircraft Establishment, Farnborough in the UK and the *Skylark Resources Rocket* (Hoare 1972) was developed. Technical design improvements made automatic stabilization of the payload possible and, a preprogrammed horizontal rotation by the electronic control unit, so that a larger area could be covered by the photographic system during the rocket trajectory. After re-entry into the earth's atmosphere a parachute recovery system permitted speedy recovery of the payload. The rocket could also be fired from short-rail transportable launchers in practically any location on earth. The prototype was successfully launched from Woomera in Australia (Ridway and Hardy 1973) and this lead to further launches from Argentina (Hardy and Ridway 1973). But, in recent years the interest in resources rockets has dropped considerably, probably because of its short-duration flight capability which makes it at best, a reconnaissance tool.

Earth satellites

On the other hand, the *earth satellite* provides an ideal platform in space for remote sensors. The earth satellite is an artificial object in space which revolves around the earth following a specific orbit. Unlike the rocket, a satellite can stay aloft for a much longer period of time, thus permitting constant surveillance of the earth. Earth satellites can be conveniently classified, according to their orbital characteristics, into three types: (1) those with near-polar *sun-synchronous* orbits; (2) those with *equatorial geosynchronous orbits*; and (3) those with *general orbits*, of which the first two are most commonly employed as sensor platforms (Cornillon 1982; Barrett and Hamilton 1982).

Sun-synchronous orbiting satellites are designed in such a way that the ascending node of each orbit of the satellite will occur at the same local time. This simply means that the angular relationship between the sun and the satellite's orbital plane is kept constant (Widger 1966; Petrie 1970). This will ensure that the angle of the sun's illumination remains constant for a given latitude over a short period. When reflected solar radiation is the major source of energy for a passive remote sensor, such as the photographic system, the sun-synchronous satellite is a more suitable platform. In order to see as large an area of the earth as possible, the *orbital inclination* of the satellite is near to 90° or near-polar so that the coverage of the earth visible from the satellite is extended nearly between the two poles. Such a satellite is about 1,000 km above the earth's surface. Good examples are the *Landsat* satellite series, designed for observation of earth resources with high-resolution imagery and the Heat Capacity Mapping Mission (HCMM) satellite, designed for high spatial resolution thermal surveys of the earth's surface.

Geosynchronous orbiting satellites are those that remain directly overhead a specific point on the earth's surface. Hence, they are also known as *geostationary*. This is achievable by increasing the orbital height sufficiently so that the *orbital period* (i.e. the time for a satellite to complete one orbit around the earth) is equal to the earth's rotation. This orbital height is found to be 35,800 km or about 5.6 times the earth's radius (*circa* 6,370 km). Such a geosynchronous orbit occurs over the equator. Examples are the communication and meteorological satellites which are required to be fixed over a particular longitude. Good examples of the meteorological satellites in this category are the USAs series of *Applications Technology Satellites* (ATS), *Geostationary Operational Environmental Satellites* (GOES) and the recent European METEOSAT System. Most of the *environmental satellites* characterized by relatively low spatial resolution data fall into this category. But data can be more frequently obtained over a single locality than in the case of earth resources satellites.

General satellites refer to those which have neither a sun-synchronous nor a geosynchronous orbit. A good example is the Seasat 1 which had a nearly circular orbit with an inclination angle of 108° and an altitude of 800 km.

Landsat resources satellites

Because of the popularity of the Landsat system for research among environmental scientists, a more detailed description of the satellite and its imaging systems is relevant at this point.

Landsat is the result of the earth resources programme developed by the National Aeronautical and Space Administration (NASA) in the USA in the early 1970s. It was originally launched on 22 July 1972 as ERTS-1 (Earth Resources Technology Satellite) and was later renamed Landsat 1. Since then, three further Landsats were successfully launched. The first three Landsats which share common orbital and imaging system characteristics can be regarded as the first generation resources satellites in the series. Landsat 4 which was successfully launched on 16 July 1982 initiates a new generation of high-resolution resources satellites, which represents an improvement over the past model. The orbits of the Landsat series are all sun-synchronous. However, for the first three Landsats, the orbit inclination angle is about 99.1° (thus providing global coverage between 81°N and 81°S), the orbital period is about 103.3 minutes, the time of crossing the equator in a north to south direction (i.e. descending node) occurs at about 9.30 a.m. every day, and the altitude of the satellite varies between 880 km and 940 km. The orbit causes the daily coverage swath to be shifted westwards in longitude at the equator by 1.43° corresponding to 159 km on ground. One complete coverage cycle consists of 251 revolutions or 18 days. In other words, it takes 18 days for the satellite to return to the same spot on earth again. With the launch of Landsat 2 and 3, the coverage cycle can be cut short to 9 or even 6 days. The coverage pattern provides 14 per cent cross-track imagery overlap at the equator, which increases as the latitude increases so that at latitude 80° the image overlap is as much as 85 per cent. This gives rise to stereoscopic viewing potential for the Landsat images (Taranik 1978).

The imaging systems onboard Landsat 1, 2 and 3 are return beam vidicon (RBV) cameras and the multispectral scanners (MSS). On Landsat 1 and 2, the RBV is a three-television camera system of the electro-optical type with a focal length of 126 mm which records the ground reflectance in three visible wavelength bands (Table 1.2). When combined together they should give a false colour composite. The system has a useful image area of 25 × 25 mm on the tube surface which carries an 81-point reseau and 4 fiducial marks. It is capable of producing high-resolution pictures consisting of 4,125 scan lines and 4,500 picture elements per scan line, which is equivalent to a ground resolution of about 80 m. The three cameras could image the 185 × 185 km area every 25 seconds (Fig. 1.25a). On Landsat 3, however, the RBV system comprises two cameras only with doubled

Table 1.2
The spectral wavelength intervals detected by the remote sensing systems on Landsats 1 and 2
(*Source*: Taranik 1978: 24)

Data Channel	System	Type of Radiation	Wavelength (μm)	NASA Code
1	RBV	Visible, yellow-green	0.475–0.575	Band 1
2	RBV	Visible, green-red	0.580–0.680	Band 2
3	RBV	Visible, red-invisible IR	0.690–0.830	Band 3
1	MSS	Visible green	0.5–0.6	Band 4
2	MSS	Visible red	0.6–0.7	Band 5
3	MSS	Invisible reflected IR	0.7–0.8	Band 6
4	MSS	Invisible reflected IR	0.8–1.1	Band 7
(5)	MSS	(Invisible thermal IR)	(10.2–12.6)	(Band 8)*

* Landsat 3 MSS, channel not on Landsats 1 and -2.

Fig. 1.25
Remote sensing systems on Landsats 1, 2 and 3: (a) Three-camera RBV system on Landsats 1 and 2, (b) Two-camera RBV system on Landsat 3 and (c) Multispectral Scanner System (MSS); (*Source*: Taranik 1978)

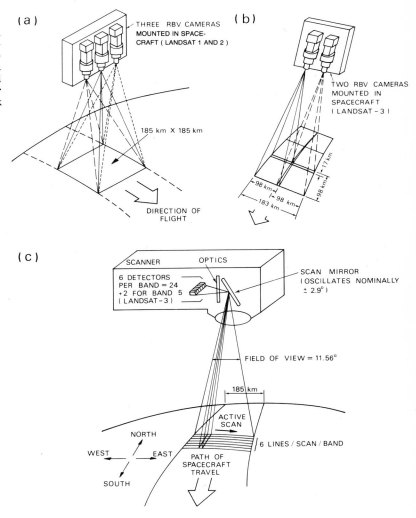

focal length in the optical system imaging only in a single spectral band of 0.505–0.750 μm (panchromatic). This resulted in reducing the areal coverage to one quarter of the area covered by a single RBV camera used on Landsat 1 and 2 but improved the spatial resolution to 40 m (Fig. 1.25(b)). It is worthy to note that the RBV system was shuttered and the imagery was obtained one frame at a time. Hence, it suffered from less geometric distortion caused by attitude changes of the sensor platform than that from a continuous scanning system. Because the RBV system was not frequently used onboard the Landsat satellites in view of some electronics and data handling problems, RBV data availability is very limited especially for Landsat 1 and 2 and is therefore less useful that the MSS data.

The MSS on the first three Landsat satellites is capable of imaging the earth's surface in four spectral bands simultaneously through a single optical system (Table 1.2). There are 6 detectors for each band so that altogether 24 detectors are used for the MSS (Slater 1980: 473). The scanning is made possible by an oscillating flat 45° mirror between the earth and the double-reflector telescope optics. It scans the earth eastwards and at right angles to the direction of the satellite motion (i.e. across-track scan) and the along-track scan is produced by the forward motion of the satellite in the southerly direction (Fig. 1.25(c)). The scan mirror rotates back and forth through ±2.9° to give a total field of view of 11.6°. Six lines of data are scanned at one time for a distance of 185 km long on the ground. During each scan, the electromagnetic radiation coming from the earth's surface and its atmosphere is reflected by the mirror into reflecting telescope optics and focused on fibre optic bundles of the detector array located in the focal plane of the telescope.

The size and arrangement pattern of these fibre optic bundles determine the instantaneous field of view (IFOV) of the detector. Each detector produces a voltage according to the amount of radiation received. The voltage, which is an analogue signal, is converted to digital values (six binary digits: 0–63) by a multiplexer. These digital values can be further scaled to wider ranges during subsequent ground processing (such as 0–127 for bands 4, 5 and 6; and 0–63 for band 7). During each scan, the voltage produced by each detector is sampled every 9.95 μs. There are approximately 3,300 samples taken along a scan line 185 km long. Thus, the IFOV is 79 × 79 m (0.62 ha), which is separated at an interval of 56 m on the ground between each sample. The radiation measurements have to be assigned to the 56 × 79 m dimension (0.44 ha) in order to maintain the spatial relationships. However, the dimension is subject to change according to variations of the spacecraft altitude. This nominal dimension of 56 × 79 m area is called a Landsat *picture element or pixel*. Therefore, the radiation measurement for a larger area is formatted to a smaller area, thus giving rise to overlaps of the areas for adjacent pixels. A geometrically corrected Landsat MSS scene of 185 × 185 km in area, nominally consists of about 2,340 scan lines and about 3,240 pixels per line (or 2,983 scan lines and 3,548 pixels per line for Landsat 3). Because of the earth's rotation during the 25 seconds for the satellite to move from the top to the bottom of the scene, the scene is a parallelogram in shape (Fig. 1.26). On Landsat 3, the MSS had an added band 8 or the thermal infrared band (10.4–12.6 μm) which employed two mercury-cadmium-telluride detectors.

A new generation of sensor system is carried onboard Landsat 4, aiming at improving the spatial resolution, spectral separation, geometric fidelity and radiometric accuracy of the data – the Thematic Mapper (TM) in addition to the four-band multispectral scanner (Salomonson and Park 1979). The orbit and coverage of Landsat 4 are somewhat different from its predecessors. Although it is still a sun-synchronous, near-polar orbiting satellite, its orbital inclination is now 98.2°; orbital period, 98.9 minutes; and nominal altitude, 705 km. It takes 16 days to cover the whole earth (except the poles). The satellite's pointing accuracy and stability have also been greatly improved. The Thematic Mapper is a scanning optical sensor operating in the visible and infrared in seven spectral bands (Table 1.3).

It works on the same basic principle as the MSS, but it provides higher spatial and radiometric resolution. The TM detector arrays are located within the primary focal plane of the telescope optics so that the incoming radiation can be transmitted directly onto the detectors. As the scan mirror sweeps back and forth in the across-track direction, it forms a raster of 16 lines in bands 1–5 and 7, and 4 lines in band 6 for each sweep direction. Data are collected in both the forward (west-to-east) and reverse (east-to-west) scans. The number of quantization levels has been increased from 0–64 to 0–256 for better radiometric accuracy. A pixel size is 30 m on ground in all except band 6 which is a thermal band with a pixel size of 120 m on ground.

The MSS on Landsat 4 is similar to those onboard Landsat 1, 2 and 3, but the pixel size has been adjusted to 80 m for compatibility with the MSS data obtained in the past. The four MSS bands which have the same spectral coverage as before have been re-numbered as bands 1, 2, 3 and 4. Finally, one should also note that the relay of Landsat 4 data to ground stations will be near real-time by using the geostationary Tracking and Data Relay Satellite (TDRS) and the Domestic

Table 1.3 Thematic Mapper spectral bands and principal applications	*Band 1 (0.45–0.52 μm)* Designed for water body penetration, making it useful for coastal water mapping. Also useful for differentiation of soil from vegetation, and deciduous from coniferous flora.
	Band 2 (0.52–0.60 μm) Designed to measure visible green reflectance peak of vegetation for vigor assessment.
	Band 3 (0.63–0.69 μm) A chlorophyll absorption band important for vegetation discrimination.
	Band 4 (0.76–0.90 μm) Useful for determining biomass content and for delineation of water bodies.
	Band 5 (1.55–1.75 μm) Indicative of vegetation moisture content and soil moisture. Also useful for differentiation of snow from clouds.
	Band 6 (10.40–12.50 μm) A thermal infrared band of use in vegetation stress analysis, soil moisture discrimination, and thermal mapping.
	Band 7 (2.08–2.35 μm) A band selected for its potential for discriminating rock types and for hydrothermal mapping.

Communication Satellite (DOMSAT) systems. This will eliminate the need to rely on onboard tape recorders to store data for transmission as practised in the past.

Seasat

It is appropriate to briefly mention here a 'general' satellite – Seasat which is the first satellite designed to observe the earth's oceans with microwave sensors. The satellite was launched into space on 28 June 1978 by NASA and Jet Propulsion Laboratory in a near-circular orbit with an inclination angle of 108° at an altitude of 800 km (Born, Dunne and Lame 1979). It circled the earth 14 times a day, covering 95 per cent of the oceans on earth every 36 hours. There were 5 sensors onboard: a radar altimeter (ALT), a Seasat-A scatterometer system (SASS), a synthetic aperture radar (SAR), a visible and infrared radiometer (VIRR) and a scanning multichannel microwave radiometer (SMMR), each of which had a specific function to perform.

The ALT operating at 13.56 Hz was for use in the measurement of the sea state, which was capable of giving an accuracy of ±0.5 m or 10 per cent (whichever was greater) for wave heights in seas less than 20 m and a root-mean-square error of 10 cm for wave heights in seas of (sea state) less than 20 m. The SASS was an active microwave wind sensor using a transmitted frequency of 14.6 GHz capable of producing an accuracy of ±2 m/sec in magnitude and ±20° in directions for winds ranging from 4 to 26 m/sec. The VIRR was a scanning radiometer operating at the visible band (0.49–0.94 μm) to provide information on cloud conditions and the infrared band (10.5–12.5 μm) to provide information on surface and cloud-top temperatures. The SAR was an active imaging system in the L-band (1.275 GHz) which looked to the right side of the satellite track with a swath width of 100 km at an incidence angle of 20°. The spatial resolution was as high as 25 m in both range and azimuth so that waves and wave spectra to oceanic wavelengths of 50 m or more could be measured. In addition, the SAR system was to help in detecting sea ice features, icebergs, water-land interfaces, and to penetrate through major storms. The SMMR operating at frequences of 6.6, 10.7, 18, 21 and 37 GHz with both vertical and horizontal polarizations was employed to observe sea-surface temperatures and to measure wind speed. The spatial resolution varied from about 100 km at 6.6 GHz to about 22 km at 37 GHz. The accuracy of the sea-surface temperature measured was about ±2 K with a relative accuracy of 0.5 K, whilst the accuracy of the wind speed measurements was about ±2 m/sec for winds ranging from 7 m/sec to about 50 m/sec.

Seasat-1, however, failed in orbit on 10 October 1978 as a result of a massive short circuit in its electrical system. But some useful data were already acquired during the 98 days of its useful life. The SAR imagery has drawn particular attention from the environmental scientists because it was the first synoptic, high-resolution radar imagery of the earth's surface including both the land and oceans for some 100 million km². A good idea of the quality of the data can be obtained from two atlases of SAR images published by the Jet Propulsion Laboratory (Ford et al 1980; Fu and Holt 1982).

Manned spacecraft

With manned spacecraft, more direct human control is possible in obtaining imagery of the earth from space. At the early stage, the astronauts onboard spacecraft such as USAs Mercury series, Gemini series and Apollo series, took photographs in black-and-white and colour of the earth. The small-format cameras of the hand-held type were generally favoured at the early stage, such as, the Maurer 220 G 70-mm camera and later the modified Hasselblad 500C 70-mm camera. The orbiting altitude was about 161 km at perigee and about 260 km at apogee. For Gemini IV and V two specific photographic experiments, namely (a) Synoptic Terrain Photography and (b) Synoptic Weather Photography Experiments, were carried out (NASA 1967; 1968). The former was specifically concerned to show major geologic structures, their form, colour and albedo with priority for photography over East Africa, the Arabian Peninsula, Mexico and the South-western United States (Gemini IV) as well as their coastal areas (Gemini V) for oceanographic and geographic investigations. The latter was intended to provide a set of higher-resolution photographs to cover a broad range of meteorological phenomena with special emphasis on cloud systems and their dynamic changes at 90-minute intervals.

In later flight missions, more advanced camera systems were used. For Apollo IX, multiband photography was attempted with four Hasselblad 500 EL cameras ($f = 80$ mm) in 0.51–0.89 μm, 0.47–0.61 μm, 0.68–0.89 μm and 0.59–0.715 μm for near-vertical stereo-coverage. More importantly, for Apollo XV, XVI, XVII missions, a totally new photographic system was assembled inside the Scientific Instrumentation Module (SIM), comprising a panoramic camera ($f = 610$ mm, aperture $f/3.5$) with a sweeping angle of 108° and a metric camera system. The metric camera system consisted of (a) a terrain camera ($f = 75$ mm, aperture $f/4.5$) equipped with forward motion compensation, (b) a stellar camera ($f = 75$ mm, aperture $f/2.8$) and (c) a laser altimeter. The exposures by the three cameras were synchronized and were all recorded on the film of the terrain camera. The panoramic camera made possible the production of high-quality stereo-photography of a large area of the lunar surface to aid in the selection of potential landing sites and exploration areas.

Following the completion of the Mercury, Gemini and Apollo programmes, a much larger spacecraft – the Skylab space station was launched on 14 May 1973 which made possible observation of the earth for more extended periods (NASA 1977). The Skylab followed a circular orbit at a nominal altitude of 435 km and an inclination angle of 50°. In mission numbers 2, 3 and 4, photography of the earth was obtained by the crew members. In particular for Skylab 2, the Earth Resources Experimental Package (EREP) was carried onboard. This included, among others, three novel imaging systems: (a) the Multispectral Photographic Camera S-190A which was an array of six matched 70 mm format cameras ($f = 152$ mm, aperture $f/2.8$) mounted to a single frame; (b) the Multispectral Scanner S-192 which operated in the visible and reflected infrared bands (in 13 separate spectral bands ranging from 0.41 μm to 12.5 μm); (c) the High Resolution Camera S-190B, a modified Hycon KA-74 reconnaissance camera ($f = 460$ mm) with an 11.5 $\times$ 11.5 cm format.

Both the S-190A and S-190B cameras which were equipped with image motion compensation were designed to provide photographs for topographic mapping applications. The S-190A produced 70 mm format photographs in six spectral bands, four of which covered the range 0.5–0.9 μm at 0.1 μm intervals on black-and-white film and the remaining two cameras exposed true- and false-colour films respectively. The resultant scale of the photography was about 1:2,900,000 and a ground resolution of 50 m was achievable. The photographic scale for the product from S-190B was about 1:1,000,000 with a ground resolution of 15–25 m, depending on the film being used (Petrie 1974). The positional accuracies of these Skylab photographs, particularly those from the S-190B camera system, has been found to be good enough (35–75 m for S-190A and 20 m for S-190B) to meet the requirements of small-scale *planimetric* mapping (Derenyi 1981). In the Skylab 4 mission, hand-held photography with Hasselblad (70 mm) and Nikon (35 mm) cameras were acquired to study the kinds of surface, air and water phenomena that could be visually identified from space. It became clear that the study of earth features was affected by the sun angle and the surface conditions (NASA Lyndon B. Johnson Space Center 1977). The Skylab space station crashed down to earth on 11 July, 1979.

The successful orbital test flight of the space shuttle in April 1981 ushered in a new era of terrestrial remote sensing. The space shuttle is a much more flexible spacecraft than the Skylab space station and provides an ideal orbital platform for collecting remote sensing data and for testing advanced sensor systems because of its ability to carry the instruments into space and return them to the ground for recalibrating and refurbishing (Taranik and Settle 1981). In the second test flight of the Space Shuttle Columbia on 12 November 1981, the Office of Space and Terrestrial Applications-1 (OSTA-1) payload was carried into orbit. The space shuttle followed a near-circular orbit at an average altitude of 262 km, an orbital period of 90 minutes and an inclination angle of 38°. The OSTA-1 payload consisted of five experiments to study the earth and its atmosphere, of which the Shuttle Imaging Radar A (SIR-A) experiment is most interesting (Settle and Taranik 1982).

SIR-A was a side-looking, synthetic aperture radar which looked at the earth's surface with horizontally polarized microwave radiation transmitted at L-band frequency (1.278 GHz or 23 cm wavelength) at an incidence angle of 50°. The ground resolution was about 40 m. About 10 million km^2 of radar images of the earth were obtained by the space shuttle. Preliminary investigations of the radar images have revealed the usefulness of the data in studying structural and geomorphic features such as faults, folds, outcrops and sand dunes in the tropical and arid regions on earth (Elachi *et al.* 1982, Cimino and Elachi 1982). The success of the SIR-A experiment will certainly lead to more interest in the use of the space shuttle as a sensor platform for remote sensing.

Approaches to image interpretation

Manual approach

We have seen that remote sensing data are basically records of reflected and emitted electromagnetic energy patterns presented as picture-like images which are highly varied in nature. In order to extract meaningful information out of these data, one has to exercise one's judgement to sieve the significant features out of the insignificance. This is the first stage of image interpretation known as *detection* (Lo 1976: 137). Detection is aided by the *spatial*, *spectral*, *radiometric* and *temporal* characteristics of the data. Spatial resolution refers to the ability of the recording system in distinguishing closely spaced objects. Larger objects tend to attract more attention than smaller ones. Thus, in Fig. 1.26 which is a Landsat picture in band 5 (0.6–0.7 μm) of the Pearl River delta area of China acquired on 25 December 1973, the field patterns are not too clearly revealed along the coast or inland. But in band 7, the reflected infrared band (0.8–1.1 μm), the strong contrast between land and water helps to pick out the larger fields on the coast and the numerous reservoirs on the hills (Fig. 1.27). On the other hand, the vegetation is more clearly shown as dark-toned patches in band 5. The sediments carried by the river into the sea are more distinctly recorded on the band 5 picture than on the band 7 picture which shows water as black. This indicates the importance of spectral resolution. Spectral resolution is the recording of the same scene in different spectral intervals; the finer the spectral interval the better is the spectral resolution which provides the rationale for multispectral or multiband remote sensing. By superimposing all four spectral bands of the Landsat image (Figs. 1.26, 1.27) together a false colour composite is formed (Pl. 1, page 36), which provides better information of the area. It is clear that the spatial and spectral resolution are interrelated. Radiometric resolution is to provide better contrast so that a greater number of discriminable grey steps between the black and white limits is achievable. Finally, temporal resolution refers to the use of imagery acquired at given time-intervals of the year to detect changes that have occurred. Figure 1.28 is a band 5 Landsat photograph of the same Pearl River delta area taken on 19 October 1979. Because harvesting has not yet taken place, the field patterns are more apparent. A comparison of Fig. 1.28 with Fig. 1.26 reveals changes in land use and coastal sedimentation patterns in the area. Another notable feature is the much lighter tone of the river water for the October scene than that for the December scene, where the river water is dark-toned. The difference is due to the higher water volume in the wet October month, which causes a greater degree of turbulence in the rivers.

The detection stage naturally leads on to the *recognition and identification* stage in which the image interpreter has to exercise general, local, as well as specific levels of reference to allocate objects into known categories. The general level is the interpreter's general knowledge of the phenomena and processes to be interpreted, the local level is the interpreter's intimacy with his own local environment, and the specific level is the interpreter's deeper understanding of the processes and phenomena he wants to interpret. In recognition and identification, the non-geometric image characteristics of *tone or colour*, *texture*, *pattern*, *shape*, *shadow*, *size* and *situation* normally give clues. Tone or colour is a record of energy reflectance or emittance which has different meanings according to the spectral sensitivity of the detector or film employed. Thus, light tone water in a thermal infrared image suggests the presence of warm water, whereas in black-and-white photography, it is the result of light reflecting to the camera lens. Similarly, the colour of an image depends on whether true colour or colour infrared film is used. As shown in Pl. 1, the false colour composite emphasizes the vegetation in red and water features in deep blue. Texture, is related the frequency of tone change, which gives one an impression of the degree of roughness or smoothness of the imaged features. Pattern, is the spatial arrangement of objects in the image. Shape, gives the general form or outline of an object. Helped by shadows, the shape factor is very useful in object identification and recognition. Long shadows tend to emphasize linear features, as is well illustrated in the case of SLAR imagery. The size or dimensions of an object is an essential clue to the identification of objects of similar shape and can be used as a standard for comparison. Finally, the situation or location of objects in relation to others is always helpful where other characteristics fail to give clues. For more detailed discussion of these various characteristics, see works by Estes and Simonett (1975), Lillesand and Kiefer (1979) and Lo (1976).

The result of identification and recognition is a list of objects and features in the area. These form the basis for the delineation of areas having homogeneous observable patterns and characteristics. This is the analysis stage. Each area so delineated has to be *classified* through a process of induction (general inference from particular cases) and deduction (particular inference from general observations). The accuracy is then evaluated by field checks. Therefore, the final stage of the interpretation is *classification*, producing spatial data which can be displayed as maps or for incorporation into a geographic information system by the computer.

Computer-assisted approach

The manual approach in image interpretation suffers from its inability to deal quickly with a large quantity of image data. This weakness is particularly evident when multispectral scanner imagery or multiband photography is to be analysed. The cross-referencing of tone values area by area and feature by feature is extremely difficult to handle manually. Analogue instruments such as the additive viewer or image enhancement methods such as density slicing can help to facilitate interpretation to a certain extent, but its speed is still not fast enough to catch up with the daily rate of remote sensing data input from space platforms. The electronic computer provides the only solution. Because image interpretation is basically a classificatory process, identification and recognition can be treated in mathematical terms, provided image data in digital form are available.

The computer-assisted approach involves a number of steps. First of all, the analogue image data have to be converted into digital form. This is done by means of a digitizing TV scanner or a microdensitometer. For image data relayed from the early satellites, the signal levels of the energy reflectance or emission have already been received in digital form. The second step is *data preprocessing*, which is a group of procedures to 'clean up' the raw input data, such as correcting geometric and radiometric distortions. This is then followed by *feature extraction*. The types of features or measurements necessary to classify the image data are selected in this stage. Possible features are spatial, spectral and temporal. For *supervised classification*, sample areas are

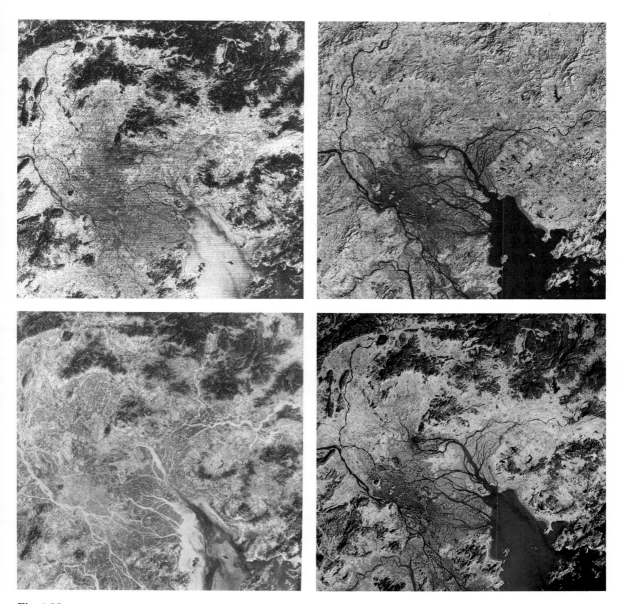

Fig. 1.26
Landsat image: Pearl River delta area of China acquired on 25 December 1973 (band 5)

Fig. 1.27
Landsat image: Pearl River delta area of China acquired on 25 December 1973 (band 7)

Fig. 1.28
Landsat image (band 5): Pearl River delta area of China acquired on 19 October 1979

Plate 1
False colour composite of Pearl River delta area, China, by combining all four Landsat bands through blue, green, yellow and red colour filters (25 December 1973)

Plate 2
A colour infrared photograph of rural Thailand showing the linear farm settlements along a canal, the major thoroughfare in the region. Each farm settlement is surrounded by trees which are displayed distinctly in red in the colour infrared

Plate 3
Gemini V photograph of the southern third of the Nile delta (*Courtesy* P.D. Lowman, National Aeronautics and Space Administration)

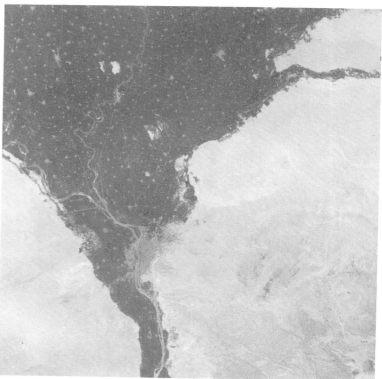

Plate 4

Colour ratio composite of Landsat-1 image of south-central Nevada (scene ID E1072–18001) consisting of the following stretched ratio image combinations using diazo colours: blue for MSS 4/5, yellow for MSS 5/6 and magenta for MSS 6/7. Mafic rocks, mainly basalt and andesite (A), are white, whereas felsic extrusive and intrusive rocks are pink (B). Playas and the two mining dumps are blue (D). Altered areas are represented by green (D) to dark-green (E) and brown (F) to red-brown (G) patterns. The green areas represent hydrothermally altered limonitic rocks, except for two areas H and I (Source: Rowan *et al.* 1976)

Plate 5

False-colour Landsat images of Mount St Helens, taken before and after the eruption on 18 May 1980: (a) taken on 11 September 1976 showing Mount St Helens as a white snow-capped feature near the centre; (b) taken on 19 August 1980 the former forest land was devastated and covered with ash which is light-coloured (compare Figs 4.6 and 4.7)

Plate 6
Landsat mosaic of temporal composites of the Indus River Valley thematically depicting the extent of flooding (*Courtesy*: M. Deutsch, *Source*: Deutsch and Ruggles 1978)

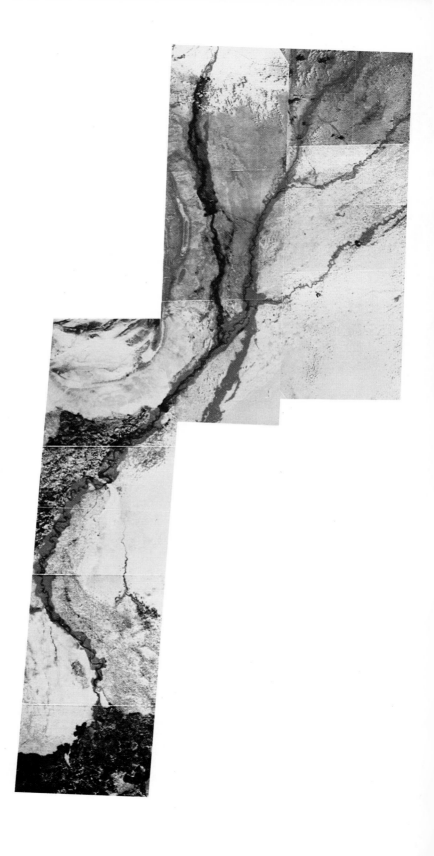

selected from the image data for more detailed examination. *Training sets* of data are then compiled. These training sets are sub-samples from the images whose identification is completely known. These will be treated as *stereotypes* of particular classes of data with which unknown areas can be compared. Statistical parameters, such as means and standard deviations, are computed for the training sets by the computer, which are employed for classifying other unknown areas. The statistical method commonly employed in supervised classification is *discriminant analysis* which specifies a set of discriminant functions to divide the feature (or measurement) space into appropriate regions. Each region should ideally contain points of only one class.

Another approach is *unsupervised classification* in which no training samples are used. The partitioning of the feature space is carried out by the method of cluster analysis which can identify natural groupings of patterns. The nature of each grouping is determined afterwards by field checks. Unsupervised classification is not generally as effective as the supervised classification because of the absence of training sets to control the results, especially when classes are only marginally separable. However, supervised classification is still too slow to handle a massive influx of satellite multispectral data. For optimal analysis, hybrid procedures incorporating the advantages of both supervised and unsupervised classification have been evolved (Swain and Davis 1978). The results of the classification from the computer-assisted approach can be output as line printer maps or cathode ray tube displays. Numerical information on the area of the mapped classes or the frequency of occurrence of each class and other useful statistical data can also be displayed by the computer if required. An example of a line printer map of land use in Hong Kong (occupying the southeast corner of the Landsat scene in Figs 1.26, 1.27, 1.28 and Pl. 1) is shown in Fig. 6.20.

Directions of application of remote sensing data

In this Chapter, a brief survey of the nature of remote sensing data and the methods of analysis has been given, which hopefully will provide the background to appreciate the many applications possible with these data. The growing activities of multispectral remote sensing of the earth from space with the use of satellites and space shuttles in recent years point to a need to constantly monitor our terrestrial environment.

Our environment which comprises man and other life forms is the *biosphere* or the life-bearing layer. It interacts with the *atmosphere* (the gaseous layer), the *hydrosphere* (the water layer) and the *lithosphere* (the solid earth crust). Remote sensing is applied to collect information needed for a better understanding of these various components of our environment. Although the diversity of applications is great, the unity of the purpose is clear: remote sensing plays primarily the role of a surveying, inventorying and mapping tool for environmental features. In subsequent chapters, an attempt will be made to illustrate the various approaches with reference to specific case studies and to discuss in some depth the problems related to these applications.

References

Barrett, E. C. and Hamilton, M. G. (1982) The use of geostationary satellite data in environmental science, *Progress in Physical Geography* **6**: 159–214.

Bird, J. B. and Morrison, A. (1964) Space photography and its geographical applications, *Geographical Review* **54**: 463–86.

Born, G. H., Dunne, J. A. and Lame, D. B. (1979) Seasat mission overview, *Science* **204**: 1405–6.

Cimino, J. B. and Elachi, C. (eds) (1982) *Shuttle Imaging Radar-A (SIR-A) Experiment.* Jet Propulsion Laboratory, California Institute of Technology: Pasadena, California.

Cooke, R. U. and Harris, D. R., (1970). Remote sensing of the terrestrial environment-principles and progress, *Transactions of the Institute of British Geographers,* No. 50, pp. 1–23.

Cornillon, P. (1982) *A Guide to Environmental Satellite Data.* Graduate School of Oceanography, University of Rhode Island.

Derenyi, E. E. (1981) Skylab in retrospect, *Photogrammetric Engineering and Remote Sensing* **47**: 495–9.

Doyle, F. J. (1972) Imaging sensors for space vehicle, Paper presented to Commission I of the 12th International Congress of International Society for Photogrammetry: Ottawa, Canada.

Elachi, C. *et al.* (1982) Shuttle imaging radar experiment, *Science* **218**: 996–1003.

Estes, J. E. and Simonett, D. S. (1975) Fundamentals of image interpretation: In Reeves, R. G. (ed.) *Manual of Remote Sensing,* Vol. II. American Society of Photogrammetry: Falls Church, Virginia, pp. 869–1076.

Ford, J. P., Blom, R. G., Bryan, M. L., Daily, M. I., Dixon, T. H., Elachi, C. and Xenos, E. C. (1980) *Seasat Views North America, the Carribean and Western Europe with Imaging Radar.* Jet Propulsion Laboratory: Pasadena, California.

Fu, L. L. and Holt, B. (1982) *Seasat Views Oceans and Sea Ice with Synthetic-Aperture Radar.* Jet Propulsion Laboratory: Pasadena, California.

Hardy, J. R. and Ridway, R. B. (1973) Skylark analyses Argentina, *Geographical Magazine* **46**: 5–6.

Hoare, D. N. (1972) The Skylark earth resource rocket, Paper presented to the 22nd International Geographical Congress, Montreal, Canada.

Hunt, G. R. (1980) Electromagnetic radiation: the communication link in remote sensing: In Siegal, B. S. and Gillespie, A. R. (eds) *Remote Sensing in Geology.* John Wiley: New York; Chichester; Brisbane; Toronto, pp. 5–45.

Hurter, F. and Driffield, V. C. (1890) Photo-chemical investigations and a new method of determination of the sensitiveness of photographic plates, *Journal of the Society of Chemical Industries* **9**: 455–69.

La Prade, G. and Leonardo, E. S. (1968) Elevation measurements from radar imagery, *Proceedings of the American Society of Photogrammetry,* 34th Annual Meeting, pp. 153–64.

Leberl, F. (1971) Metric properties of imagery produced by side-looking airborne radar and infrared linescan system: In Kure, J. (ed.) *Proceedings of the ISP Commission IV symposum.* ITC Publications A/50: Delft, pp. 125–48.

Lillesand, T. M. and Kiefer, R. W. (1979) *Remote Sensing and Image Interpretation.* John Wiley: New York; Chichester; Brisbane; Toronto.

Lo, C. P. (1976) *Geographical Applications of Aerial Photography.* Crane Russak: New York; David and Charles: Newton Abbot; London; Vancouver.

Loor, G. P. de (1969) Possibilities and uses of radar and thermal infrared systems, *Photogrammetria* **24**: 43–58.

Lowe, D. S. (1980). Acquisition of remotely sensed data, in Siegal, B. S. and Gillespie, A. R. (eds), *Remote Sensing in Geology,* John Wiley, New York, Chichester, Brisbane and Toronto, p. 59.

McDonald, H. C. and Lewis, A. J. (1976) Operation and characteristics of imaging radar, in Lewis, A. J. (ed.) *Geoscience Applications of Imaging Radar Systems, Remote Sensing of the Electromagnetic Spectrum* **3**: 23–45.

Moffitt, F. H. and Mikhail, E. M. (1980). *Photogrammetry,* Harper and Row, New York, p. 76.

NASA (1967) *Earth Photographs from Gemini III, IV and V.* US Government Printing Office: Washington, DC.

NASA (1968) *Earth Photographs from Gemini VI through XII.* US Government Printing Office: Washington, DC.

NASA Ames Research Center (1978) *High Altitude Perspective.* US Government Printing Office: Washington, DC.

NASA Lyndon B. Johnson Space Center (1977) *Skylab Eyplores the Earth.* US Government Printing Office: Washington, DC.

Petrie, G. (1970) Some considerations regarding mapping from earth satellites, *Photogrammetric Record* **6**: 590–624.

Petrie, G. (1974) Mapping from earth satellites, *Proceedings of P.T.R.C. Annual Symposium*, paper no. 11.

Ridway, R. B. and Hardy, J. R. (1973) Skylark over Woomera, *Geographical Magazine* **45**: 289–97.

Sabins, F. F. (1978) *Remote Sensing: Principles and Interpretation*, W. H. Freeman, San Francisco, p. 134.

Salomonson, V. V. and Park, A. B. (1979) An overview of the Landsat-D project with emphasis on the flight segment: In Tendam, I.M. and Morrison, D. B. (eds), *Proceedings of 1979 Machine Processing of Remotely Sensed Data Symposium*, pp. 2–11.

Settle, M. and Taranik, J. V. (1982) Use of the space shuttle for remote sensing research: recent results and future prospects, *Science* **218**: 993–5.

Slater, P. N. (1980) *Remote Sensing: Optics and Optical Systems.* Addison-Wesley: Reading, Mass.; London; Amsterdam; Don Mills, Ont.; Sydney; Tokyo.

Swain, P. H. and Davis, S. M. (1978) *Remote Sensing: the Quantitative Approach.* McGraw-Hill: New York; London; Paris; Singapore; Sydney; Tokyo; Toronto.

Taranik, J. V. (1978) *Characteristics of the Landsat Multispectral Data System*, Open File Report 78–187. US Geological Survey.

Taranik, J. V. and Settle, M. (1981) Space shuttle: a new era in terrestrial remote sensing, *Science* **214**: 619–26.

Thompson, L. L. (1979) Remote sensing using solid-state array technology, *Photogrammetric Engineering and Remote Sensing* **45**: 47–55.

Tomlinson, R. F. (1972) *Geographical Data Handling*, Vol. 1. International Geographical Union Commission on Geographical Data Sensing and Processing for the UNESCO/IGU Second Symposium on Geographical Information Systems: Ottawa, Canada.

Ulaby, F. T., Moore, R. K. and Fung, A. K. (1981) *Microwave Remote Sensing, Active and Passive, Volume 1, Fundamentals and Radiometry*, Addison-Wesley, Reading, Massachusetts; London, p. 423.

Vozikis, E. (1983) Analytical methods and instruments for mapping from balloon photography, *Photogrammetric Record* **11**: 83–92.

Wester-Ebbinghaus, W. (1980) Aerial photography by radio controlled model helicopter, *Photogrammetric Record* **10**: 85–92.

Whittlesey, J. H. (1970) Tethered balloon for archaeological photos, *Photogrammetric Engineering* **36**: 181–6.

Widger, W. K. (1966) Orbits, altitude, viewing geometry, coverage and resolution pertinent to satellite observations of the earth and its atmosphere, *Proceedings of the 4th Symposium on Remote Sensing of Environment*, pp. 484–537.

Chapter 2 The human population

Population: the focus of environmental study

In the twentieth century human population has grown more rapidly than ever before. The peak of its growth occurred in the 1970s when an annual average growth rate of about 2 per cent was recorded and was regarded as the highest growth rate in human history (Freedman and Berelson 1974). The growth was the result of falling death rates in response to the provision of better medical facilities and disease eradication programmes as well as the increasing affluence of the population brought about by industrialization and urbanization. Although world population has shown signs of a lowering growth rate as a result of declining global fertility since 1981, the net addition still amounts to 80 million per year. It was estimated by the United Nations that by the year 2110 the world population could stabilize itself at 10.5 billion, which will be nearly two and half times larger than the present population of 4.4 billion (Salas 1981). The impact of population growth on global resources, the global environment and global development remains a major challenge to be faced by mankind. Our study of the terrestrial environment derives significance and meaning only when its various elements are observed from the viewpoint of human utilization (Trewartha 1953). It is therefore appropriate to start our application of remote sensing to the population field.

General areas of applications

Because remote sensing data provide a synoptic view of the terrestrial environment, an obvious application is to map the spatial distribution of population and to examine the impact the human presence has brought to the living environment. In other words, remote sensing data faithfully record the interactions of man with his environment at different levels of sophistication according to the scale and the sensor type employed. An outgrowth from this application is to devise methods to estimate population number in an area, which is essential information required to estimate the spatial distribution of population when planning the economic development of a region or a country.

The extraction of information in these two areas of application involves two fundamental approaches of remote sensing: (a) the visual or manual approach of image interpretation and (b) the formulation of appropriate mathematical models to yield the data desired.

Population distribution

People are so small that they cannot be easily picked out from aerial photographs or from other forms of remotely sensed imagery. The spatial distribution pattern of population can only be inferred indirectly from the settlement pattern or from other visible evidence. Thus, from Pl. 2, page 36, which is a high oblique colour infrared photograph of rural Thailand, one can readily see the evenly spaced farm settlements marked by dense clusters of tree groves in bright purplish red along the banks of the straight canal which functions as the major thoroughfare in the region. Large-scale aerial photographs are useful in giving insights into the characteristics of population distribution in a habitat for the present or in the past at micro-scale. For a macroscopic view, imagery acquired by sensors from space platforms reveals distinct spatial patterns which can be treated analytically in a modelling attempt for explanation. The following case studies will serve to bring out the complementary nature of these two different approaches.

Study of the past population: the search for evidence

The use of aerial photography with a view to discover man-made structures of the past on the earth's surface is by now a well-established practice among archaeologists. A significant contribution of photographic data is the comprehensive view they give to a site which makes the strategic assessment of a region of interest possible. The ultimate aim of archaeology is to reconstruct the overall picture of a human community, its economy and setting (Bradford 1957). For example, if a site of an abandoned medieval village is discovered, one should, with the aid of the aerial photographs, locate and map the remains of the ancilliary features such as its fields, trackways, kilns or fishponds and then interpret how the people of the past made use of such natural resources as grassland, forests, minerals, etc. found in their region. The use of large-scale vertical and oblique aerial photographs provides enough details for such a kind of in-depth photo-interpretation to be made (St. Joseph 1977).

The discovery of archaeological sites from aerial photographs is based on evidence produced by tone or colour changes. The most important evidence is provided by *vegetation markings* caused by differential growth of vegetation responding to hidden differences in the soil beneath. The increased depth of soil filling holes and hollows may produce a denser and taller growth of vegetation, whereas certain materials underneath the surface that are impenetrable to roots will cause poor stunted growth. The occurrence of these vegetation markings is affected by season, weather and moisture conditions. Long-rooted cereal crops such as wheat, oats or barley produce the best effect of crop markings particularly in a dry season (Fig. 2.1). For vegetation that fails to grow properly, for example above buried roads or along wall-foundations *parch marks* are produced. On an aerial photograph they provide important evidence for revealing urban sites. It is interesting to note that in the tropics where irrigation and rice cultivation are practised, crop marks are not visible because crop sites are covered by a dense growth of trees and scrubs.

Another important source of evidence are *soil markings* which occur when earth works have been nearly levelled and become visible or when the soil is bare after ploughing or has only the thinnest covering of vegetation. The colour and tone contrast depend on the sub-soil. The contrast is particularly strong in the UK on chalk and limestone but in general, the contrast shows best after a long period of rain and wind in the area.

Other considerations include the effect of sunlight and shadows which can emphasize subtle differences in relief. The time of photography and direction of orientation in the case of oblique photography

Fig. 2.1
Crop marks, Orton Waterville, Huntingdonshire, UK. The photograph was taken on the evening of 21 June 1966 when the corn was reflecting buried features in the gravel. Note the long shadows cast by the taller crops which emphasized the effect (*Cambridge University Collection*: copyright reserved)

all help in the detection of complicated archaeological sites.

The great amount of air reconnaissance work carried out by Professor J. K. S. St Joseph of the University of Cambridge reported in the British archaeological journal, *Antiquity*, from 1964 to 1980, provides ample evidence to the usefulness of aerial photography in reconstructing settlements and human activities of the past. In Fig. 2.1, crop marks reveal a complex archaeological site on the south side of the River Nene in the parishes of Alwalton and Orton Waterville in Huntingdonshire, England (St. Joseph, 1969: 314–5). The site is located within a meander of the River Nene on a gravel river terrace standing high enough to be free of flood. The aerial photograph which was taken in the evening as indicated by the long shadows of trees reveals very well the buried features in the gravel. The crop marks resulted from the difference in height of the corn crop are further emphasized by the shadows. The main group of crop marks suggests a maze of sub-rectangular and less regular enclosures occupying the same ground successively. The small marks are probably indications of pits, pot-holes and even graves, and a long line of pits across the top of the photograph can be seen. The two roughly circular enclosures in the foreground suggest the possibility of the foundation-trenches of large native huts, or drainage ditches around such huts. This site is one of the many discovered in the region designated for the Peterborough New Town which covers an area of about 5,200 ha. A study of archaeological sites such as this can reveal social development of the inhabitants and changing land use patterns over some three millennia.

The use of multispectral aerial photography has also been tested for the purpose of archaeological research with a view to determining whether slight differences in the tone or the colours of growing crops can be detected (Hampton 1974). Four Vintern 70-mm F95 cameras ($f = 100$ mm; aperture size $f/2$) were used to fly photography over known archaeological sites of different geological setting in central southern England. The films employed were black-and-white panchromatic, true and false colour, each being combined with a number of different filters. Each site was photographed four times, from spring to late summer, so that the major crop growth changes were covered. A visual approach in evaluating the resultant multispectral photography was adopted, which involved tracing archaeological details (such as crop marks) from negatives or positives of each spectral band over a light table. The investigation has pointed towards some advantages in the use of multispectral aerial photography with two or at most three cameras. Under dry weather conditions, the panchromatic film with a yellow or orange filter gave its best results in late June and July. False colour infrared film also could produce good results in early June and late July. Thus, by using two cameras loaded with black-and-white panchromatic and false colour infrared films, most of the anomalies of vegetation growth could be recorded.

So far, the application of remote sensing to archaeology has been confined to the use of large-scale vertical or oblique aerial photography because of the need to emphasize vegetation and soil markings – the basic clues for archaeological site discovery. However, in recent years, non-photographic remote sensors have begun to be employed for application in archaeology. An interesting example is the use of synthetic aperture radar in resolving the problem of the population size of the Maya civilization in Central America (Adams 1980; Adams *et al.* 1981). The Maya area is a lowland region all below 1,000 metres and covers some 250,000 km². It receives heavy rainfall which results in luxuriant growth of vegetation which hampered the use of traditional aerial photography. At present, over 300 Maya centres with formal architecture which functioned as true cities are known. The size of these centres is judged by the number of courtyards and associated buildings they possess. The largest centre, called Tikal, has the equivalent of 85 courtyards and hundreds of associated buildings. These suggest population densities ranging up to 600 people per km² in the most densely populated zones. An urban population of 50,000 was thought to have been reached in the case of Tikal, and between 5–10,000 in the case of smaller centres during the Classic period (AD 250–900). However, all these population figures appeared to be incompatible to the sustaining capacity based on the traditional pattern of extensive slash-and-burn agriculture, which could only give a maximum calculated carrying capacity of 77 persons per km² of cultivated land. In the Rio Bec region in the north increased productivity due to hillside terracing which was widespread, could explain the high population density there. But as hillside terracing is not found in the southern regions, other forms of intensive cultivation have to be looked for. In view of the climatic conditions in the study area, the imaging synthetic aperture radar, originally designed for scanning the surface of Venus, was employed. The wavelength used was 25 cm. The slant range resolution and azimuth resolution of the system were respectively 15 m and 8 m. The radar was mounted on an aircraft flown at an altitude of 7.3 km. The resultant scale of the radar imagery was therefore about 1:250,000 (Fig. 2.2). Supporting black-and-white infrared and panchromatic photography was also flown simultaneously. The radar imagery was found to be capable of detecting such archaeological features as ancient cities, raised roads, edges of extensive paved zones and canals, which all exhibited radar

Fig. 2.2
Synthetic aperture radar imagery of the Pasion River (Seibal) region. Swamps and canals are revealed (N to left) (*Source*: Jet Propulsion Laboratory)

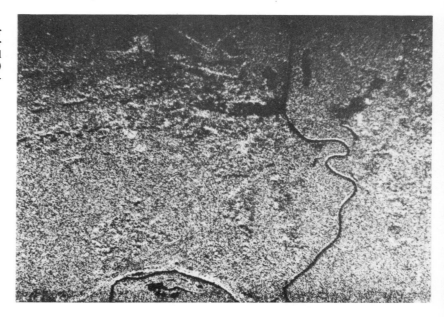

signatures. The small scale of the radar imagery, however, would not permit 'points' data to be detected. But small grey networks of lines (linear features) could be detected within the swamps, which suggest the occurrence of extensive canal systems (Fig. 2.2). By filtering out the non-archaeological features through an enhancement process, a 20 per cent residue of grid patterns was left. These were ground checked and confirmed to be canals (Fig. 2.3). Obviously, the grid pattern detected is that of the largest canals and their maximum extents. The radar imagery revealed that an area between 1,250 and 2,500 km² in the southern lowland was modified by the canals which probably served as drainage channels to drain off excessive water in the swamps and so could be utilized for cultivation (the so-called 'drained field' agriculture). This system of agriculture could support a minimum of 10 persons per ha of land. With this discovery it is now possible to explain the high population density of 300 to 600 persons per km² in the Tikal and Rio Bec regions, because the cultivation of the drained-field areas

Fig. 2.3
Map of the Maya lowlands showing drained field zones detected by radar in relation to known Maya cities of the southern and intermediate areas (*Source*: Adams *et al.*, 1981)

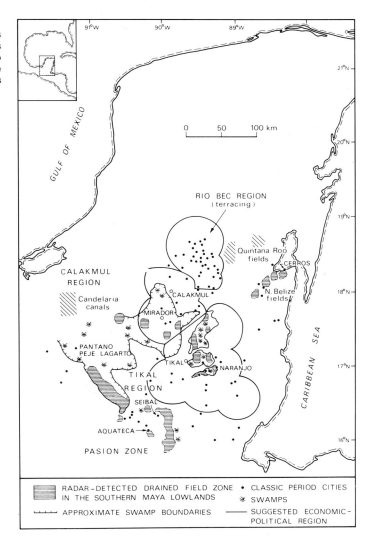

45

alone would support a population of 1.25 million. It is logical to assume that in Late Classic time (AD 600–900) the drained-field zones were in maximum use to support peak populations. The radar data also give new perspectives on Maya civilization. One important contribution is that locational analysis of Maya settlements becomes simpler. Swamps in the area are believed to be assets rather than wasteland and may have been the most productive land available, as evidence by the location of large urban centres on the edges of large swamps. In addition, these canals served the dual purpose of water control and transportation so that bulk commodities could easily be transported, thus enabling a larger population than usual to be supported at a site. Such an intensive and sophisticated system of agriculture is also very vulnerable to collapse and this might explain the decline of the Maya civilization in AD 900. The poor spatial resolution of the radar data restricts the accuracy of the estimate that one can make of the areal extent of the swamps and the intricate canal system which in turn affects the estimation of the population density in the area. But the usefulness of radar imagery in resolving archaeological problems in climatically difficult environment is confirmed.

Present population distribution and mathematical modelling

The present population distribution pattern can be inferred from the settlement pattern in macro-scale revealed by imagery acquired from space platforms. The Gemini V photograph of a portion of the Nile Delta in Egypt has been employed by Tobler (1969) to study the relationship between population size and land area (Pl. 3, page 36). The individual settlements of varying sizes on the delta are very distinctly identified and the regularity of spacing of the smaller settlements in relation to larger ones is noteworthy. This seems to suggest the applicability of Walter Christaller's 'Central Place Theory' in this area, for the hierarchical arrangement of settlements in the landscape. The Central Place Theory attempts to explain systematically the size, frequency and spacing of settlements from the economic point of view (Alao et al. 1977). It postulates that settlements with the lowest level of specialization would be equally spaced and surrounded by hexagonally shaped hinterlands. For every six of these settlements there would be a larger, more specialized settlement, which in turn would be spaced equally from other settlements with the same level of specialization as itself. It is thought that each settlement at the next higher level of specialization will serve three times the area and population of that at the immediate lower level. This kind of arrangement of settlements is called the 'marketing principle' or a k = 3 hierarchy, implying that the number of settlements at successively less specialized levels in the hierarchy increases geometrically in the order of 1, 3, 9, 27 and so on. This model provides the basis for the prediction of population in the settlements. Thus, if one can measure the built-up areas of the settlements in the hierarchy to provide the basis for ranking, a knowledge of the distances of separation of these settlements will permit prediction to be made of the population size of individual settlements (Holz et al. 1973).

In the case of the Nile Delta area, Tobler (1969) further observed that the built-up area of a settlement should be proportional to the population according to the following equation:

$$r = aP^b \qquad\qquad [2.1]$$

where r is the radius of a circle of the same area of the settlement, P is the population, a is a coefficient and b is an exponent. This formula is based on the law of allometric growth put forth by Julian S. Huxley in 1932 to describe the relationship of the growth of an organ or part of an organ to that of the whole organism (Huxley 1932; Reeve and Huxley 1945; Nordbeck, 1965). It was found that the settlements in the Nile Delta area exhibited much higher a and b values than those for settlements in the USA and Europe. By dividing all the radii by four in the Nile Delta, a proportional circle map of population distribution (based on 1960 population census data) was drawn by the computer using equation [2.1] and the latitude-longitude positions of settlements in 1:50,000 scale (Fig. 2.4). When compared with the Gemini photograph (Pl. 3), a good agreement between settlement built-up area and population size is seen, thus proving the validity of the theory. It is suggested from this study that the coefficient a provides a useful indication of settlement packing which reflects the characteristics of the individual society in spatial organisation. Clearly, in the case of the Nile Delta, the settlements are 16

Fig. 2.4
Computer plot of towns using radii calculated from the 1960 Census of Egyptian populations and latitudes and longitudes from 1:50,000 topographic maps (cf. Pl 3; *Source*: Tobler 1969)

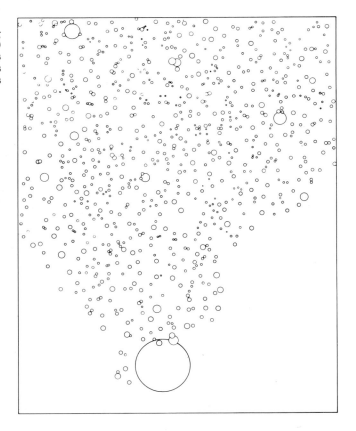

times more compact than settlements in the USA. Obviously, equation [2.1] can be employed to compute the population of a settlement with the input of the built-up area measured directly from imagery or maps. Further discussion of such an application will be found in a later section (pp. 53–57).

Another useful source of remote sensing data for the study of the settlement is the Shuttle Imaging Radar-A (SIR-A) data obtained with a synthetic aperture imaging radar (in L-band frequency or 23 cm wavelength) from the Space Shuttle Columbia in November 1981. The radar image has a ground resolution of about 40 m and a scale of 1:500,000. As radar scattering is affected by the degree of surface roughness, vegetation and man-made structures, particularly settlements are most distinctive, because the walls of buildings form corner reflectors with the surface or because of the abundance of metallic structures or both (Elachi *et al.* 1982; Cimino and Elachi 1982). From Fig. 2.5 which shows a part of the North China Plain to the north of Chinan, the hamlets, villages, towns and cities as well as the road network are clearly delineated (Fig. 2.6). Lo (1984) has made use of these data to test the validity of Walter Christaller's Central Place Theory in China and found that the settlements exhibited a mixed random-clustering pattern, thus revealing a lack of conformity to the Christallerian structure of central places.

The thermal infrared data provide insights into another aspect of spatial population distribution, i.e. the heat energy pattern of the population. The Defense Meteorological Satellite Program (DMSP) of the United States Air Force collects night-time images of the world in the thermal infrared band of 8–13 μm at an altitude of about 830 km in a near polar, sun-synchronous orbit (Croft 1978; Cornillon, 1982). A comparison of a DMSP night-time image of eastern USA (Fig. 2.7) with a population map of the same area (Fig. 2.8) produced by the US Bureau of Census (1973) clearly reveals the close correlation between the two. Welch and Zupko (1980) made use of these

Fig. 2.5
Shuttle Imaging Radar-A data of Dezhou area, China, taken on November 12, 1981. Note that each settlement is displayed as a bright spot (*Courtesy*: Jet Propulsion Laboratory and National Aeronautics and Space Administration)

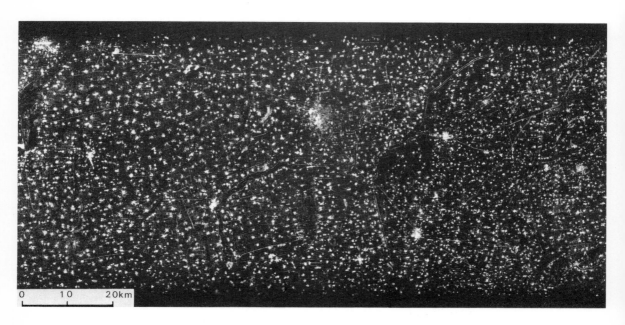

Fig. 2.6
Major communication lines, canals, rivers and settlement in Dezhou area, China (cf. Fig. 2.5)

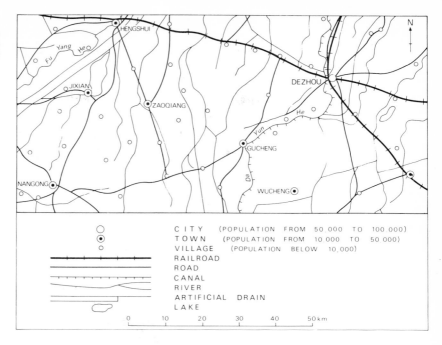

Fig. 2.7
DMSP thermal infrared image of Eastern USA produced from USAF DMSP film transparencies archived for NOAA/NESDIS at the University of Colorado (IRES/National Snow and Ice Data Center)

Fig. 2.8
Night-time population distribution map of eastern USA in 1970 produced by the US Bureau of Census (cf. Fig. 2.7)

DMSP data to study the relationship between population and energy utilization pattern in USAs cities. Because built-up area is proportional to population size, it would be possible to think of the energy distribution pattern of a settlement as a dome in which the x, y plane represents the illuminated urban area (IUA) and the z axis the energy utilization (Fig. 2.9). In the DMSP images, the settlements appear as bright spots of varying sizes and shapes (Fig. 2.7). Welch and Zupko selected 35 cities including 31 urbanized areas with a population of 50,000 or more and 4 with a population of less than 50,000 in the east and west of the DMSP images of the USA on 15 February 1975. Contact duplicate positive film transparencies were used and with the aid of a Bausch and Lomb Zoom Transfer Scope, the IUAs of the selected cities were matched with the built-up area of the cities delineated on a base map of 1:750,000. A systematic correspondence between the two was observed, with the IUA always being slightly larger than the built-up area as depicted on the base map, largely because of the effects of halation and system spread function at the small image scales. The DMSP images were then placed in a Joyce Loebl Mark III CS microdensitometer and a series of image density profiles were generated in the x and y directions at 0.4 mm intervals across each IUA. The profiles were plotted at the same 1:750,000 scale and superimposed on the IUA's xy grid. The density of each grid

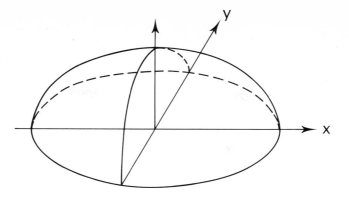

Fig. 2.9
The Illuminated Urban Area (IUA) as a three-dimensional domed figure in which the z-coordinates are proportional to energy consumption (*Source*: Welch 1980)

intersection was determined by interpolation. These IUA density patterns of individual cities can be plotted by a computer as three-dimensional perspective drawings (Fig. 2.10), The mean volumes of these IUA domes were computed using the x and y profiles with the aid of Simpson's Rule:

$$V = h/3 \; (A_0 + 4A_1 + 2A_2 + 4A_3 + 2A_4 + \ldots + 2A_{n-2} + 4A_{n-1} + A_n) \qquad [2.2]$$

where V is the mean volume in mm³, h is the distance between microdensitometer profiles at the map scale in mm, and A is the area under each microdensitometer profile above the midpoints of the profile edges in mm². The mean volumes of IUA domes of individual cities were compared with the population and energy consumption (kwh) data, and strong correlations were observed between energy consumption and volume as well as between energy consumption and population (Fig. 2.11). A regression line in the form of $y = ax^b$ could be generally fitted to each of the relationships, where y is the dependent variable such as population or energy consumption in kwh, a is the coefficient (the y intercept), x is the independent variable such as energy consumption or volumes of the IUA dome in mm³, and b is the exponent (the slope of regression line). This equation is similar in form to the allometric growth formula mentioned before (equation

Fig. 2.10
Three-dimensional perspectives of the energy consumption patterns represented by the IUA domes of four cities in eastern USA (*Source*: Welch 1980)

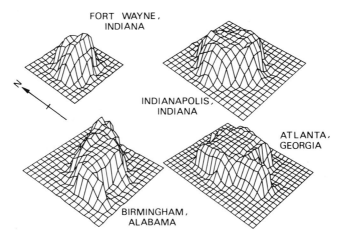

Fig. 2.11
Regression equations between population and IUA dome volume (a) and energy utilization and IUA dome volume (b) for the urbanized areas recorded on the DMSP images of the east and west study areas (*Source*: Welch and Zupko 1980)

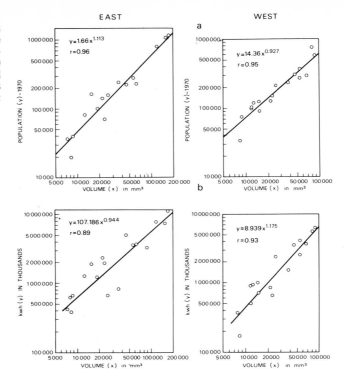

[2.1]. It represents a way whereby urban energy consumption pattern can be quantitatively measured from remotely sensed imagery.

Population estimation

Although many countries in the world already have regular censuses, the accuracy of these census data varies from country to country. In fact, aerial photography has been used in checking the reliability of the population census data in Jamaica (Eyre, Aldophus and Amiel 1970) and the USA (Clayton and Estes 1980). In both cases, the census data were found to be deficient. In Jamaica it was observed that there were three types of errors in the census: (1) omission of existent population; (2) inclusion of non-existent population; (3) inappropriate definition of urban population. In the USA, checking with high-altitude colour infrared photography pinpointed an overcount error of the residential structures in the Federal Census statistics. Apart from the problem of accuracy, population changes so quickly that many of the census data become out-of-date as soon as they are published. Therefore there is a need to regularly update the census data so that accurate population statistics are available for use in planning the economic development of a country. Already a great interest has been shown by research workers in both developed and developing countries in using remote sensing data for the purpose of population estimation and the method has been refined to such an extent that it becomes practically viable at different levels of sophistication. One should note here that population estimation from remote sensing data in no way replaces censuses conducted at regular ten-year intervals by countries. The value of the method is its low cost and efficiency in providing accurate and up-to-date estimates of population for the planners at any point in time.

There are different approaches in the use of remote sensing data for population estimation according to the type of population to be dealt with and the technological level available. The choice also depends on whether a knowledge of the distribution of the population by districts is required. In general, four types of approaches are identifiable: (1) estimation based on measured land areas; (2) estimation based on counts of dwelling units; (3) estimation based on measured land use areas; (4) estimation based on spectral radiance characteristics by individual pixels. Each approach has its merits and demerits as detailed below.

Estimation based on measured land areas

As observed in previous sections, the allometric growth model in the form of $A = aP^b$ can easily be applied in reverse to compute an estimate of the population (P) provided the value of A (the built-up area of a settlement) is known. The accuracy of the estimated population figure obviously depends on the validity of the mathematical model employed. According to Nordbeck (1965), the model developed for cities of the USA was $A = 0.00151\ P^{0.88}$ where A is the area measured in square miles. Wellar (1969) measured the areas of ten settlements in the area of Houston and San Antonio, Texas from Gemini XII photographs and input them individually in the model. He discovered that the estimated populations of several cities differed considerably from the actual population. It appeared that smaller settlements with a population below 10,000 were very closely approximated by the estimation population. For larger settlements, under- and over-estimation of the actual population were observed. However, it was not clear whether the discrepancy was caused by the built-up area measurements or by the inapplicability of the allometric growth model.

In applying the allometric growth model to estimate the population of Chinese cities, Lo and Welch (1977) modified the equation [2.1] to

$$P = aA^b \qquad\qquad [2.3]$$

$$\text{or } \log P = \log a + b \log A \qquad\qquad [2.4]$$

where P is the estimated population and A is the built-up area of the settlement. In addition, a linear regression in the form of

$$P = a + b A \qquad\qquad [2.5]$$

was introduced to explain the more rapid growth of larger cities. Based on 1953 census data and the built-up areas of 124 cities as measured from maps at a scale of 1:250,000, produced in 1951–56, equation [2.4] was found to be:

$$\log P = 4.8733 + 0.7246 \log A \qquad\qquad [2.6]$$

$$\text{or } P = 74696\ A^{0.7246} \qquad\qquad [2.7]$$

The correlation coefficient for the model was found to be $+0.75$ (Fig. 2.12). This model was employed to estimate the population of 13 cities in 1972–74 based on area input from Landsat images. Both band 5 and band 7 of the Landsat 70 mm positive film transparencies of these cities were placed in an I^2S 6040 PT additive viewer and projected onto the screen of the viewer as colour composite at a scale of 1:500,000. The boundaries of these cities were delineated on plastic overlays registered to the viewing screen and traced with a planimeter to obtain area measurements. These cities had populations within the range of 500,000–3,000,000. The outlines of individual cities are particularly well defined in band 7, the infrared band (Fig. 2.13).

As equation [2.6] was derived from 124 cities with population ranging from 25,000 to over 2,000,000 its application to the Landsat area data resulted in over- and under-estimation. It was particularly noteworthy that for extremely large cities with populations in excess of 2,500,000 the allometric growth model tended to underestimate their populations. To refine this approach in the case of China, cities were grouped together according to regions and a separate allometric growth model was computed for each region by means of the 1951–56 data. In this way, the number of sampled cities employed for computation became smaller and the range of populations involved was less great. Thus, the models for Northeast and Central China were respectively $\log P = 4.8611 + 0.6312 \log A$ (or $P = 72627 \, A^{0.6312}$) and $\log P = 4.9087 + 0.8071 \log A$ (or $P = 81040 \, A^{0.8071}$). All these models improved the accuracy of the population estimate of the cities in their proper regions. As for the excessively large cities with populations over 2,500,000, the linear growth model represented by equa-

Fig. 2.12
A regression equation between population and land area for Chinese cities based on 1951–56 data (*Source*: Lo and Welch 1977)

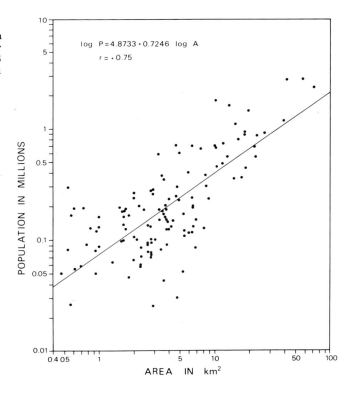

$\log P = 4.8733 + 0.7246 \log A$

$r = +0.75$

POPULATION IN MILLIONS

AREA IN km²

Fig. 2.13
Enlarged Landsat image (band 7) of the city of Shenyang taken on 21 November 1972. Note airfields to the north, south, east and west of the city. Also note that railway lines are clearly displayed. Freshly ploughed fields are displayed in dark tone in this infrared band (0.8–1.1 μm)

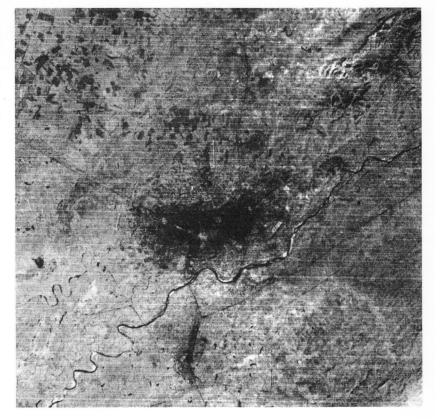

tion [2.5] was found to be more appropriate probably because of their faster rates of growth. The models computed for Northeast and Central China were respectively P = 21989 + 30452A and P = 14915 + 57119A. Based on the 1970 population statistics and Landsat area measurements, one could arrive at some new figures for predicting population in Chinese cities:

$$\log P = 5.3304 + 0.4137 \log A \qquad [2.8]$$

$$\text{or } P = 213993 \, A^{0.4137} \qquad [2.9]$$

which should be the more accurate model to use if Landsat data are to be used.

It is obviously possible to formulate a more complex regression model to explain the relationship between population and land area. This has been done, notably by Holz *et al.* (1973) and Ogrosky (1975). The theoretical basis is Christaller's Central Place Theory mentioned before. Holz *et al.* (1973) developed the following model:

$$P_i = a + b_1L_i + b_2P_j - b_3D_{ij} + b_4A_i \qquad [2.10]$$

where P_i is the population of urban area i,

L_i is the number of direct lines L between i and the other urban area,

P_j is the population of the nearest larger urban area j,

D_{ij} is the highway distance between urban area i and the nearest larger urban area j, and

A_i is the observable occupied dwelling area of urban area i.

This model was applied to estimate the population of 40 urban centres in the Tennessee Valley in 1953 and 1963. The population of these centres varied from 2,500 to 20,000. Large-scale aerial photography was used to extract data for variables L_i and A_i. By using a stepwise linear regression the two models obtained for 1953 and 1963 were respectively:

$$P_i \ (1953) = -1711.896 + 174.997 \ L_i + 0.027 \ P_j - 8.650 \ D_{ij} + 0.142 \ A_i \qquad [2.11]$$

and $P_i \ (1963) = -4628.394 + 484.192 \ L_i + 0.016 P_j + 122.846 \ D_{ij} + 0.079 \ A_i \qquad [2.12]$

They gave a multiple correlation coefficient for 1953 and 1963 of 0.95 and 0.88 respectively, thus indicating that over 91 per cent in 1953 and over 77 per cent in 1963 of the variations in population of urban areas could be explained. Therefore, the 1953 equation resulted in a better estimation of population than the 1963 equation. They also noted that the smaller urban areas exhibited less variation between population and land area than that for the larger urban areas, thus suggesting that for larger urban centres more variables other than land areas should be added.

On this basis, Ogrosky (1975) attempted an improvement of the regression model by refining the variables in equation [2.10]. For A_i, the urban area variable, allowance was made for major parks, water bodies and undeveloped military reservations in urban areas, and for L_i, surface links, arbitrary values of 1, 2 and 4 were assigned to railroads, secondary roads and multiple-lane highways respectively to differentiate the importance of each link. The variable P_j, the population of the nearest larger urban area j, was replaced by a new variable A_j or the land area of the nearest larger urban area j. Finally, the variable of D_{ij}, the highway distance between urban area i and the nearest larger urban area j, was unchanged but was measured directly from the imagery, with ferry routes also included where appropriate. High-altitude colour infrared aerial photography at a scale of 1:135,000 was employed to test the applicability of the regression model to 18 urban areas with populations between 11,000 to over 500,000 in the Puget Sound Region of the USA, centred around Seattle. By applying a stepwise regression procedure, the following model was obtained:

$$\log P_i = 0.7724 + 0.0106 \ L_i + 0.0004 \ A_j + 0.0143 \log D_{ij} + 0.6896 \log A_i \qquad [2.13]$$

It is noteworthy that a logarithmic transformation of the variables was used. The resultant correlation coefficient was +0.986 or a coefficent of determination of +0.973, thus representing an improvement over the work of Holz and others (1973).

The approach illustrated above is only suited to estimate the overall population of a settlement and appears to be more suited for use with small-scale aerial photography or even imagery acquired from space platforms such as Landsat. The measurement of settlement area is greatly facilitated by the availability of electronic digitizers and planimeters. The use of image inhancement techniques to emphasize the settlements for automatic measurement by a computer-controlled unit will speed up the process, but the difficulty of delineating accurately the settlement's built-up area has given rise to errors in the estimates (Anderson and Anderson 1973). This approach is most interesting in the sense that one can formulate and test new mathematical models with a view to arrive at a more reliable estimation of population from remote sensing data.

Estimation based on counts of dwelling units

This approach is more suitable when large scale aerial photography is available because it requires that individual dwelling types be identified and counted. If we know the number of persons normally resident in each dwelling unit, then multiplying this factor by the total number of dwelling units counted in a region will give us an estimation of the total population. The factor of resident density or number of persons per dwelling unit is obtained from census statistics if available or from sample surveys on ground. A classic example of the application of this method of population estimation is Hsu's work on Atlanta, Georgia, USA (1971). He made use of aerial photography at the scale of 1:5,000 to estimate and map the 1963 population distribution of the Atlanta area. A transparent grid covering a photographic area of 1/4 square mile size was constructed as the base for obtaining housing counts comparable to those on a 1:24,000 scale topographic map grid cell. By moving this transparent grid on the photograph and registering it with the corresponding map grid, the number of dwelling units were counted and recorded. Since it was assumed that a dwelling unit was occupied by one household, the number of persons per household which could be obtained from census statistics was employed in computing a population density for each grid cell as follows:

$$\frac{\text{Population}}{\text{density}} = \frac{(\text{persons per household}) \times (\text{number of dwelling units})}{\text{grid cell area}} \quad [2.14]$$

In this way, a choropleth map of population density of the Atlanta area in 1968 was obtained. The accuracy of the estimation depended heavily on the skill of the photo-interpreter in identifying and counting the dwelling units. It was found in a field check that where evergreen vegetation and dark-coloured roofs and driveways occurred together, dwelling units were missed out in the count. There were also some difficulties in distinguishing high-rise apartment buildings from multistorey office buildings. But in most residential areas, the random error was less than 3 per cent.

Apart from the practical difficulty of identifying and counting the dwelling units correctly, one should also note that the assumption of one single household occupying one dwelling unit may not be valid in cities of many developing countries. It is not uncommon to have two or even three households sharing the same dwelling units, and it is not surprising to find this method usually gives rise to underestimation of population, especially if it is applied to the urban area (Clayton and Estes 1980). Even in a rural area such as the New Territories of Hong Kong, the Chinese villagers traditionally live in extended families, and a village house may contain more than one household. To avoid this difficulty in an application carried out by Lo and Chan (1980), the household concept was abandoned and the number of persons actually living in the village house was determined in the field. This application also required initially classifying the rural dwelling units in the study area into four types, namely, (a) traditional village houses, (b) temporary shacks or huts, (c) new village houses type A and (d) new village house type B. It was noted that the residential density varied from 4.4 persons per dwelling unit of the new village house type to 3.1 persons per dwelling unit of the temporary shack type. The result of this application revealed again an underestimation of the rural population in the study area with an overall accuracy of about 6 per cent.

Another application to estimate the rural population made by Allan and Alemayehu (1975) employed an ingenious method of sampling to combine with the counting of dwelling units from the aerial photographs at a scale of 1:20,000. The study area was the Awurajah (sub-provincial district) of Wolamo which lies 400 km south of Addis Ababa in Ethiopia and is a densely populated rural area. The block of photography comprised 73 overlapping vertical photographs and covered an area of 199.7 km². No population data existed for the area, and one had to devise a sample scheme to obtain population density data. A transact of 1 km wide by 13 km long was first run in the field to obtain some preliminary ideas of the population density and distri-

Fig. 2.14
Location of plot samples and transect line, Wolamo, Ethiopia (*Source*: Allan and Alemayehu 1978)

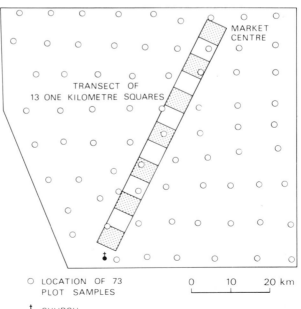

Particulars	One-kilometre-square plots in the transect													Totals
	1	2	3	4	5	6	7	8	9	10	11	12	13	
per cent arable	83	68	53	23	34	53	82	79	80	63	58	59	74	62
per cent pasture	0	21	37	77	64	47	18	5	11	36	22	27	18	29
estimated households	77	64	61	33	27	61	101	98	47	43	91	92	98	893
estimated population (households ×5)	385	320	305	165	135	305	505	490	235	215	455	460	490	4465

Table 2.1
Population and land use in the transect aligned south-west to north-east in the Wolamo area in Ethiopia
(*Source*: Allan and Alemayehu 1975)

bution (Fig. 2.14). The transect was deliberately aligned to include areas of both high and low population densities. Along the transect the number of households were counted and the area of arable and pasture land was also measured. A household size of 5 persons was arrived at, giving an estimated population of 4,465 (Table 2.1). A close relationship was found to exist between arable land and population density in the area (Fig. 2.15). The transect study also revealed a great variation of population density in the area from 135 to 490 persons per square kilometre.

After this transect, population estimation was to be applied to the area covered by the whole block of aerial photographs. The method of plot sampling normally employed by vegetation scientists was adopted in conjunction with the aerial photography. The principal point of the photograph was used to locate the centre of the sample plot because this is geometrically the least distorted part of the photograph. In this way, a non-aligned systematic sample was obtained without too much trouble (Fig. 2.14). The sample plot is circular in shape, which has the

Fig. 2.15
Relationship between proportion of arable land and population density as revealed by Table 2.1 (*Source*: Allan and Alemayehu 1975)

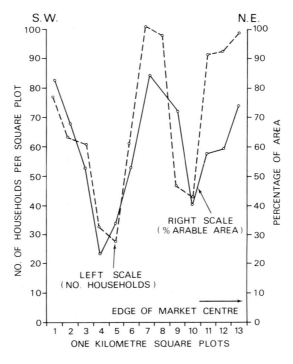

advantage of minimizing the edge-length of the plot, thus reducing the problem of deciding whether dwellings are inside or outside of the plots. The circular plot was initially drawn with a diameter of 1 cm which was later doubled to 2 cm, 4 cm, 8 cm, etc. progressively. The aerial photographs were interpreted with the households identified and counted. For an estimate of households with a sample error of 10 and 15 per cent, at the 0.05 significance level, a sample plot with a diameter of over 2 cm was needed. It was noteworthy that the estimates for the total population declined as the diameter of the plot increased. The edge decision problem would be more significant in the small sized plots than the large ones. This error was found to result in a 10 per cent excess in the count estimate each time as the sample size was halved. Thus, by progressive doubling of the sample size according to such a 10 per cent increment to the largest sample size possible, a final estimate of population of 47,700 for the study area was arrived at, giving an overall population density of 250 persons per square kilometre. Unfortunately, no confirmation of the accuracy of the final result could be made because of the inavailability of census data. However, this method is most interesting and demonstrates clearly the usefulness of statistical sampling in aiding population estimation from aerial photographs for a developing rural environment.

Clearly, the approach of counting dwelling units would face more problems in an urban environment where high-rise residential buildings abound. Lindgren (1971) who applied this method to the metropolitan area of Boston USA recommended the use of colour infrared photographs at a medium scale (say, 1:20,000) to improve the results of the dwelling units estimation. By making use of such criteria as roof type, relative size of structure, number of storeys and division of buildings, availability of parking and amount and quality of vegetation, the estimates of dwelling units could be easily made from the colour infrared photography because of the sharp contrast between buildings and vegetation as well as between built-up and non-built-up areas. A correct identification of 99.5 per cent of the residential structures was achieved. However, the study area happened to comprise primarily of multi-family structures as opposed to single-family ones. This caused the estimates of dwelling units to be less accurate than the estimates of residential structures. It was also pointed out that a familiarity with the study area could significantly improve the accuracy of the results.

Finally, it is also possible to replace the dwelling unit count with a dwelling unit length measurement such as has been practised by Collins and El-Beik (1971). In estimating the population in a residential area in the city of Leeds in England, they measured from the aerial photographs ('split-vertical' type with a scale of 1:10,000) the strip lengths of back-to-back and terraced houses to the nearest millimetre but counted the semi-detached houses. Where the terraced houses were three storeys high, the measured length was multiplied by a factor of 1.5. Population figures were then related to these measurements or counts to give population density figures. However, it appears that length measurement is only practicable in the case of a homogeneous terraced or back-to-back residential housing area. There is also the added difficulty of scale distortion caused by relief displacement and tilts of aircraft. It is therefore more common to rely on dwelling unit counts.

Estimation based on measured land use areas

This approach dispenses with the need for tedious and the time-consuming counting of individual dwelling units, thus eliminating the reliance on large-scale aerial photography. In essence, it involves interpreting and measuring the land use of an area with special attention given to the residential use which is further subdivided into distinctive classes according to the cultural characteristics of the study area. The population density associated with each residential land use category is obtained by sample survey in the field or from existing census data. The population (P) of the area can then be determined by the following equation:

$$P = \sum_{i=1}^{n} (A_i D_i) \qquad [2.15]$$

where A_1, A_2, ... A_n are the areas devoted to residential land use categories 1 to n, and D_1, D_2, ... D_n are the corresponding population densities associated with residential land use categories 1 to n. A typical example of such an approach is the work by Kraus, Senger and Ryerson (1974). They made use of 70 mm black-and-white panchromatic photography at an approximate scale of 1:600,000 and 23 × 23 cm format colour infrared photography at approximate scales of 1:120,000 and 1:60,000 all taken on 4 April 1971 to estimate the population of four cities in California (Fresno, Bakersfield, Santa Barbara and Salinas) which represented a range of sizes and cultural or environmental situations. They enlarged the 70 mm photography to a scale of 1:40,000 and mapped the land use of these four cities according to the following scheme: (a) single family residence; (b) multi-family residence; (c) trailer park residence; (d) commercial or industrial uses. The area devoted to each type of land use was measured with the aid of a polar planimeter. The population densities associated with the three residential land use categories in (a), (b) and (c) were determined with the aid of 1970 US Census Block Data. Areas of each residential land use category were identified on the Census Block data maps and random samples of blocks within each residential land use category were then obtained to determine the population density for that land use in each city.

The results of the land use maps were found to be generally quite accurate except for small, isolated apartment units and older homes which tended to be obscured or misclassified as single family residences. The population estimates obtained for the four cities in this way exhibited an error ranging from an underestimate of 9.17 per cent in Fresno to an overestimate of 7.00 per cent in Santa Barbara. It was noteworthy that the population was underestimated in three of the four cities, probably because of the difficulty in identifying residences in the older central business districts. On the other hand, the overestimation in Santa Barbara was due to great variations in single family residential lot sizes, causing an overestimation of population density for single family residential land use. But on the whole the mean error for all four cities combined was an under−estimation of 4.51 per cent which compared favourably with works by Hsu (1971) and Collins and El-Beïk (1971) discussed before. One major criticism of the work by Kraus *et al.* was the use of small scale 70 mm photography which was blown up to an excessive extent (15 times). There appeared to be no check on the accuracy of the area measurement because relief displacements and tilts of the aircraft could have caused significant scale distortions. But on the whole, this approach is much more efficient than the house count approach. In an area where dwelling units are extremely mixed up, this approach of population estimation is particularly valuable.

In developing countries where illegal squatter structures usually occur in the rural-urban fringes of the cities, an estimation of the squatter population can be most efficiently obtained from aerial photographs with this method. A knowledge of the squatter population figure and their spatial distribution is indispensable for the development of public housing in these countries. In a study of squatter settlements in the Kai Tak district of Hong Kong, Lo (1979) made use of large scale (1:10,000) true colour and colour infrared photography to estimate the squatter population using the equation [2.15]. To eliminate distortion caused by relief displacements and tilts of aircraft, the Bausch and Lomb Zoom Transferscope was used to transfer the delineated squatter area from aerial photographs to a topographic map at the scale of 1:10,000. With the superimposition of a dot grid, the areal extent of the squatter settlement can be measured within an accuracy of ±5 per cent. The estimated squatter population was found to be accurate within ±2 per cent.

This approach of population estimation has been improved upon by Thompson (1975) who advocated the use of a base population count obtained from the census. Any residential land use changes (new addition and demolition) since the base year is then mapped from the current aerial photographs. The population density for the residential land use is obtained. From the measured area of residential land use changes, an estimation of population change since the base year can be obtained. By adding to or subtracting from the base year population, an accurate estimation of population for the current year can be obtained. In this approach, it is necessary to have aerial photographs or land use maps for the base year with which comparison with the current aerial photographs or land use maps can be made. He demonstrated his approach with the use of high-altitude colour infrared photography at the scale of 1:100,000 obtained with a Wild RC-8 camera for 573 urban census tracts in the Washington, DC area of the USA for June 1970 and July 1972. The US Geological Survey (USGS) land use map was employed, which classified urban land use into 11 categories, namely, industrial, transportation, commercial and services, strip and cluster development, single-family residential, multi-family residential, improved open space, unimproved open space, water, agriculture with residence, and wetland. Because of the 4 ha minimum area for mapping adopted by the USGS, small residential units may have been omitted. Also, because these land use data do not indicate residential multistoreyed buildings found in the inner city area, the area measurements obtained from these maps will under-estimate the amount of residential usage. As great difficulties arose in differentiating between 'single-family residential use' and 'multi-family residential use' in terms of their characteristic population densities, a single residential land use class was employed as the basis for population estimation while other land uses containing population were ignored. The residential land use added to each census tract since 1970 was interpreted from the aerial photographs for 1972. The population density for the residential land use was obtained from the 1970 census data. The product of the net increase in residential land in the area since 1970 and the population density for that land gave an estimate of additional household population in the two year time period. Addition of this figure to 1970 household population enumerated by the census gave the estimate of population for 1972 with a 5 per cent underestimation. Thompson (1975) indicated certain limitations of his method, which included the underestimation of multi–family residential land use in both inner and outer urban areas; inaccurate population density value produced for the residential land; the exclusion of other non-residential population, containing urban fringe populations associated with agricultural land uses and inner city populations associated with commercial or services uses; the lack of correspondence in time between the population census data and the aerial 'photography employed. He concluded that this land use method was less accurate than the dwelling unit count method, but it provided a simple way to give a fairly good estimate of population for the whole urban area from high-altitude photography. In addition, the method was best applied to a large area below the county level, to suburban locations rather than inner cities, to rapidly changing rather than slowly changing or unchanging environments, to areas with little multi-family development, to homogeneous rather than highly mixed land use, to predominently single-storeyed rather than multistoreyed housing developments with a relatively long, about four or five year, time span.

Estimation based on spectral radiance characteristics by individual pixels

With the increasing availability of multispectral data from satellites such as Landsat, population estimation can be carried out with a higher degree of automation. Unfortunately, the rather low spatial resolution (250 m) of the MSS data for the first three Landsats tended to restrict the accuracy of the result. Hsu (1973) suggested the idea of the use of selected Landsat 1 multispectral radiance data cell by cell (1×1 km) in combination with corresponding ground truth data and low altitude aerial photography to develop a multiple regression model to predict the population of a study area. He believed that the radiance data would reflect the roofs of houses differently from their surroundings. Unfortunately, his idea was not pursued further until ten years later when Iisaka and Hegedus (1982) presented a more rigorous account of this method in an application to estimate the population distribution in Tokyo, Japan. It was observed that the radiance in the four spectral bands of the Landsat MSS data depended on the ground covering materials or the land use of the area. An area with low population density was assumed to have more natural components i.e. vegetation cover or bare soil, than another area of high population density where artificial structures are dominant. By using a grid size of 500×500 m area, the relationship between population and spectral radiance characteristic was expressed as follows:

$$P = Ax_1 + Bx_2 + Cx_3 + Dx_4 + E \qquad [2.16]$$

where P is the population in a 500 × 500 m area, and x_1, x_2, x_3, x_4 are the spectral reflectance values of the four bands of the Landsat MSS data. To calibrate the model, reference data of population distribution in the Tokyo ward area for 1970 and 1975 were obtained to compute the population density in all the 500 × 500 m grids. Landsat 1 and Landsat 3 MSS CCT data for the Kanto area in Japan acquired on 26 November 1972 and 24 January 1979 respectively were obtained. A sample of 88 grid cells representing a wide range of population density values was selected from the reference data. No samples were selected in the central business district (CBD) of Tokyo where zoning, building systems and materials were rather different from those of suburban areas. There was no correlation between building density and population density in this CBD. The two sets of Landsat CCT data were geometrically corrected and matched to a common scale. After resampling, the 500 × 500 m grid cell corresponded to 10 pixels of the MSS data. Statistical means of the radiance values (in digital counts) for each of four spectral bands for each grid area were computed. It was found that the data of bands 6 and 7 showed a strong linear regression correlation with the population density (Fig. 2.16).

A cluster analysis was then carried out for all the sampled grid cells in the study area with the mean reflectance values for each band as inputs. Four and three clusters of grid cells were identified respectively for the 1972 and 1979 Landsat MSS data, each of the clusters being associated with one typical population density. When these clusters of cells were mapped separately for 1972 and 1979, it became clear that the sample grid cells in cluster 4 in 1972, which were all located in the suburban residential area of Tokyo, moved one cluster higher because rapid housing development had taken place. On the other hand, grid cells located near the CBD seldom changed their cluster classes. After these preliminary investigations, the spectral and population data were combined together for stepwise linear regression analysis. It was found that only three Landsat MSS spectral bands (bands 4, 6 and 7) were strongly correlated with population. The specific regression equations for the 1972 and 1979 data were respectively:

$$P_{(1972)} = 2.196x_1 - 7.648x_3 - 3.167x_4 + 215.857 \qquad [2.17]$$

$$P_{(1979)} = 7.971x_1 - 13.681x_3 - 0.297x_4 + 54.819 \qquad [2.18]$$

In order to verify these two regression models, they were applied to compute the population in the study area first with a pixel size resolution and then with 500 × 500 m grid cells. Comparison of the estimated population was then made cell by cell with that obtained by the population census. The differences between the estimated values and the original census data of population densities (ΔP) were plotted against the original census population densities for the 88 sampled cells for 1972 and 1979 (Fig. 2.17). It was seen that most of the grid cells were within the zone of bounds of the standard error of estimation, i.e. quite close to zero error. Those grid cells which showed a great divergence from the original value were found to have non-residential land uses such as schools, railway stations, churches, etc. which were misclassified as residential in nature. If these cells were excluded, the multiple regression coefficients for the 1972 and 1979

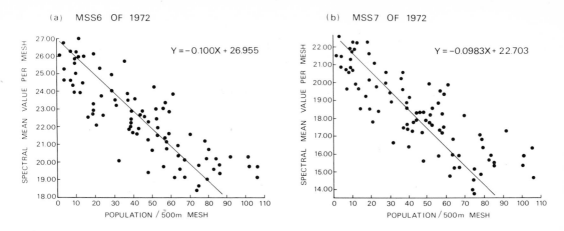

$Y = -0.100X + 26.955$

$Y = -0.0983X + 22.703$

Fig. 2.16
Regression equations between population and spectral mean values of (a) MSS 6 and (b) MSS 7 Landsat data (*Source*: Ilisaka and Hegedus 1982)

data would be increased from 0.836 to 0.939 and from 0.770 to 0.899 respectively. Clearly, the major weakness of this method is the presence of the non-residential structures which have no direct relationship with population density. But this method does permit population density maps or population distribution maps of a region to be displayed by the computer with ease and is useful for geographic information system development. Herein lies the potentially fruitful research area for applied remote sensing.

Inference of socio-economic characteristics of the human population

Fig. 2.17
Verification of the model by comparing the differences between the estimated values and the original census data of population densities (ΔP) with the original census population densities for the 88 sampled cells for (a) 1972 and (b) 1979 (*Source*: Iisaka and Hegedus 1982)

Another type of demographic data which are useful to planners are socio-economic characteristics of the population. These, however, can only be inferred through indirect evidence of housing types and environmental characteristics from aerial photography. In a pioneering paper by Green (1957), a sociologist, it was argued that human groups occupied physical space and facilities as a result of their interaction and adjustment to the environment. Thus, the directly observable physical data from the aerial photographs have meaningful socio-

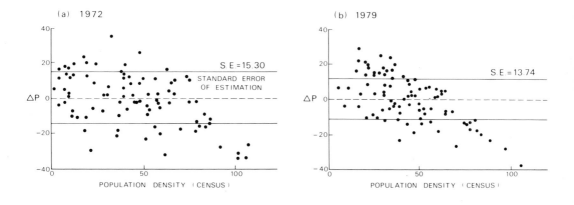

logical correlations. In this way, a method of correlative photo—interpretation or 'inventory-by-surrogate' can be developed. Green's work is a good example of this approach. He was interested in studying the social structure of the city of Birmingham, Alabama in the USA. He extracted the following types of data from aerial photographs on each residential sub-area of the city: (1) the location of residential areas relative to three concentric zones with their midpoint in the Central Business District (CBD); (2) the description of the residential areas in terms of the internal and adjacent land usage from which an ordinal scale of 'residential desirability' (namely, favourable, neutral, unfavourable) was established; (3) the prevalence of single-family homes, i.e. a count of the single-unit, detached type dwellings, thus yielding second ordinal scale of high, medium and low occurrence; (4) the density of housing in average numbers of dwellings per block, giving rise to a third ordinal scale: high, medium and low density.

Ground checks were then conducted to determine the accuracy of these data extracted from the aerial photographs. It was revealed that: (a) up to 99 per cent of the residential structures were correctly identified; (b) the total number of individual dwelling units in all types of structures was underestimated by 7 per cent; (c) the overall average density of dwelling units per block was underestimated by 1.7 per cent; (d) the detached, single-dwelling units were overestimated by 5.3 per cent. These results were regarded to be sufficiently accurate. The residential sub-areas in the city were ranked according to the two indices: (a) the prevalence of single dwelling units; (b) the density of dwelling units per block, first with the photo data and then with the ground data. These two sets of ranking were statistically compared with Spearman's Rank Correlation. Near perfect correlations of 0.98 and 0.99 respectively were obtained, thus proving that the photo data were as valid as the ground data.

The sociological characteristics of these residential sub-areas from other statistical sources were studied to see whether they were related to the photo data. It was confirmed that significant differences between zones in the city existed in educational status, average monthly rentals and adult crime rates. Another observation was the consistency in correlation between the prevalence of single-family houses and socio-economic status of the urban sub-areas in terms of occupational status, education, income, rental values and ethnic composition.

Finally, these social and physical data were combined by means of the Guttman Scaling Technique to form a single continuum defining the exact position or rank of each sub-area in relation to every other sub-area in the sample. A scale of 'residential desirability' of USAs cities was constructed, based on the four photo-data categories. Another scale – the 'socio-economic status scale' – could also be constructed using the five social data items of: (a) median annual income; (b) prevalence of within-dwelling crowding; (c) prevalence of home ownership; (d) prevalence of social disorganization; (e) educational achievement. These two scales were found to be strongly correlated. Thus, physical data extracted from aerial photographs can be employed to predict the social structure of the city.

The same approach has been followed by Mumbower and Donoghue (1967) using large-scale (1:10,000) aerial photography to delineate urban poverty areas by correlating it with such environmental features as structural deterioration of the houses, debris, clutter, lack of vegetation, walks and paved streets. They discovered that urban poverty was closely associated with residential areas located adjacent to the CBD, industry and major urban arteries. These were also found to be strongly correlated with low income, unemployment, low education achievement, family crowding, crime, low health status and lack of community facilities. Metivier and McCoy (1971) made use of large-scale (1:6,000) black-and-white aerial photography to map the urban poverty areas in Lexington Kentucky, USA. Their map was found to be about 80 per cent accurate. They also discovered that house density was the most significant criterion for the identification of sub-standard housing on the photographs, which served also as an index of socio-economic conditions. House density which was expressed as the number of houses per ha for each block was strongly and negatively correlated with median family income ($R = -0.875$), with average house value ($R = -0.787$), with average rent ($R = -0.801$) and with high percentage of non-white population ($R = 0.829$) (McCoy and Metivier 1973). An extension of this work was carried out in Albany, New York, USA, using smaller scale aerial photography and measuring average lot size of tracts instead of houses per block to determine density (Henderson and Utano 1975).

In the case of a city in a developing country, where a great influx of population from a rural area has severely strained the housing situation, the urban poverty region lies not only in the old slum area of the city centre with deteriorating structure but also in the flimsy squatter structures on the periphery of the city. Aerial photography can be usefully employed to detect these poverty-stricken regions. The availability of sequential aerial photographs could reveal clearly the pattern of growth or decline of these squatter areas and provide a basis to evaluate the government's land use policy, as demonstrated by Lo (1979) in his study of Hong Kong. The environmental variables, notably those relating to vegetation and housing conditions, can be more conspicuously revealed by colour infrared photography (Marble and Horton, 1969). One should note, however, that much ground survey work is required before accurate inference can be made concerning the socio-economic characteristics of population from aerial photographs.

Conclusions

In this chapter, the application of remote sensing data to the study of past and present population of our earth has been examined. It is interesting to note the widespread use of aerial photography by archaeologists to discover sites of human settlements with the aid of tonal change or shadowing effect. Such an application is largely descriptive in nature, but in recent years non-photographic remote sensing systems such as radar has been used by archaeologists to solve problems in a climatically difficult environment where aerial photography is less suited. As for the study of present population, there is a greater variety of remotely sensed aerospace imagery being used. Along with the conventional aerial photography, thermal infrared line scanner as well as synthetic aperture radar data have been employed. The space platforms have provided data which are suitable for use in testing macro-scale models of population distribution on the earth, such as Christaller's Central Place Theory and the allometric growth model. A great deal of attention has been given to population estimation from remotely sensed imagery acquired from both aircraft and spacecraft. This indicates that an accurate knowledge of the population number at any moment is still lacking in many countries. The four different approaches to population estimation have by now been well tested and represent a whole range of scales of application.

In general, the dwelling unit count approach based on large-scale aerial photographs gives the most accurate result and is best used for selected small areas. On the other hand, the land area approach can give quick estimates of population for entire cities and is more suited for use in conjunction with satellite image data. The land use area approach is most efficient in obtaining population estimates of large areas from medium to small scale aerial photography. The estimation of population based on spectral reflectance at the pixel level using Landsat MSS digital data opens the way for computer-assisted estimation and mapping of population which needs geographic information systems. With the improvement in the spatial and spectral resolution of the Thematic Mapper data acquired by Landsat 4, one will expect better results to emerge from this approach in population estimation. Finally, socio-economic characteristics of the human population can be inferred indirectly from aerial photographs through correlating them with the directly observable physical structures. This approach depends heavily on field work and is characterized by a high degree of generalizations. Much investigation is needed before this approach is operationally useful.

References

Adams, R. E. W. (1980) Swamps, canals, and the locations of ancient Maya cities, *Antiquity* **54**: 206–14.

Adams, R. E. W., Brown, Jr., W. E. and Culbert, T. P. (1981) Radar mapping, archaeology and ancient Maya land use, *Science* **213**: 1457–63.

Alao, N., Dacey, M. F., Davies, O., Denike, K. G., Huff, J., Parr, J. B. and Webber, M. J. (1977) *Christaller Central Place Structures: An Introductory Statement*, Department of Geography, Northwestern University, Evanston, Illinois.

Allan, J. A. and Alemayehu, T. (1975) Rural population estimates from air photographs: an example from Wolamo, Ethiopia, *The ITC–Journal* no.1: 85–100.

Anderson, D. E. and Anderson, P. N. (1973) Population estimates by humans and machines, *Photogrammetric Engineering* **39**: 147–54.

Bradford, J. (1957) *Ancient Landscape: Studies in Field Archaeology*, G. Bell, London.

Cimino, J. B. and Elachi, C. (eds) (1982) *Shuttle Imaging Radar-A (SIR-A) Experiment*, Jet Propulsion Laboratory, California Institute of Technology: Pasadena, California.

Clayton, C. and Estes, J. E. (1980) Image analysis as a check on census enumeration accuracy, *Photogrammetric Engineering and Remote Sensing* **46**: 757–64.

Collins, W. G. and El-Beik, A. H. A. (1971) Population census with the aid of aerial photographs: an experiment in the city of Leeds, *Photogrammetric Record* **7**: 16–26.

Cornillon, P. (1982) *A Guide to Environment Satellite Data*, Graduate School of Oceanography, University of Rhode Island.

Croft, T. A. (1978) Night-time images of the earth from space, *Scientific American* **239**: 69–79.

Elachi, C. *et al.* (1982) Shuttle imaging radar experiment, *Science* **218**: 996–1003.

Eyre, L. A., Adolphus, B. and Amiel, M. (1970) Census analysis and population studies, *Photogrammetric Engineering* **36**: 460–6.

Fredman, R. and Berelson, B. (1974) The human population, *Scientific American* **231**: 31–9.

Green, N. E. (1957) Aerial photographic interpretation and the social structure of the city, *Photogrammetric Engineering* **23**: 89–99.

Hampton, J. N. (1974) An experiment in multispectral air photography for archaeological research, *Photogrammetric Record* **8**: 37–64.

Henderson, F. M. and Utano, J. (1975) Assessing urban socio-economic conditions with convertional air photography, *Photogrammetria* **31**: 81–89.

Holz, R., Huff, D. L. and Mayfield, R. C. (1973) Urban spatial structure based on remote sensing imagery, in Holz, R. K. (ed). *The Surveillant Science: Remote Sensing of the Environment.* Houghton Mifflin: Boston; Atlanta: Dallas; Geneva; Illinois; Hopewell; New Jersey, Palo Alto; pp. 375–80.

Hsu, S. Y. (1971) Population estimation, *Photogrammetric Engineering* **37**: 449–54.

Hsu, S. Y. (1973) Population estimation from ERTS imagery: methodology and evaluation, *Proceedings of the American Society of Photogrammetry 39th Annual Meeting,* pp. 583–91.

Huxley, J. S. (1932) *Problems of Relative Growth*, Methuen, London.

Iisaka, J. and Hegedus, E. (1982) Population estimation from Landsat imagery, *Remote Sensing of Environment* **12**: 259–72.

Kraus, S. P., Senger, L. W. and Ryerson, J. M. (1974) Estimating population from photographically determined residential land use types, *Remote Sensing of Environment* **3**: 35–42.

Lindgren, D. T. (1971) Dwelling unit estimation with color-IR photos, *Photogrammetric Engineering* **37**: 373–8.

Lo, C. P. (1979) Surveys of squatter settlements with sequential aerial photography – a case study in Hong Kong, *Photogrammetria* **35**: 45–63.

Lo, C. P. (1984) Chinese settlement pattern analysis using Shuttle Imaging Radar-A data, *International Journal of Remote Sensing* **5**: 959–67.

Lo, C. P. and Chan, H. F. (1980) Rural population estimation from aerial photographs, *Photogrammetric Engineering and Remote Sensing* **46**: 337–45.

Lo, C. P. and Welch, R. (1977) Chinese urban population estimates, *Annals of the Association of American Geographers* **67**: 246–53.

McCoy, R. M. and Metivier, E. D. (1973) House density vs socio-economic conditions. *Photogrammetric Engineering* **39**: 43–7.

Marble, D. F. and Horton, F. E. (1969) Extraction of urban data from high and low resolution images, *Proceedings of the Sixth International Symposium on Remote Sensing of Environment*, University of Michigan, Ann Arbor, Michigan, pp. 807–17.

Metivier, E. D. and McCoy, R. M. (1971) Mapping urban poverty housing from aerial photographs, *Proceedings of the Seventh International Symposium on Remote Sensing of Environment*, University of Michigan, pp. 1563–69.

Mumbower, L. E. and Donoghue, J. (1967) Urban poverty study, *Photogrammetric Engineering* **33**: 610–18.

Nordbeck, S. (1965) *The Law of Allometric Growth*, Discussion Paper No. 7, Michigan Inter-University Community of Mathematical Geographers: Department of Geography, University of Michigan.

Ogrosky, C. E. (1975) Population estimates from satellite imagery, *Photogrammetric Engineering and Remote Sensing* **41**: 707–12.

Reeve, E. C. R. and Huxley, J. S. (1945) Some problems in the study of allometric growth, in Le Gros Clark, W. E. and Medawar, P. B. (eds) *Essays on Growth and Form Presented to D'Arcy Wentworth Thompson*, Oxford University Press, Oxford, pp. 120–51.

St. Joseph, J. K. S. (1969) Air reconnaissance: recent results, 18, *Antiquity* **43**: 314–5.

St. Joseph, J. K. S. (1977) Air photography and archaeology, in St. Joseph, J. K. S. (ed.) *The Uses of Air Photography*, John Baker, London, pp. 135–53.

Salas, R. M. (1981) *The State of World Population, 1981*, United Nations Fund for Population Activities.

Thompson, D. (1975) Small area population estimation using land use data derived from high altitude aircraft photography, *Proceedings of the American Society of Photogrammetry* (Fall Convention), pp. 673–96.

Tobler, W. (1969) Satellite confirmation of settlement size coefficients, *Area* **1**: 30–4.

Trewartha, G. T. (1953) A case for population geography, *Annals of the Association of American Geographers* **43**: 71–97.

US Bureau of the Census (1973) *Population Distribution, Urban and Rural in the United States: 1970*, United States Maps, GE-70, No. 1, US Government Printing Office; Washington, DC.

Welch, R. (1980) Monitoring urban population and energy utilization patterns from satellite data, *Remote Sensing of Environment* **9**: 109.

Welch, R. and Zupko, S. (1980) Urbanized area energy utilization patterns from DMSP data, *Photogrammetric Engineering and Remote Sensing* **46**: 201–7.

Wellar, B. S. (1969) The role of space photography in urban and transportation data series, *Proceedings of the Sixth International Symposium on Remote Sensing of Environment*, Vol. II, pp. 831–54.

Chapter 3 The atmosphere

Characteristics of the atmospheric environment

The atmosphere or the gaseous realm which surrounds our earth exercises a great impact on the activities of mankind. In the lowest 80 km of the atmosphere, there is a homogenized mixture of gases composed predominantly of nitrogen (78.08 per cent by volume), oxygen (20.95 per cent), argon (0.93 per cent) and carbon dioxide (0.033 per cent). The remaining gases, in decreasing order of percentage by volume, are neon, helium, krypton, xenon, hydrogen, methane and nitrous oxide, which add up to less than 0.003 per cent of the total volume. However, the lower atmosphere is never completely dry and water vapour which is generally about 1 per cent but can be as high as 4 per cent is present. Another variable gas, ozone, is also present at the surface of the earth and may contribute up to 0.00007 per cent by volume. Carbon dioxide is a gas of great importance, despite its small percentage, because of its ability to absorb heat, thus permitting the lower atmosphere to be warmed by

Fig. 3.1
Characteristics of the thermal structure of the atmosphere up to the lower part of the ionosphere (*Source*: Petterssen 1969)

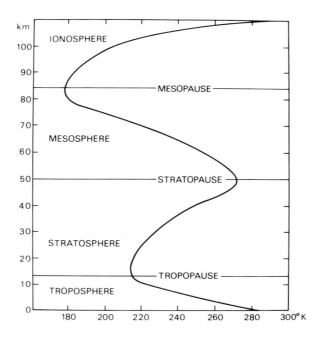

the incoming solar radiation and the emitted longwave radiation from the earth's surface. Carbon dioxide is also an effective emitter of radiation which helps to cool the upper atmosphere.

The atmosphere is held to the earth by gravitational pull. Thus, the air thins out and the pressure decreases as you go up the atmosphere. A vertical subdivision of the atmosphere into four layers, *viz., troposphere* (0–12.5 km), *stratosphere* (12.5–50 km), *mesosphere* (50–80 km), and *thermosphere* or *ionosphere* (beyond 80 km), according to the thermal structure, is evident (Fig. 3.1). Each layer is clearly demarcated by a upper boundary called a *pause*, hence the names *tropopause, stratopause* and *mesopause*. The human population lives in the tropopause, the lowest layer of the atmosphere, where virtually all the clouds and weather occur (Petterssen 1969, p. 36). This will form the focus of our discussion on the role played by remote sensing in collecting data to better understand the dynamic circulatory processes of the atmospheric system. The data required are basically three types: temperature, pressure and humidity.

The role of satellite platforms in collecting meteorological data

The collection of meteorological data has been traditionally ground based with the use of such standard instruments as barometers, thermometers, anemometers, rain-gauges and sunshine recorders distributed in a dense network of stations over a country (Sutton 1967). To collect pressure, temperature and humidity data in the upper atmosphere, a *radiosonde* is used. This consists of a small barometer, a thermometer and a hygrometer built into a cage which is carried by a balloon up to a height of about 36 km. A tiny radio, powered by batteries is also contained in the cage to telemeter the information back to earth. After the Second World War, ground-based military radar systems were used to determine winds in the upper atmosphere by tracking a radar target attached to a balloon. From the track of the balloon, wind speeds and directions can be computed. Radars have also been used to measure the intensity of rain and snow. In recent years, *Doppler radars* have been developed specifically to detect tornadoes and other forms of storms. These radars are designed on the Doppler shift principle which permits frequency changes of radar waves caused by the movement of the target to be measured (Battan 1973). In this way, Doppler radar can measure the speed at which a target moves towards or away from the radar set. Doppler radar systems can also be carried on board aircraft for collecting rainfall data associated with severe weather systems and active mid-latitude fronts for atmospheric research (Ray *et al.* 1980).

Specialized sounding rockets have also been employed to obtain pressure, temperature, density and wind data of the uppermost layers of the atmosphere for meteorological and geophysical studies. But the most significant development in recent years is the great variety of meteorological satellites which provides ideal platforms for remote sensing of meteorological phenomena in both the lower and upper atmosphere of the earth. These platforms make regular monitoring of a large part of the earth's atmospheric environment possible with suit-ably designed remote sensing instruments. In particular, the geosta-tionary satellites which can be placed at the desired longitude to orbit the equatorial plane of the earth at an altitude of 35,800 km in an east-to-west motion to match the earth's rotation give global views of the earth over the same subsatellite point every day and permit continuous collection of data (Fig. 3.33). However, they cannot see the poles because of the earth's curvature. The other type of meteorolog-ical satellite is the sun-synchronous or polar orbiting kind which are launched to an altitude about 1,100 km and permit viewing the same spot on the earth's surface twice a day. The data collected are of much higher resolution (both spatially and radiometrically) than those from the geostationary satellites, thus complementing the macroscale study of the earth's atmosphere with a mesoscale one. Together they all contribute towards an understanding of the microclimatology or micro-meteorology of an area based on ground measurements (Barrett 1974, pp. 8–9; Barry, 1970). To illustrate the usefulness of these satellite platforms in remote sensing of the atmosphere, the meteorological satellite systems developed by the USA will be briefly examined and employed as examples in subsequent discussions in this Chapter because their data are more generally available to the public.

The series of US meteorological satellites was begun with the successful launching of TIROS-I (Television and Infrared Observation Satellite) on 1 April 1960 (Fig. 3.2). Ever since then ten TIROS have been put into orbit at an average altitude of 700 km. These represented the first generation of the research and development polar

Fig. 3.2
Research and operational me-teorological satellites (*Source*: Cornillon 1982)

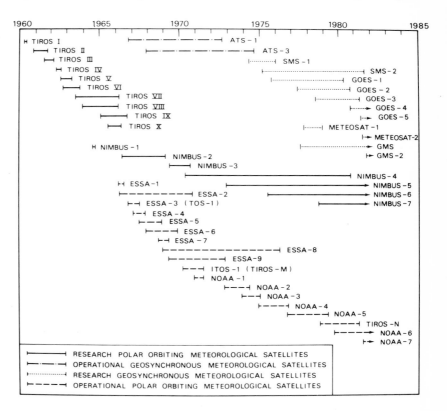

orbiting satellites devoted to meteorology. The Nimbus series from No. 1 to No. 7 initiated on 28 August 1964 was the second generation of the research and development polar orbiting satellites which orbited the earth at an average altitude of 1,000 km. They carried more advanced remote sensing instruments and permitted an improved pointing accuracy for the sensors. There was also a generation of research and development geosynchronous meteorological satellites known as Application Technology Satellites (ATS) that have been orbiting the equatorial plane of the earth since 7 December 1966 at an average altitude of 35,800 km. Six satellites in this series had been launched by the end of 1979.

Parallel to these research and development satellites were the operational meteorological satellites for civilian applications. The first generation of the polar orbiting and sun-synchronous operational satellites were the ESSA (Environmental Science Services Administration) series which was first launched on 3 February 1966 to orbit the earth at an average altitude of 1,500 km. Altogether there were nine satellites in this series. To provide improved operational infrared and visual observations of the earth's cloud cover, a second generation of sun-synchronous, polar-orbiting meteorological satellites known as the ITOS/NOAA (National Oceanic and Atmospheric Administration) satellite series were developed. The prototype ITOS-I (Improved TIROS – Operational Series) which was also known as TIROS-M was launched on 23 January 1970. The subsequent satellites in this series were re-designated NOAA (1 through 5) and flew at an average height of 1,450 km. The satellites were placed in one of two different orbital configurations: one with an ascending node (from South to North) at 1500 hours and the other with a descending node (from North to South) at 0900 hours local solar time. More recently, the prototype of the third generation of operational polar-orbiting, sun-synchronous meteorological satellites known as TIROS-N was launched, on 13 October 1978, with the specific objectives of measuring the earth's atmosphere, surface and cloud cover, and near surface environment. Two more satellites in this series designated as NOAA-6 and NOAA-7 have been successfully launched since. The altitudes of these satellites were much lower, about 850 km on average, thus resulting in faster moving satellites (about 6.6 km/sec as compared with 5.8 km/sec of the previous series). However, the attitude stabilization system has been designed to maintain the pointing accuracy of the sensors to ±0.2° of the local geographic reference.

One may also note that there is a series of military operational sun-synchronous polar-orbiting meteorological satellites under the Defense Meteorological Satellite Program (DMSP) which was made available to civilian use in February 1973. The first series of satellites known as DMSP Block 5B/C consisted of six satellites designated as P, Q, R, S, T and Y which flew at an altitude of about 850 km and was designed in such a way that two satellites collected data simultaneously, one at about noon (ascending) and midnight (descending), and the other at about dawn (ascending) and dusk (descending). In this way, the sensors could collect data over the entire earth four times a day. The high resolution sensors (two scanning radiometers) collected data in the visible (0.4–1.1 μm) and the thermal infrared (8–13 μm) channels. The Block 5D series introduced in September 1976 was a follow-on to the Block 5B/C series. It consisted of four satellites designated F1 to F4 collecting data of the earth four times a day using two satellites in the series. However, the F4 satellite collected data at different times with a mid-morning (1000 hours local solar time) descending node. It is noteworthy that the DMSP data were collected to meet Department of Defense requirements.

There are corresponding operational geosynchronous meteorological satellites for civilian applications. This is known as the SMS/GOES satellite series as a follow-up to ATS. SMS-1 (Synchronous Meteorological Satellite) was launched on 17 May 1974, and the third satellite in the series, SMS-C, was renamed GOES-1 (Geosynchronous Operational Environment Satellite), after its successful launch on 16 October 1975. GOES-2 was also launched on 16 June 1977 to complement GOES-1 for a full coverage of the USA. Altogether five GOES satellites have been launched at the time of writing. These satellites carry sensors to provide day and night observations of temperature and humidity of the earth.

The GOES series of satellites can be made geostationary at any desired longitude and can be moved forward or backward in their orbits to replace the one that is failing to function or to supplement the data collection effort. Hence, the terms GOES-East and GOES-West are now used for the satellites located at longitudes 75°W and 135°W respectively rather than the actual name of the satellite. Together with the European geosynchronous satellite called METEOSAT and the Japanese Geostationary Meteorological Satellite (GMS) the whole globe can be monitored between 70°N and 70°S (Fig. 3.3) and the system of five geostationary satellites has contributed significantly to the success of the First Global Experiment (FGGE) of the Global Atmospheric Research Programme (GARP) mounted in 1978–79 as well as the World Weather Watch (Barrett and Hamilton 1982).

Meteorological satellites carry a wide variety of sensors to monitor the earth-atmosphere system. Basically, they can be classified into *imaging* and *non-imaging sensors*. The *imaging sensors* include vidicon cameras and electro-optical scanners. Vidicon cameras (VCS) of wide, medium and narrow angle types have been used on the TIROS series of satellites, sensing in the visible channel of 0.55–0.75 μm and producing photographs with rather coarse spatial resolutions of 2.5–3 km, 2 km and 0.3–0.8 km respectively as the field of view narrowed down. The ESSA, Nimbus and NOAA satellite series made

Fig. 3.3
The positions of the current system of geostationary meteorological satellites. The USSR satellite nominally over the Indian Ocean has never been flown. The position was temporarily filled during 1979 by a USA GOES. Local coverage is now provided by an Indian satellite (INSAT). The rings show the extent of high quality imagery (*Source*: Barrett and Hamilton 1982)

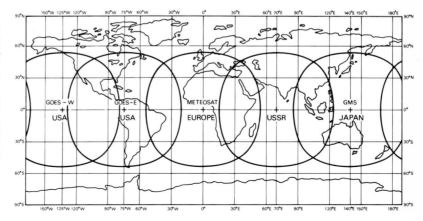

use of more advanced vidicon cameras (AVCS) with a doubled surface diameter of the image tube and consequently improved spatial resolutions. The Spin-Scan Cloud Cameras (SSCC) were carried on the geosychronous ATS satellites which utilized the spinning motion of the satellite and the small mechanical motion of an optical telescope to scan the earth. ATS-3 carried a Multicolor Spin-Scan Cloud Camera (MSSCC) which was capable of responding to colour signals of blue, green and red, which could then be combined on the ground to give colour pictures of the earth-atmosphere system.

The electro-optical scanners that were also used on meteorological satellites for the collection of multispectral imaging data, were ones such as the two-channel Temperature-Humidity Infrared Radiometer (THIR) in Nimbus 5 through 7, the two-channel Very High Resolution Radiometer (VHRR) in the NOAA satellites 2 through 5, the four-channel Advanced Very High Resolution Radiometer (AVHRR) in the TIROS-N and NOAA 6 and 7 series, the five-channel Coastal Zone Color Scanner (CZCS) in Nimbus–7, and the two-channel Visible Infrared Spin Scan Radiometer (VISSR) in the geostationary SMS/GOES satellites. These scanners most commonly record radiation in the visible (0.55–0.70 μm) and thermal infrared (10.5–12.6 μm) spectral bands.

The *non-imaging sensors* that were widely employed on meteorological satellites included all types of *radiometer sounders*, designed to operate at or beyond the wavelengths in the infrared, i.e. greater than 1 μm. A good example is the Selective Chopper Radiometer (SCR) onboard Nimbus 5 with 16 spectral channels in the infrared portion of the spectrum, capable of giving a thermal resolution of 0.1 K. Eight of the channels were within the 15 μm carbon dioxide (CO_2) band. All the 16 channels viewed the same column of the atmosphere vertically at slightly different times with a spatial resolution of 28 km. The radiation from the earth reaches the detector either through a cell containing CO_2 or through an empty cell depending on the position of the mirror chopper. The incoming radiation is therefore modulated only at frequencies at which the gas absorbs, which is detected with a phase-sensitive detector. The phase-sensitive detector gives a signal proportional to the radiation from the earth absorbed by the CO_2 cell. The empty cell permits space to be viewed all the time to provide a zero reference (Houghton and Smith, 1970). Another example is the Vertical Temperature Profile Radiometer (VTPR) in use onboard NOAA 2 through 5. This was a scanning radiometer employing interference filters to measure atmospheric temperature from the earth's surface to about 30,500 m using eight spectral channels in the infrared portion of the spectrum. Scanning was achieved by a stepping mirror. Each channel records the radiation emitted from a layer of specific height in the sub-satellite atmospheric column.

Other non-imaging sensors of note are the passive scanning microwave radiometer, incorporating antennae. An example was the Scanning Microwave Spectrometer (SCAMS) on Nimbus 6 with five spectral channels, selected near the water vapour line at the frequency of 22 GHz (wavelength 1.36 cm), near the atmospheric window at 32 GHz (1 cm) and near the oxygen band at 54 GHz (5 mm), which were aimed at providing global maps of tropospheric temperature profiles and, over oceans, the liquid and water vapour content of the atmosphere every 12 hours. An improved version is the Scanning Multichannel Microwave Radiometer (SMMR) on Nimbus 7. Another example, the Electrically Scanning Microwave Radiometer (ESMR) on Nimbus 5 and 6, performed scanning by electronically steering the beam of the radiometer with a phased array. No moving parts were involved and the angle of the antenna in this microwave radiometer is stepped across a scan line in discrete intervals. The microwave receiver was sensitive to a centre frequency of about 19.4 GHz (or 1.6 cm wavelength). These microwave radiometers have the advantage that surface and near-surface winds can be sensed in all but the most severe weather. Because of the much lower magnitude of the terrestrial emittance or solar reflectance in microwave frequencies than in the visible or thermal infrared, the instantaneous field of view (IFOV) of the radiometer has to cover a large enough area on the earth to detect the radiant energy by the receiving antenna. Hence, spatial resolutions coarser than 25 km are quite usual.

From the above review, one sees that sensor systems employed by the various meteorological satellites are highly sophisticated and are capable of collecting basic data on the temperature, pressure and humidity of the atmosphere at frequent intervals. On the other hand, this massive in-flow of data from a great diversity of sensor systems requires the research worker to select wisely those data relevant to his problems. Cornillon (1982), has compiled a reference guide for more detailed descriptions of these satellites and their sensors.

Directions of application of satellite remote sensing to the atmospheric environment

Remote sensing applications to the atmospheric environment have been characterized by an orientation towards the analysis of the large volume of meteorological satellite data. Clearly, the scale of the investigation has been both global and regional. The global scale is concerned with the general circulation within the earth-atmosphere system whereas the regional scale focuses on phenomena that occur within a small area of the earth, of about 10 to 100 km across, such as, studies of tornadoes, thunderstorms or urban heat islands (Barrett 1974). Apart from scale considerations, availability of the satellite data in image and non-image forms obtained at frequent time intervals permits three approaches to processing the data for various applications: (1) the visual or computer-assisted interpretation of the spatial patterns of the original two-dimensional imagery; (2) the mathematical analysis of the one-dimensional non-image radiation measurements to derive useful meteorological parameters; (3) the detection of trends or patterns of changes of weather phenomena by comparing one-dimensional or two-dimensional meteorological data acquired at different times. These approaches in combination with the scale factor lead to remote sensing applications in (1) applied macroclimatology with special emphasis on the radiation budget of the earth-atmosphere system, atmospheric moisture and circulation patterns and (2) mesoscale and synoptic scale analysis of weather phenomena for forecasting purposes. These will be examined in greater detail below.

The radiation budget of the earth-atmosphere system

The earth-atmosphere system must operate to maintain a balanced budget of heat. Heat and moisture are continuously being exchanged between the earth and the atmosphere (Fig. 3.4). Any transfer of moisture from one place to another is equivalent to a transfer of heat. In the long run, the earth-atmosphere system must return to space as much energy as it receives from the sun, the major energy source. The mean value of input from the sun which is generally regarded as constant is $1.95 \text{ cal/cm}^2 \text{ min}$ or 1.38 kW/m^2, hence its name *solar constant*. Therefore, the solar constant measures the rate at which solar radiation is received at the top of the atmosphere on a unit surface normal to the incident radiation. The radiation budget describes the energy exchange between the earth and space up to about 50 km altitude and can be defined as

$$Q_n = W_s - W_r - W_e \qquad [3.1]$$

where Q_n is the radiance balance (net radiant flux density), W_s is the incoming solar radiation, W_r and W_e are radiant flux densities reflected and eimitted from the earth to space respectively. Normally, a quantity called *planetary albedo* (α_p) which is determined by the ratio W_r/W_s, i.e. the proportion of solar energy reflected back to space by the earth's surface. Therefore, equation [3.1] can also be written as

$$Q_n = W_s (1 - \alpha_p) - W_e \qquad [3.2]$$

All these parameters can be determined from remote sensing by meteorological satellites, described in works by Vonder Haar and Suomi (1971), Raschke *et al.* (1973) and Gube (1982), using polar-orbiting sun–synchronous TIROS, Nimbus, ESSA as well as geosynchronous METEOSAT data. Onboard the Nimbus-6 satellite, an Earth Radiation Budget (ERB) experiment was carried out (Smith *et al.* 1977; Jacobowitz, 1979). Previous efforts in radiation balance studies

Fig. 3.4
Basic components of the radiation budget of the earth–atmosphere system (*Source*: Houghton 1979)

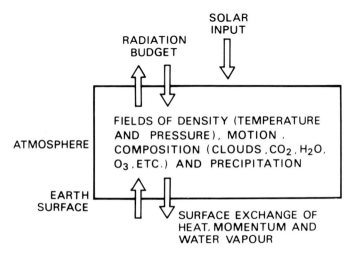

have been reviewed by Kondrat'ev, Borisenkov and Morozkin (1970, pp. 56–86). The earth's radiation balance is an important subject of research because our weather and climate are determined by the geographical distribution of the radiative energy. To illustrate this application, the ERB experiment of Nimbus 6 is examined here in greater detail.

The ERB experiment was equipped with a 22-channel radiometer (Table 3.1). The first ten channels were selected to measure incoming solar radiation over the entire shortwave and longwave spectrum. These solar channels were calibrated with reference to a standard pyrheliometer. The next four channels were fixed, wide-angle (130°) channels designed to measure the earth radiation flux in the short-wave region (0.2–3.8 μm), and in total ($\lambda > 0.2$ μm), leaving the earth-atmosphere system. The final eight channels were narrow-angle bi-directional scanning telescopes that observed the earth-reflected solar radiation (0.2–4.8 μm) and the earth-emitted longwave radiation (4–50 μm) for a local area of 250–500 km linear dimensions. The four wide-angle earth flux channels provided a direct measure of the total earth flux passing through a unit area at the satellite altitude of 1,100 km. Together with the solar constant measured by the ten solar channels, the planetary albedo and the net radiation balance of the earth-atmosphere system could be determined according to equation [3.1]. It is noteworthy that the earth radiation fluxes measured by the wide-angle sensors have to be converted to an irradiance (W/m²) by means of a calibration equation which was developed during an extensive prelaunch thermal vacuum calibration. Since the irradiance is the energy leaving the earth that reaches each unit area at the satellite elevation, it can be expressed in terms of a unit area located at the top of the atmosphere (about 15 km above the earth's surface) by applying an inverse square transformation which assumes that radiation is scattered isotropically from all surfaces. For the ERB wide-angle channel measurements of earth radiation fluxes, they were found to be somewhat low, possibly caused by the effective field-of-view of the channels being larger than originally supposed, thus enabling more

Table 3.1
Characteristics of ERB sensing channels
(*Source*: Jacobowitz *et al.* 1979)

	Solar channels									
Channel number	1S	2S	3S	4S	5S	6S	7S	8S	9S	10S
Spectral interval (μm)	0.2–3.8	0.2–3.8	0.2–50+	0.53–2.8	0.7–2.8	0.40–0.51	0.34–0.46	0.30–0.41	0.28–0.36	0.25–0.30
Fields-of-view (degrees)										
Maximum	26	26	26	26	26	26	26	26	28	28
full response	10	10	10	10	10	10	10	10	10	10

	Earth flux channels			
Channel number	11E	12E	13E	14E
Spectral interval (μm)	0.2–50+	0.2–50+	0.2–3.8	0.7–2.8
Fields-of-view (degrees)				
Maximum	133.3	133.3	133.3	133.3
With stop		112.4		

	Scanning channels							
Channel number	15	16	17	18	19	20	21	22
Spectral interval (μm)	0.2–4.8	0.2–4.8	0.2–4.8	0.2–4.8	4.5–50	4.5–50	4.5–50	4.5–50
Fields-of-view (degrees)	ALL 0.25 × 5.14 DEGREES							

thermal radiation to exit through the aperture of the sensor (Jacobowitz *et al.* 1979). The solar channels were also found to be in error due to problems with the instrument's calibration. The value of the solar constant given by the ERB measurements was 1.392 kW/m² which is nearly 2 per cent higher than the generally accepted value of 1.37 kW/m². However, despite these problems, the ERB experiment has succeeded in showing that the earth-atmosphere system is close to radiational balance over an entire year, with a strong seasonal variation (Fig. 3.5). It is noteworthy that the albedo and longwave radiation cycles did not match and were out-of-phase by nearly 180° (6 months), apparently being caused by the occurrence of maximum snow and ice cover (with the highest albedo) during the period of minimum longwave radiation as well as the higher degree of cloudiness in the Northern Hemisphere winter than in the Southern Hemisphere winter. The absorbed solar radiation (i.e. incoming minus reflected solar radiation) exhibited a variation with time dependent upon the variation of the incoming solar radiation. The net radiation which is the difference between the absorbed solar radiation and the outgoing longwave radiation showed correspondence in phase to the absorbed solar radiation (Fig. 3.5). Figure 3.5(b) shows radiative heating dominating over radiative cooling. This annual imbalance of the net radiation is most likely the result of the overestimation of the solar constant (1.391 kW/m²). By using a lower value of 1.368 kW/m² as indicated by the dashed line in Fig. 3.5(b), a more balanced picture results.

The ERB experiment has also been able to map the geographical distribution of the outgoing radiation fluxes from the earth's surface in both the coarser (from the wide-angle measurements) and the finer (from the narrow-angle measurements) resolutions. Figure 3.6 is an example of the former, showing net radiation of the world for August 1975. Clearly, one can see that the earth cooled down throughout most

Fig. 3.5
Monthly mean radiation budget parameters July 1975-December 1976. Dashed line in lower figure corresponds to the location of the zero line if a solar constant of 1368 W/m² were used in place of the value of 1391 W/m² determined from the ERB observations (*Source*: Jacobowitz *et al.* 1979)

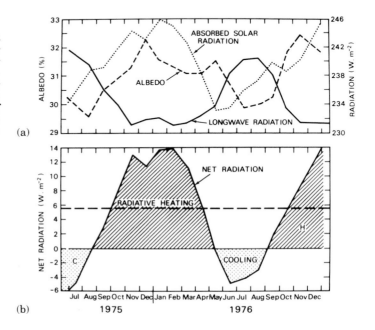

Fig. 3.6
Map of the net radiation for August 1975 from Nimbus-6 ERB wide-angle earth-flux observations (*Source*: Smith *et al.* 1977)

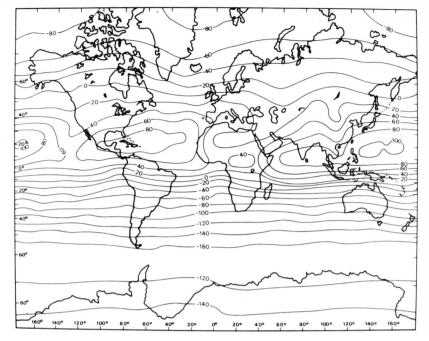

of the Southern Hemisphere and that maximum heating occurred over the Carribean and western North Atlantic as well as over the eastern North Pacific. A heating minimum occurred over North Africa. Figure 3.7 showed the same net radiation in greater detail because the narrow-angle scanning channel measurements were used to prepare the map. The monthly sample of radiance measurements for each target area were integrated over azimuth and zenith angle. The net radiation map (Fig. 3.7) resolves more details and reveals a radiation deficit of -20 to -40 W/m^2 over the Sahara Desert as compared with 20 W/m^2 over the same area in the wide-angle data (Fig. 3.6). On the whole, the ERB experiment indicated that the planetary albedos, long-wave radiation fluxes and net radiation of the earth-atmosphere system were about 30 per cent, 240 W/m^2 and -4 W/m^2 respectively for the months of July and August 1975 which corresponded well with the Nimbus 3 estimates given by Raschke *et al.* (1973). The satellite data tended to reveal a darker and radiatively warmer planet earth than was believed earlier on the basis of calculations using climatological data.

The availability of geostationary satellite data permits the radiation budget of the earth-atmosphere system to be determined with greater accuracy. The polar-orbiting satellites such as Nimbus 6 cross the equator twice, one at local noon (ascending node) and one at midnight (descending node). Only two daily measurements can be made over a certain location which may cause a biasing error in computing the daily mean radiation fluxes (Saunders and Hunt 1980). However, the use of a network of geostationary satellites to cover the globe (except the polar regions) as shown in Fig. 3.3 permits measurements to be made at intervals of 30 minutes. Thus, the radiation budget parameters can be adequately sampled to take into account

Map of the net radiation for August 1975 from Nimbus-6 ERB narrow-angle scanning channel observations (*Source*: Smith *et al.* 1977)

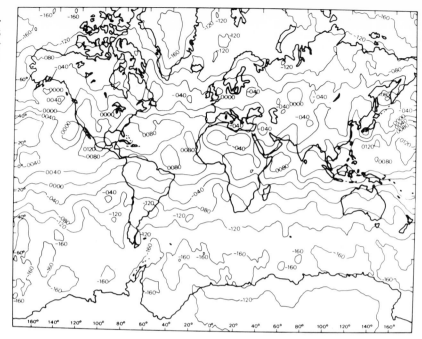

their diurnal variation. Gube (1982) presented a method to derive radiation budget parameters, i.e. planetary albedo and emitted long-wave radiation, at the top of the atmosphere from the visible (VIS 0.4–1.1 μm), thermal infrared 'window' (IR 10.5–12.5 μm) and infrared water vapour absorption (WV 5.7–7.1 μm) bands of the METEOSAT images. In her models of spectral radiation fluxes at the top of atmosphere (from spectral radiance measured by METEOSAT), Gube included the correction factors for the anisotropic nature of radiation in the form of regression equations. These were the factors governing the viewing geometry, i.e. the zenith angle (z) between the satellite and the point of observation as well as the azimuth angle (ϕ) of the satellite. The direction of solar illumination or the solar zenith angle z_0 was also incorporated (Fig. 3.8). By using the spectral response of 0.4–1.1 μm in the VIS channel, reflected shortwave radiation from the earth was deduced. The emitted longwave radiation from the earth (4–100 μm) was derived from the IR channel which gave the temperature of the radiating surface together with lower tropospheric humidity and the WV channel which gave the upper tropospheric water vapour content. This investigation has succeeded in not only mapping the daily mean albedo of the African continent in the Inter-Tropical Convergence Zone (ITCZ) along the equator but also drawing attention to the geographical regions with a pronounced diurnal cycle of radiation. It was clear that an increase in longwave radiation occurred during sunshine hours (Fig. 3.9a). This was particularly severe over the desert areas (reaching 50–60 W/m^2) but less so over moist forest surfaces (only about 20 W/m^2). Even in areas with cloud cover, greater temperature changes occurred over high clouds of convective origin such as the cirrus and cumulonimbus clouds as the cloud top height increases (Fig. 3.9b). Over the African continent there was an afternoon minimum in longwave

Fig. 3.8
Geometric relationship between the sun, the satellite and the point of observation. z and ϕ are respectively zenith angle and azimuth angle of the satellite and z_0 is zenith angle of the sun (*Source*: Gube 1982)

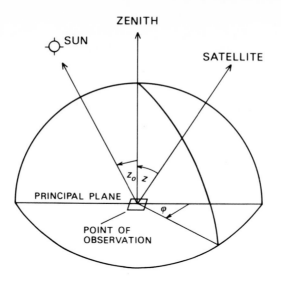

Fig. 3.9
Time series of albedo and long-wave flux determined from METEOSAT VIS, IR and WV image data (23/24 November 1979): (a) reflector type 'land'. (b) reflector type 'cloud' and (c) reflector type 'sea' (*Source*: Gube 1982)

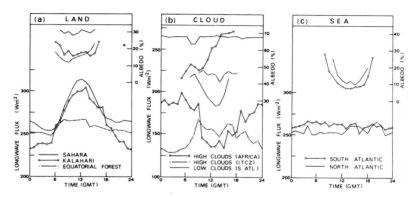

flux. On the other hand, high clouds over sea (the ITCZ region in Fig. 3.9b) showed minimum longwave flux values at night. Practically constant temperatures of the emitting surfaces exhibited no longwave flux change, as over low clouds (Fig. 3.9b) and sea (Fig. 3.9c). The albedo variations over the same locations could be explained by solar zenith angle changes which modified cloud and atmosphere scattering processes and surface reflection. Thus, the albedo varied symmetrically for the desert surface (Fig. 3.9a), low cloud (Fig. 3.9b) and sea (Fig. 3.9c). However, the albedo change over high clouds of convective orgin (Fig. 3.9b) was large, increasing from 45 per cent at 600 hours GMT to 70 per cent at 18 hours GMT, this increase being accompanied by a decrease in longwave radiation. One major problem to note with the use of METEOSAT data is the uncertainty about the calibration factors in the IR, VIS and WV channels.

Atmospheric temperature structure

Closely associated with the radiation budget investigation is the remote sounding of atmospheric temperature structure from the satellite platform. Temperature measurements can be made by using the emission spectrum of carbon dioxide centered at a wavelength of 15 μm in the infrared (Houghton and Smith 1970). Carbon dioxide is present in the atmosphere, uniformly mixed to within 1 or 2 per cent. Its emission spectrum when viewed from above will be dependent only on the vertical distribution of the atmospheric temperature. These temperature measurements are particularly affected by the presence of clouds. The first remote temperature sounding of the atmosphere from a satellite was made from TIROS VII using a simple filter radiometer with a spectral band pass covering the whole of the 15 μm carbon dioxide band which observed emission in the lower stratosphere. This was followed by the Satellite Infrared Spectrometer (SIRS) and the Infrared Interferometer Spectrometer (IRIS), carried onboard Nimbus 3 launched on 14 April 1969. In particular, the SIRS measured the radiance of the earth and the atmosphere in seven narrow spectral intervals in the 15 μm carbon dioxide band and in one interval of minimum absorption at 11.1 μm, the water vapour window (Wark and Hilleary 1969). The field of view was a 200 km square area and the radiances were sampled every 8 seconds. The instrument, which is basically a diffraction-grating spectrometer, has a means of checking its calibration in orbit. The deviation of a vertical temperature profile involved the solution of seven simultaneous radiative transfer equations which took the following form:

$$I(v_i) = B[v_i, T(p_s)]\tau\ (v_i,p_s) - \int_1^{\tau(v_i,p_s)} B[v_i, T(p)]\ d\tau(v_i,p)$$
$$i = 1, \ldots . 7 \tag{3.3}$$

where I is the measured radiance at frequency v_i; B is the Planck radiance at v_i, pressure p and temperature T; τ is the fractional transmittance from any pressure level p to the top of the atmosphere, and p_s refers to surface pressure.

An example of the vertical temperature profile over Kingston, Jamaica obtained from SIRS data in comparison with radiosonde data is shown in Fig. 3.10. A very close fit between the two sets of data was seen. The main difference occurred in the middle troposphere where the SIRS sounding was warmer. The temperature structure in the atmosphere up to 30 km altitude can be measured with an accuracy such that the mean temperature of 200 mb thick layers can be determined to better than 2 K, under cloudless conditions. Improved versions of the infrared instruments were carried onboard subsequent Nimbus satellites, such as, the Selective Chopper Radiometer (SCR), the Infrared Temperature Profile Radiometer (ITPR) on Nimbus 4; the High Resolution Infrared Radiometer (HIRS/1) which possessed spectral channels in both the 4.3 μm and 15 μm bands onboard Nimbus 6; the Vertical Temperature Profile Radiometer (VTPR) onboard the operational NOAA series of satellites from 2 through 5 and the High Resolution Infrared Sounder (HIRS/2) for lower atmosphere temperature sounding and the Stratospheric Sounding Unit (SSU) for upper atmosphere temperature sounding onboard TIROS-N, NOAA 6 and 7 (Brownscombe and Nash 1982).

Microwave spectrometers such as the Nimbus E Microwave Spectrometer (NEMS) onboard Nimbus 5 have also been used to measure atmospheric temperature from satellite platforms; advanced versions are the Scanning Microwave Spectrometer (SCAMS) onboard Nimbus

Fig. 3.10
Comparison of temperature profiles derived from SIRS data with radiosonde data on 14 April 1969 in Kingston, Jamaica (1200 GMT). SIRS, 16°N, 73°W, 1612 GMT, clear sky. (*Source*: Wark and Hilleary 1969)

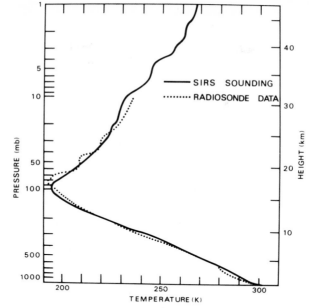

6 and the operational Microwave Sounding Unit (MSU) onboard TIROS-N and NOAA 6 and 7. These passive microwave sensors have the advantage of being able to measure the temperature profile under cloud-covered conditions. Mostly they are used to measure not only temperature but also water vapour and liquid water. Therefore the five spectral channels of the SCAMS were selected in the water vapour band (22.235 GHz), the atmospheric window at 31.650 GHz, the oxygen band at 52.85 GHz, 53.85 GHz and 55.45 GHz (Staelin *et al.* 1977). The primary surface parameters determining the microwave emission are surface temperature and surface emissivity (or dielectric properties of the surface). The product of these two parameters gives the brightness temperature of the surface. The last three channels of SCAMS were selected at the oxygen absorption band which measured the average temperature of layers about 10 km thick, centered near altitudes of 2, 6 and 13 km respectively. The accuracy of the temperature measurements were found to give root mean square errors of 1.7 K.

Because radiation sources occur at different heights in the atmosphere, gradients of transmission with respect to the logarithms of pressure known as *weighting functions* can be computed for each carbon dioxide channel of the instrument. In other words, the weighting functions describe the region of atmosphere from which the radiation detected in the different spectral channels originates. Figure 3.11 illustrates the weighting functions for different channels of the HIRS/1 and SCAMS instruments on Nimbus 6. The vertical resolution of these remote sounding measurements is not high compared with radiosonde measurements, being at best about 5 km whereas their horizontal resolution is much better, being less than 100 km for the infrared and about 250 km for the microwave observations. There are problems of interpretation of microwave measurements caused by variations of surface reflectivity and clouds can adversely affect the

Fig. 3.11
Weighting functions (gradient of transmission with respect to logarithm pressure) for instruments sounding the temperature of the lower atmosphere on the Nimbus 6 satellite for (a) 15 μm channels of HIRS, (b) 4.3 μm channels of HIRS and (c) channels of SCAMS (*Source*: Houghton 1979)

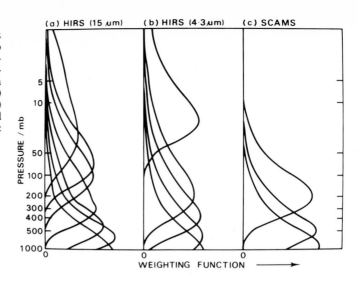

accuracy of infrared observations. Therefore, for best results both infrared and microwave observations should be employed together.

Surface radiation studies

The radiation and temperature measurements from the satellite platforms can be extended to study earth surface radiation characteristics in the rural and urban environments. An important area of investigation is desertification caused possibly by increased surface albedoes which produce subsequent rainfall deficits. The increased albedo results in a net radiative loss which gives rise to general subsidence and drying over the area, thereby inhibiting or reducing the convection necessary for rain. The increased surface albedoes may be natural or man-made as a result of improper land management. Rockwood and Cox (1978) tested such a hypothesis with geostationary satellite data obtained from the visible band (0.55–0.75 μm) of the SMS-1 over the Sahel region in northwestern Africa (Fig. 3.12). They combined simultaneous satellite and aircraft data together to map the earth's surface albedo which is different from the planetary albedo in terms of radiation budget. The top of the atmosphere is referred to as planetary albedo. To infer the surface albedo, they established first an empirical relationship between the mean SMS-1 brightness count (B) on a 20 × 20 nautical miles (37 × 37 km) grid and the mean surface albedo (α_{sfc}) using a second order polynomial equation of the form:

$$\alpha_{\text{sfc}} = C_o + C_1 B + C_2 B^2 \qquad [3.4]$$

From this it was then possible to estimate the mean surface albedo values for input to the radiation budget equation below to obtain the effective reflectivity of the earth-atmosphere system, (ρ_{sys}):

$$\rho_{\text{sys}} = 1 - \alpha_{at} - \tau_{at}(1 - \alpha_{\text{sfc}}) \qquad [3.5]$$

Fig. 3.12
Visible photograph from SMS-1 satellite at 1200 GMT on 20 September 1974 (*Courtesy*: Rockwood and Cox 1978)

where α_{at} is the effective atmospheric absorptivity and τ_{at} is the effective atmospheric transmissivity. τ_{at} had been calculated from near-surface aircraft data whereas α_{at} was estimated to be in the range of 22–23 per cent for the conditions lying between deserts and tropical rainforests. Then, an empirical relationship was established between the effective reflectivity (ρ_{sys}) and the mean brightness count (B) using a similar second-order polynominal equation as shown in [3.4]. The surface albedo (α_{sfc}) was then computed by applying the equation:

$$\alpha_{sfc} = 1 - [(1 - \rho_{sys} - \alpha_{at})/\tau_{at}] \qquad [3.6]$$

This study assumed the reflected radiance field at the top of the atmosphere to be isotropic, an assumption not unreasonable for low latitudes and small solar zenith angles. The land surface in the study area was classified into 8 classes based on soil, vegetation and moisture characteristics which all have an impact on the surface albedoes (Table 3.2). This was done with the aid of the aerial photographs of the study area. Each class of surface was also related to the inferred surface albedo. This method was applied to the area for 1200 GMT 2 July 1974, 1200 GMT 10 August 1974 and 1200 GMT 20 September 1974 of the SMS-1 data. Surface albedo maps were produced for these three dates which covered the wet and dry seasons in the area (Fig. 3.13). It was observed that there was a change in the surface albedo north-south gradient from July to September. At about 15°N

Class	Surface Type	Characteristics	Surface Albedo (α_{sfc})	
			Mean	*Range*
0	Swampland, oceans	coastal swamps, rivers, smooth oceans; >50% water with small solar zenith angle	0.09	0.0 –0.10
1	Dense forest	uniform dark vegetation; 10% light vegetation or soils	0.15	0.10–0.16
2	Moderate forest	mostly dark vegetation; 30% light vegetation or soils	0.17	0.16–0.21
3	Mixed vegetation	evenly mixed surface; 50% dark vegetation; 50% light vegetation or soils	0.22	0.21–0.26
4	Savanna	mostly low, light grasses cultivated fields, some dark scrub; 70% light dry grasses; 30% dark scrub or rock	0.28	0.26–0.31
5	Mixed desert	dry, light surface, coloured soils and rock outcroppings; 50% light soils; 50% scrub, light grasses	0.33	0.31–0.36
6	Moderate desert	sparse vegetation, mostly light sands, scrub; 30% low scrub, light grasses, rock	0.39	0.36–0.42
7	Desert	uniform sand surface; little variation in surface over large areas; 10% low scrub or coloured rock	0.43	0.42–1.0

Table 3.2
Surface classification system as applied to NorthWestern Africa (*Source*: Rockwood and Cox 1978)

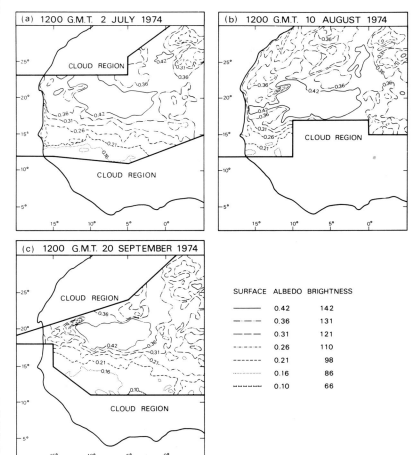

Fig. 3.13
Surface albedo analysis for (a) 1200 GMT 2 July 1974, (b) 1200 GMT 10 August 1974 and (c) 1200 GMT 20 September 1974 (*Source*: Rockwood and Cox 1978)

there was a change in albedo from 25–19 per cent, representing a relative 25 per cent decrease. At 14°N and 17°N relative changes of 15 and 5 per cent in albedo respectively were observed. North of 18°N, changes from 0–15 per cent were inferred. It was also noteworthy that surface class 7, corresponding to a desert surface having less than 10 per cent scrub vegetation, had albedo values greater than 42 per cent and was nearly unaffected by the seasonal variation of moisture. Surface class 3 (even mixture of vegetation and light soils), which was the dominant class at 15°N during July, had an albedo range of 21–26 per cent, appeared to change to surface class 2 (moderate forest) with a lower range of albedoes (16–21 per cent). Surface class 6 or the moderate desert (albedo range 36–42 per cent) was reduced from 24 per cent in area on 2 July to 13 per cent in area by 20 September. On the whole, the Sahara region north of 18°N was found to be a comparatively stable region with respect to large changes in albedo, but smaller scale variations could occur over a very large area, whilst south of 18°N larger scale changes appeared to be more common. In a subsequent study over the same area by Norton, Mosher and Hinton (1979) using cloud-free ATS-3 green sensor data (0.48–0.58 μm), surface albedo maps with similar patterns of variation were produced. They made use of normalized modal brightness values as surface albedo values in the form:

$$A = I_0 D_m / \cos \theta \qquad [3.7]$$

where A was the surface albedo, I_0 was the normalizing factor from the Sahara normalization point (a bright area located northwest of Timbuktu centered at 19°N, 05°W), D_m was the modal brightness value over each 2° grid, and θ was the solar zenith angle. They concluded that seasonal reflectance variations of as much as 80 per cent were found in the central Sahel during years of normal precipitation while the variation for drought years generally amounted to less than 15 per cent. A surface reflectance cycle coincident with the drought intensity seemed to exist. However, the hypothesis that increased albedoes would produce rainfall deficits remained unproved.

Another interesting application of radiation budget and temperature measurements from satellite remote sensing is the urban heat island study undertaken by Carlson *et al.* (1981), which illustrates the usefulness of polar-orbiting thermal infrared satellite data in microclimatological investigations. The application was based on a four-layer surface model to infer the surface energy fluxes, moisture availability and thermal inertia from HCMM radiance data whose noise equivalent temperature error (NEΔT) was 0.3 K. The four layers of the models were a mixing layer, a 50 m surface turbulent layer, a thin transition layer between the air and the surface interface, and a 1 m ground layer (Fig. 3.14). The daytime model gave the following energy balance equation:

$$R_n = G_o + H_o + E_o$$
$$= (1 - A_o)S - \varepsilon_g \sigma T_o^4 + \varepsilon_a \sigma T_a^4 \qquad [3.8]$$

where R_n was the net radiation, G_o the sensible heat flux into the grounds, H_o the sensible heat flux into the atmosphere, E_o the latent

Fig. 3.14
Basic structure of the four-layer
surface model employed in the
urban heat island study (*Source*:
Carlson *et al.* 1981)

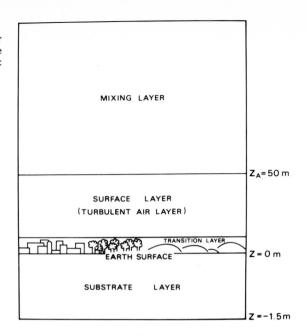

heat flux into the atmosphere, A_o the surface albedo, T_o the temperature at the ground-air interface, σ the Stefan-Boltzmann constant, T_a
the temperature at the top of the surface layer (50 m high), S the net
solar flux, ε_g the ground emissivity and ε_a the emissivity of the atmosphere. From the latent heat flux E_o, the moisture availability (M)
which measured water saturation at ground surface was determined.
The thermal inertia which measured the rate of heat transfer at the
ground-air interface was determined by

$$P = \lambda K^{-\frac{1}{2}} \qquad [3.9]$$

where λ was the thermal conductivity of the substrate layer determined from the ground surface flux G_o in the standard conduction
equation, and K was the thermal diffusivity of the substrate identical
to λ/C_g, C_g being ground capacity. During night-time, solar heating of
the surface is absent and H_o becomes negative and tubulence dimin-
ishes. Heat flux is no longer determined directly by net radiation but
dependent passively on the *lapse rate*, i.e. temperature decreasing with
height. The model was solved with reference to the conditions of the
near-surface stability. After the solution of the daytime and
night-time surface layer models, the satellite radiance data were
input into these models to produce a series of maps for surface
temperature, sensible heat flux, moisture availability and thermal
inertia (Fig. 3.15). When the technique was applied to Los Angeles and
St. Louis, temperature maxima were observed to be related to
commercial and industrial centres with minima in the mountains and
vegetated areas during daytimes whilst the night-time patterns were
less well-defined with temperatures inland being warmer than those
near the coast, probably a direct reflection of urban rather than any
marine influences (Figs. 3.16a and b). The temperature range was also

Fig. 3.15
Flow diagram for inferring surface parameters from satellite observations (*Source*: Carlson *et al.* 1981)

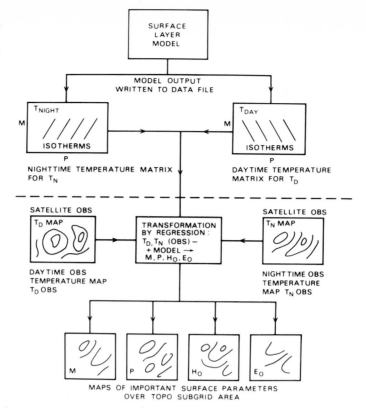

much smaller at night indicating a smaller rural-urban difference than in the daytime. The minima in moisture availability corresponded closely with the areas of daytime temperature maxima (Fig. 3.16c). On the other hand, thermal inertia patterns were ill-defined and not useful (Fig. 3.16d). It was concluded that the heating up of the urban area caused differential heating which resulted in small-scale circulations. The urban heat island was therefore attributable to the presence of surfaces that could not retain moisture – a result of inadvertent human modifications of the land cover.

Cloud classifications

Satellite images record cloud patterns from which one can deduce the three-dimensional structure of wind and pressure fields. This makes use of four factors. First, clouds are suggestive of the condition of stability in the atmosphere. Thus, cloud-free skies are associated with subsidence under anticyclonic conditions or the influence of very dry continental air whilst deep cloudiness over a wide area is indicative of instability. Secondly, the relative importance of vertical or horizontal motion in the troposphere can affect cloud patterns: small horizontal motion is associated with mottled, cellular cloud patterns whilst strong vertical motion is indicated by elongated cloud cells or

Fig. 3.16
Los Angeles (a) surface temperature analysis (31 May 1978 day time ~ 1330 LST); (b) surface temperature analysis (30 May 1978 night-time ~ 0230 LST, units in °C); (c) moisture availability analysis (30–31 May 1978); (d) thermal inertia analysis (30–31 May 1978; units in cal cm^{-2} k^{-1} s$^{-\frac{1}{2}}$; (*Source*: Carlson *et al.* 1981)

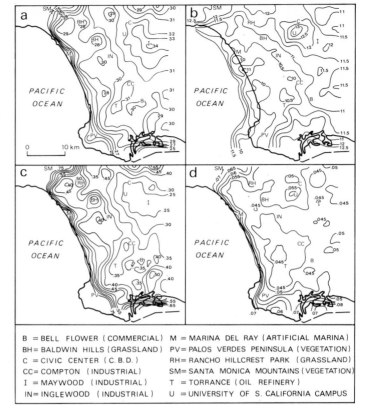

B = BELL FLOWER (COMMERCIAL)   M = MARINA DEL RAY (ARTIFICIAL MARINA)
BH = BALDWIN HILLS (GRASSLAND)   PV = PALOS VERDES PENINSULA (VEGETATION)
C = CIVIC CENTER (C.B.D.)   RH = RANCHO HILLCREST PARK (GRASSLAND)
CC = COMPTON (INDUSTRIAL)   SM = SANTA MONICA MOUNTAINS (VEGETATION)
I = MAYWOOD (INDUSTRIAL)   T = TORRANCE (OIL REFINERY)
IN = INGLEWOOD (INDUSTRIAL)   U = UNIVERSITY OF S. CALIFORNIA CAMPUS

cloud lines. Thirdly, strongly and complexly layered clouds are frequently generated along zones of contact and mixing between air streams possessing contrasting temperature and, or humidity characteristics. If clouds at different levels can be distinguished, variations in the direction and speed of the air flow at different levels in the troposphere can be detected. Finally, one should take note of the modifications in air masses and air streams caused by the underlying topography which creates the difference in cloud patterns over land and sea surfaces.

A useful first step in using satellite images is therefore to identify the different types of clouds. Clouds have been classified into three major types based on their forms: (1) *cirrus* or curly clouds; (2) *stratus* or layered clouds; (3) *cumulus* or lumpy clouds. They can be further distinguished by their heights above the ground into: (1) high clouds of over 6 km high (prefix cirro); (2) middle clouds of between 2–6 km (prefix alto); (3) low clouds of below 2 km; (4) clouds of vertical development (Table 3.3). Cloud types were conventionally identified visually from fixed positions on the ground. With the use of satellite images, the viewing is vertically, from above. Without stereoscopic viewing capability, it is extremely difficult to differentiate the cloud's altitude of occurrence. The lower spatial resolution of the satellite images also means that microscopic elements of the clouds cannot be depicted. However, Conover (1963) developed a generic classification of clouds for use in the recognition of clouds from satellite images. His

Form	High	Middle	Low	Vertical Development
Curly	Cirrus (Ci)	–	–	–
Layered	Cirrostratus (Cs)	Altostratus (As)	Stratus (S), nimbostratus (Ns)	–
Lumpy	–	–	Cumulus (Cu)	Cumulonimbus (Cb)
Partly layered, partly lumpy	Cirrocumulus (Cc)	Altocumulus (Ac)	Stratocumulus (Sc)	–

Table 3.3
A simple classification of clouds based on form, altitude and development
(*Source*: Gedzelman 1980)

classification scheme was based on six characteristics: (1) *cloud brightness* relating to the depth of cloud, the nature of cloud constituents and the angle of illumination; (2) *cloud texture* relating to the degrees of smoothness of the clouds; (3) *vertical structure* relating to the structure and heights of the clouds deduced from cloud shadows; (4) *forms* of cloud elements, relating to their degree of regularity; (5) *patterns* of cloud elements, relating to their degree of organization; (6) the *size* the elements and patterns. Conover's scheme has been widely accepted. Barrett (1970, 1974) proposed a genetic cloud classification scheme for meso- and macro-scale arrangements of clouds. The genetic scheme involved both cloud arrangements and the meteorological processes that give rise to them. Basically, the scheme separated those cloud arrangements caused by meteorological processes in the free atmosphere from those that were governed more strongly by the earth's surface patterns which included the nature (i.e. land or water), temperature and topography of these surfaces.

Meteorological satellite images have been used for the preparation of cloud charts or *nephanalyses* which show the distribution of cloudiness over an area. Normally, four categories of cloudiness, *viz.*, open (less than 20 per cent cloud cover), mostly open (20–50 per cent cloud cover), mostly covered (50–80 per cent cloud cover) and covered (over 80 per cent cloud cover) are used visually to compile these charts. Better cloud information will be particularly useful in energy and moisture budget modelling, rainfall monitoring and sea surface temperature mapping. Recently, some progress has been made towards computer-assisted nephanalysis. Basically, this involves the use of computer algorithms to identify *cloud types* and to estimate *cloud amount*.

Most workers made use of spectral features (i.e. cloud brightness) in the visible, water vapour and thermal infrared channels (singly or jointly) to classify cloud types with the computer (Greaves and Chang 1970; Shenk, Holub and Neff 1976). Textural features have also been used in combination with cloud brightness to analyse cloud type and cloud amount (Parikh 1977; 1978; Parikh and Ball 1980; Harris and Barrett 1978). Parikh's work (1977) provided a good example of the usual approach. She made use of 25 spectral and 144 textural features extracted from polar-orbiting NOAA-1 visible (0.52–0.72 μm) and infrared (10.5–12.5 μm) tropical cloud data. The spectral features measured quantities such as brightness, cloud top temperature and variation in brightness and temperature. The textural features measured factors such as the amount of local variation within the cloud sample and the overall homogeneity of the sample data. These textural features were defined from grey level difference histogram frequencies such as mean, contrast, angular second moment and entropy for four directions and four distances on both visible and infrared data arrays. Both a four-class problem, in which cloud types were grouped into (1) 'low', (2) 'mix', (3) 'cirrus' and (4) 'cumulonimbus' classes, and a three-class problem, in which the 'mix' class was excluded, were attempted. A supervised approach making use of training sets was adopted and four types of classifiers – maximum likelihood, one-against-the-rest, voting and Fisher, using sample *a piori* probabilities – were all tried. It was concluded that the maximum likelihood classifier produced the best results with a classification accuracy of 91 per cent for the four-class problem and 98 per cent for the three-class problem. This type of classification was extended to the geosynchronous SMS-1 data collected by the Visible and Infrared Spin Scan Radiometer (VISSR) (Parikh 1978). Decreases in classification accuracy ranging from 4–11 per cent were observed in the four-class problem but were not found to occur in the well-defined three-class problem where the 'mix' class was excluded. Therefore, the inaccuracy seemed to be caused by the 'mixed clouds' class.

Harris and Barrett (1978), using cloud brightness and cloud texture data, extracted from Defense Meteorological Satellite Program (DMSP) high resolution (0.61 km), visible (0.4–1.1 μm) and thermal infrared (8–13μm) imagery (already converted into a two-dimensional digital array by a microdensitometer) have been able to classify clouds into three categories, *viz.*, (1) sheet and, or layered cloud ('stratiform'), (2) cellular and, or tower cloud ('cumuliform') and (3) broken and, or mottled cloud ('stratocumuliform' and 'mixed'), with an accuracy of over 72 per cent. Cloud texture was evaluated by measuring the standard deviation and vector dispersion of density values within small grid squares (5×5 pixels) of the satellite image. A supervised classification approach using linear discriminant analysis was adopted to allocate each grid square to one of the three cloud categories and a no-cloud category. A cloud-type and cloud-amount nephanalysis was constructed by the computer (Fig. 3.17 and Fig. 3.18).

Desbois *et al.* (1982) presented an unsupervised approach in the automatic classification of clouds using the geostationary METEOSAT data in the visible (VIS), infrared window (IR) and infrared water vapour (WV) channels with the method of dynamic cluster analysis. By this method, natural groupings were formed without *a priori* classification. These individual clusters were later visually identified from the imagery into cloud types or terrain features. Unfortunately, no accuracy figure was given. The major problems of automatic cloud classification are the lack of agreement in the cloud classification scheme and the inavailability of a standard, for accuracy evaluation. It appears possible to include cloud size, form, organization and shadows as specified by Conover (1963) into the computer algorithm as contextual information to improve classification accuracy (Harris 1982). Also, the use of multispectral data such as those provided by Landsat may prove a valuable approach for improving cloud identification by machines (Barrett *et al.* 1976; Harris and Barrett 1978).

Apart from its role in cloud classification, cloud brightness is useful in estimating cloud-top temperature as demonstrated by Park *et al.* (1974) who made use of the polar orbiting Nimbus 4 data (11.5 μm channel) and the geosynchronous ATS 3 digitized data, with an emphasis on clouds of great depths, such as cumulonimbus. The 11.5 μm channel of the Temperature-Humidity Infrared Radiometer

Fig. 3.17
The objective analysis computed from the DMSP image shown in Fig. 3.18. The spatial resolution of this map is about 40 km (*Source*: Harris and Barrett 1978)

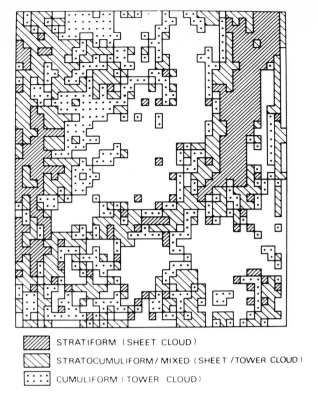

STRATIFORM (SHEET CLOUD)

STRATOCUMULIFORM / MIXED (SHEET / TOWER CLOUD)

CUMULIFORM (TOWER CLOUD)

Fig. 3.18
Defense Meteorological Satellite Program (DMSP) high-resolution (0.61 km) visible (0.4–1.1 μm) image of a part of western Europe and the North Atlantic, 0804 GMT, 30 April 1975. (*Courtesy*: SSEC, Madison, WI)

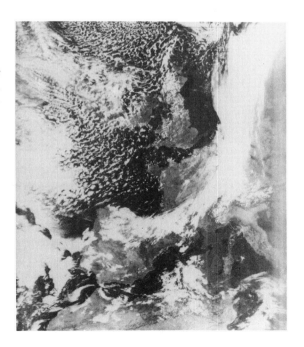

(THIR) of the Nimbus 4 does record energy originating from the cloud-top if the field of view is filled with clouds, because the absorption by water vapour, carbon dioxide and ozone in this spectral region is very minimal. The cloud brightness values were obtained from the ATS 3 digitized data for the corresponding date and time over the same study areas in the USA (latitude 35°N–40°N; longitude 75°W–105°W). By using a simple regression equation of the form $Y = aX^b$ where a and b are the coefficients to be determined, a relationship between temperature (X) and brightness (Y) can be obtained, revealing a strong negative correlation (over -0.9) between them (Fig. 3.19). If the cloud-top temperature can be estimated in this way, a knowledge of the cloud types, surface temperature as well as vertical distribution of water vapour and temperature (lapse rate) would allow the height of cloud tops to be determined. However, a correction temperature of 5 K had to be introduced to the cloud-top temperature because of the 3–7 K discrepancy between the satellite measured and the radiosonde measured cloud-top temperatures. It was also found that brighter clouds were thicker clouds, as depicted in Fig. 3.20. Desbois *et al.* (1982) have also demonstrated a graphical method of estimating cloud-top temperature from the two-dimensional histogram of the IR and WV channels of METEOSAT. This was deduced from the intersection between the principal axis of the histogram for a particular type of cloud and a superimposed curve of the form $F(T) = (x,y)$ so that $x = F_{ir}(T)$ and $y = F_{wv}(T)$ where F_{ir} and F_{wv} were functions associated with the temperature T of a blackbody having a radiometric value x in the IR channel and y in the WV channel (Fig. 3.21).

Fig. 3.19
The relationship between temperature and brightness (*Source*: Park *et al* 1974)

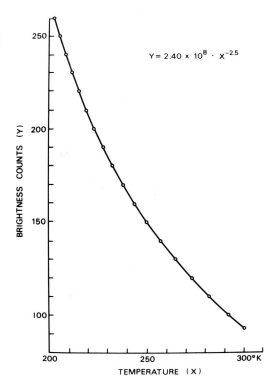

$Y = 2.40 \times 10^8 \cdot X^{-2.5}$

BRIGHTNESS COUNTS (Y)

TEMPERATURE (X)

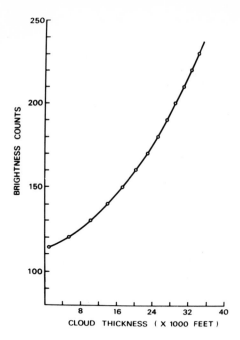

Fig. 3.20
The relationship between brightness and cloud thickness (*Source*: Park *et al.* 1974)

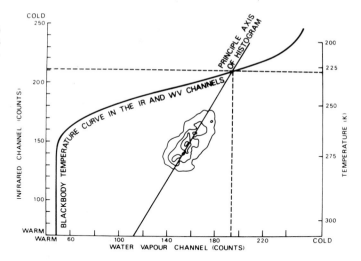

Fig. 3.21
Determination of the cirrus top temperature for the three classes of cirrus (*Source*: Desbois *et al.* 1982)

Photogrammetry has also been applied to determine cloud heights accurately by using stereographic pairs of geosynchronous SMS/GOES images (Minzner *et al.* 1978; Hasler 1981). This has proved to give a better estimate of cloud heights than that inferred from thermal infrared data. Stereographic pairs of images were obtained by synchronizing the SMS-2 spin-scan camera with that of SMS-1 so that the northern most limb of the earth was viewed by both satellites within a few seconds of each other. With this configuration, both cameras continued to view the same latitude band of the earth within a few seconds of each other over the entire north-to-south stepping of the cameras. The stereomodel of the cloud fields is then measured by

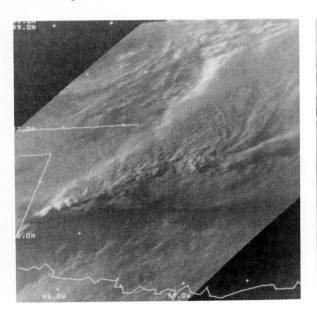 

Fig. 3.22
(a) GOES-West image of tornadic thunderstorms in Oklahoma on 3 May 1979. The storms were scanned by the satellite at 0051:19 GMT. This image has been digitally remapped into the image coordinate system of the GOES-East image (b) to form a stereopair.
(b) GOES-East image of the same storms scanned at 0051:27 GMT. Both (a) and (b) images can be viewed under a stereoscope to produce a three-dimensional model (*Courtesy*: Hasler 1981)

means of a special stereoscope and a floating dot with a parallax bar. The stereoscope permits image tilt corrections to be made. The measured parallax is employed in a parallax equation to compute heights. Crude cloud heights data obtained were then corrected for curvature and related to the sea-level reference point. The height can be measured consistently within ±500 m. Minzer *et al*. (1978) found the stereographic cloud top heights to be consistently about 1 km higher than the radar heights and reported height accuracies of ±0.1–0.2 km with a maximum of ±0.5 km. This approach permits the contouring of the cloud fields at 5 km intervals over an extensive area. Hasler (1981) reported more recent developments involving the use of the computer image analysis technique to digitally remap the GOES-West (135°W) image into the image coordinate system of the GOES-East (75°W) image to form a stereopair (Fig. 3.22 and Fig. 3.23). He confirmed that absolute heights accurate to about 0.5 km were obtainable from stereo measurements.

Stereographic observations have also been applied to other meteorological problems such as mapping the three-dimensional cloud geometry associated with severe storms; measuring cloud top and base heights; estimating winds from cloud motions; examining the dynamics of convective clouds from a time sequence of cloud top height maps; determining atmospheric temperature from stereo heights and infrared cloud top temperatures; and estimating cloud emissivity. Clearly, stereographic observations from geosynchronous satellites have provided most valuable aids in meteorological research.

Rainfall estimation

Visible and thermal infrared satellite images of clouds provide a means to estimate rainfall. This is based on the assumptions that high

Fig. 3.23
Stereo cloud top height contour analysis (in kilometres above sea-level) made from a stereo image pair (Fig. 3.22) of an Oklahoma tornadic thunderstorm complex at 0051 GMT 3 May 1979 (*Source*: Hasler 1981)

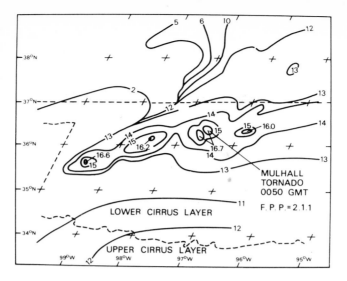

cloud brightness as observed from visible images normally implies a greater probability of rain and low cloud-top temperatures as detected from infrared images also means a greater chance of rain. All these characteristics indicate the presence of large thick clouds. In this way, precipitating clouds can be distinguished from others on both these images. However, there are also other factors determining the production of rain in clouds so that in some cases these assumptions do not hold true. Cloud brightness, for example, is not only governed by cloud thickness but also by other factors such as the area-to-volume ratio and the orientation of the solar beam. All these geometric factors determine the optical path of the light within the cloud (Barrett and Martin 1981).

Methods of rainfall estimation using the visible or infrared images fall into two groups: (1) the time-independent cloud indexing methods; (2) the time-dependent life-history methods. The *cloud indexing methods* require identification of cloud types and areas to estimate the amount of rainfall, assuming a certain intensity or probability of rainfall associated with each cloud type. A good example is the so-called Bristol method developed by Barrett (Barrett and Martin 1981). Fundamentally, it was argued that

$$R = f (C_a, C_t, S_w, A) \qquad [3.10]$$

where R is monthly rainfall, C_a is cloud cover in grid squares of selected size, C_t is cloud type with an associated probability and intensity of rainfall, S_w is a synoptic weighting factor and A is the altitude. S_w is to account for the more intense rains anticipated in the conventional clouds of the tropics. This method attempts to relate rain gauge data with polar-orbiting satellite cloud data. The grid squares ($1/6°–1°$ in size) are first identified as 'gauge cells' and 'satellite cells'. Then the study area is subdivided morphoclimatically, and regression models are established to relate cloud index data from the satellite images with the observed rainfall data from the rain gauges for an historic period of time. Both visible and infrared images are used to identify cloud types during daytime, but at night only infrared images are used. Where a grid cell contains more than one type of precipitable clouds, the two most significant types are considered separately, and the resulting cloud index for the cell is taken to be their sum. Based on these ground truths, regression models can be employed to estimate rainfall for other data sparse cells (Fig. 3.24). Because clouds with similar appearances do not always rain equally, guidelines have been established for individual rainfall estimates to be floated upwards or downwards according to the observed cloud index and rainfall relationships at gauge cells affected by the same rain-cloud system, thus eliminating the need to employ the synoptic weighting factor S_w. This method of rainfall estimation has been successfully applied to northwest Africa in 1977 for the desert locust monitoring project (Barrett 1981). On the whole, the method was found to be giving twice-daily rainfall amounts correctly on 94 per cent of all occasions, with rain estimated for rain cases on 76 per cent of such occasions in northwest Africa. It was clear that the accuracy of rainfall estimation was affected by the type of climatic regions involved and the densities of ground observation (rain gauge). However, the use of geostationary satellite image data will help to improve the accuracy of these estimations by giving information on the development or dissipation of rain-cloud systems in the atmosphere. As a result of the difficulties in automatic cloud classification discussed in the previous section, computer-assisted rainfall estimation using this method does not appear to be feasible at the present stage. There are other cloud index methods of rainfall estimation, notably those developed by National Earth Satellite Service (NESS) in the USA (Follansbee 1973; 1976).

Fig. 3.24
Flow diagram of the cloud indexing method in rainfall estimation (*Source*: Barrett 1981)

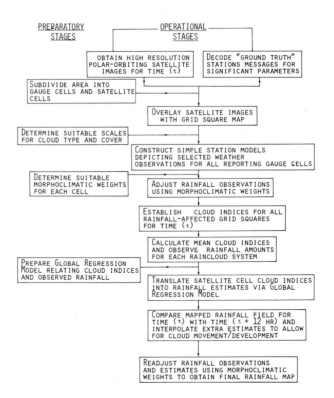

The first method estimated rainfall using NOAA polar-orbiting satellite data and the following equation:

$$R = (K_1A_1 + K_2A_2 + K_3A_3)/A_o \qquad [3.11]$$

where R is average rainfall across the broad study area for each 24 hour period; A_o is the area under study; A_1, A_2 and A_3 are areas of A_o covered by the three most important types of rain-producing clouds (cumulonimbus, cumulocongestus and nimbostratus); K_1, K_2 and K_3 are empirical coefficients. This method therefore only considered convectional clouds in the low latitudes. Later, the equation was simplified to

$$R = K_1A_1/A_o \qquad [3.12]$$

to take into account the diurnal variability of cloud in rain in these low latitudes. NESS later extended the method to estimate rainfall in higher latitudes using the following equation:

$$P_{\frac{1}{2}} = 0.09 \ r_{\frac{1}{2}} \ E(P_{30}) \qquad [3.13]$$

where r is the fraction of the estimated period that a point is covered by precipitating clouds and $E(P)$ is the climatic normal (or expected) precipitation for the point and the subcripts refer to periods in days. For a more detailed discussion of these methods, one is referred to an excellent review by Barrett and Martin (1981).

The *life-history method* examines the development of convective clouds from a time-sequence of geostationary satellite images over individual grid cells. The interval between consecutive pictures must be short. A good example is the Wisconsin method developed by Stout *et al.* (1979) who related volumetric rain rate (R_v) to cumulonimbus cloud area and areal change for estimation of tropical oceanic convective rainfall by

$$R_v = a_1A + a_2 \ dA/dt \qquad [3.14]$$

where A is the cloud area at time t, dA/dt is cloud area change, and a_1 and a_2 are empirical coefficients. They made use of the geostationary SMS-1 visible and infrared data at intervals of 30 minutes or less to test their model. A cumulonimbus cloud was first identified based on its growth, movement, texture, brightness and temperature. Its area and location were measured in the first picture. The cloud area was approximated by a single threshold contour of temperature or brightness. The thresholds were 200 W/m² and 245 K for visible and infrared data respectively. This cumulonimbus cloud was followed in each picture in the sequence until it disappeared. This was repeated for all cumulonimbus clouds in the first picture. After this, new cumulonimbus clouds were searched for in the second and succeeding pictures. Volumetric rain rate (R_v) was calculated step by step for each cloud. They could be summed to give rainfall amount for any time interval or mapped to show rainfall distribution. Comparison with hourly radar rain rates gave a strong correlation of 0.84 (Fig. 3.25). The figure also indicated that satellite rainfall tended to be smoother than radar rainfall. A comparison between the maps of satellite rain-

Fig. 3.25
Satellite and radar hourly rain rates averaged over a disc 204 km in radius centred at 8°30'N, 23°30'W (*Source*: Martin 1981)

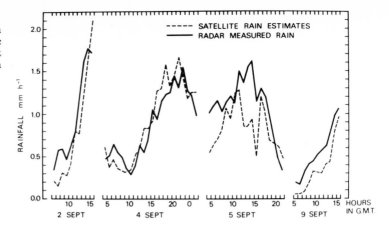

fall and radar rainfall also indicated a slight displacement towards the north and northwest (Fig. 3.26). Otherwise, the 24-hour rainfall pattern showed remarkable agreement. Obviously, the method was inflexible and quite tedious to apply in practice. It also demanded much skill. However, this example shows very well how the life-history method of rainfall estimation works. Griffith and Woodley (Griffith *et al.* 1978) estimated summer time convective rainfall from geosynchronous satellite (SMS-1 and ATS-3) thermal infrared images calibrated by a combined system of rain gauge and radar data over south Florida. The cloud area of the cumulonimbus was found to be related directly to the radar echo area which in turn was linearly related to volumetric rainfall rate. By measuring the cloud areas from the infrared satellite images and obtaining the appropriate echo areas from the relationship:

$$A_e = c\,(t, A_{max}).A \qquad [3.15]$$

Where A_e is the equivalent echo area, A is the cloud area, A_{max} is the maximum satellite cloud area and $c(t)$ is the growth or decay coefficient, it is possible to estimate the volumetric rainrate (R_v) from:

$$R_v = a(t).b(T).A_e \qquad [3.16]$$

Fig. 3
Maps of (a) satellite rainfall and (b) radar rainfall (*Source*: Martin 1981)

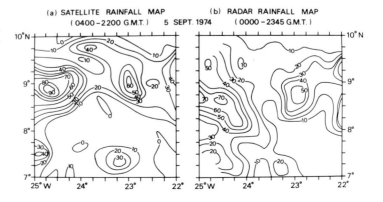

where *a(t)* is the growth or decay coefficient and *b(T)* is the coefficient for the apportionment of rain by cloud top temperature structure. The major advantage of this method is that it can be automated. The method was found to give an over-estimation of rainfall and performed better on heavier rain than on light rains (Griffith and Woodley 1981).

Another approach is by Scofield and Oliver (1981) who estimated rainfall at half-hourly or hourly intervals from convective systems (such as thunderstorms) by using GOES infrared and high-resolution visible images by a decision-tree technique. The thermal infrared images of a thunderstorm in two consecutive pictures were enhanced digitally. They were then compared to detect any changes. The active portion of the thunderstorm was identified. An estimate of the rainfall (R) for the active portion of the thunderstorm could be obtained from the meteorological factors expressed in the following equation:

$$R = [a_1 (T.dA/dt) + a_2 + a_3 + a_4]. a_5(dW) \qquad [3.17]$$

where *dA/dt* is the expansion of the coldest contour in the enhanced thermal infrared images; *T* is the cloud top temperature; *dW* is the departure of precipitable water from a summertime normal, and other parameters relate to the occurrence of overshooting tops (a_2), merging thunderstorms and convective cloud lines (a_3) and the duration of the storm (a_4). The application of this method to flash flood producing thunderstorm over Chicago and to Hurricane Allen crossing southern Texas has been found to be successful (Scofield 1981).

Apart from the use of satellite visible and infrared data, one should note the emergence of the *passive microwave* method in rainfall estimation. As has been outlined in Chapter 1, the passive microwave sensor detects thermal emission the intensity of which is linearly related to the blackbody temperature corresponding to the *brightness temperature*. However, the brightness temperature is also affected by the direction of polarization of the electromagnetic wave. At 37 GHz the calculated brightness temperature difference of horizontally and vertically polarized radiation emerging through a standard atmosphere exceeds 30 K (Rodgers *et al.* 1979). Three constituents of the atmosphere are: water vapour, oxygen and liquid water droplets. Water vapour has a weak absorption line of microwaves at 22.235 GHz and a sequence of very strong lines at 183 GHz and higher. Oxygen has two major absorption lines, one near 60 GHz and one at 118.75 GHz. Thus, by selecting the appropriate operating frequencies, the microwave sensor can either look through or look at the atmosphere. It has been found that the frequency intervals best suited for measurement of rain are less than 20 GHz and close to 30 GHz (Barrett and Martin 1981, p. 137). Microwave radiometry from satellites has been used to remotely measure rainfall from Nimbus 55 (using ESMR-5 at 19.35 GHz frequency or 1.55 cm wavelength with a horizontal polarization), Nimbus 6 (using ESMR-6 at 37 GHz frequency or 0.81 cm wavelength with horizontal and vertical polarization) and Nimbus 7 (using Scanning Multichannel Microwave Radiometer SMMR at 6.6, 10.7, 18, 21, 37 GHz frequencies or respectively 4.6, 2.8, 1.7, 1.4, 0.8 cm wavelengths). With the use of 19.35 GHz frequency for the ESMR-5, rainfall estimation over the oceans has been quite successful because the emissivity of the ocean surface is small and not highly variable which thus provides a good background for observing rain

(Wilheit *et al.* 1977). However, it would be less successful over land because of the poor contrast between rain clouds and background as a result of the large and highly variable emissivity of land surfaces. The ESMR-6 which is tuned to a higher frequency of 37.0 GHz is more sensitive to liquid water droplets and, when measuring in both polarization components, is capable of giving a more accurate rainfall estimation over land (Rodgers *et al.* 1979). It has been shown that while both water on the ground and precipitation have low brightness temperatures, the brightness temperature emerging from wet ground is polarized in contrast to the unpolarized one from the precipitation. This provides a means to distinguish wet soil from dry ground and rain, especially if the surfaces are warm. By applying a supervised approach and a Bayesian classifier in a computer, Rodgers and his collaborators have been able to use the ESMR-6 data to generate maps of rain and wet ground over the southeastern USA at 70 and 80 per cent confidence levels (Fig. 3.27). A general agreement in the rain area was confirmed, but much shortage of rain in areas to the east and southwest was also noted.

It is worthy to note that the brightness temperature at 37 GHz becomes insensitive to changes in rainfall rate once rainfall rates exceed 8 mm/hour. On the whole, it seems that the microwave method tends to underestimate the rain amount. One major problem that remains to be solved is the effect of *non-uniform beam filling* caused by large spatial variation in the rain rate within the fields of view of the microwave antenna. This would cause a bias in the rainfall estimate. Another problem is the physical rain model which relates rain layer thickness with the rainfall rate and climate region to be adopted (Burke *et al.* 1982). Thus there are still many problems waiting to be

Fig. 3.27
Rainfall pattern over the southeast USA, near 1630 GMT on 14 September 1976 (*Source*: Rodgers *et al.* 1979)

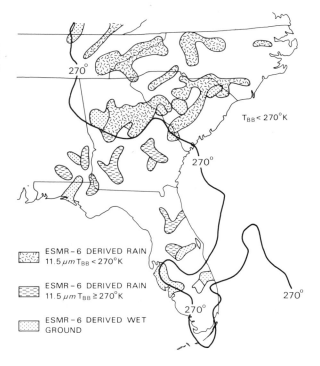

Fig. 3.28
Nimbus 4 THIR photofacsimile imagery in the (a) 6.7 μm and (b) 11.5 μm spectral regions at local midnight over the western USA on orbit 2565, 16 October 1970. Area A is cloud mass visible in both spectral regions. Areas B and C are regions that are clear in the 11.5 μm image but show very noticeable detail in the 6.7 μm image (*Source*: Steranka *et al*. 1973)

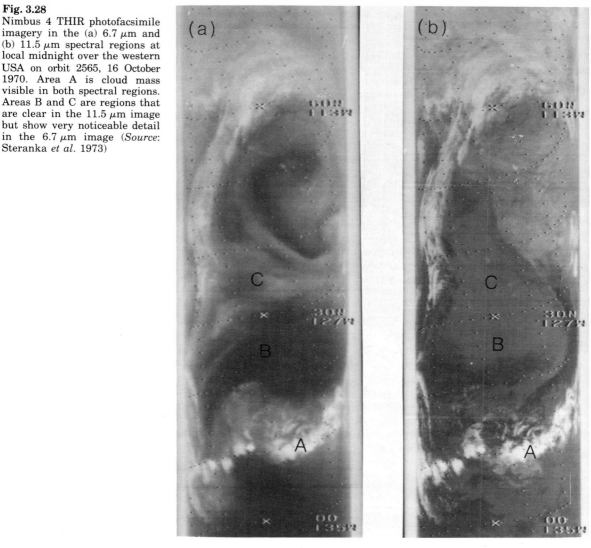

solved before this new remote sensing tool can be applied operationally to collect rainfall data from the atmosphere.

Water vapour analysis

The moisture content of the atmosphere is another important parameter which can be measured directly from satellite images. The Nimbus 4 satellite carried the Temperature-Humidity Infrared Radiometer (THIR) which monitored radiation in the 6.5–7.2 μm water vapour absorption region with a 23 km spatial resolution. The radiation

Fig. 3.29
Enhanced 400 mb moisture analysis over the USA at 1200 GMT on 16 October 1970. The moisture patterns were obtained from the 6.7 μm data in Fig. 3.28(a) (*Source*: Steranka *et al.* 1973)

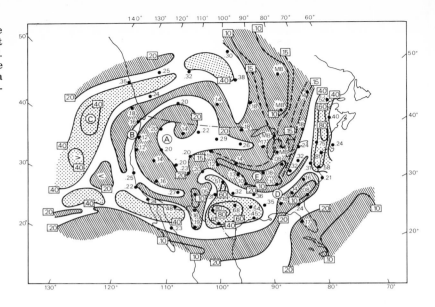

so monitored principally came from emission in the 500–250 mb region of the upper troposphere. By examining the photofacsimile imagery in the 6.7 μm channel, moisture-rich or cloudy regions can be detected by the different levels of grey shading. The drier areas exhibit dark grey to black tones while the wet regions are white or light grey (Fig. 3.28). By associating these grey shades with the radiosonde soundings, an experienced analyst can obtain a good idea of the vertical stratification of the moisture content (Steranka *et al.* 1973). From the same imagery, it would be possible to estimate the *mixing ratio*, i.e. the mass of water vapour compared to the mass of dry air within which the water vapour is contained, in the atmosphere over the USA. The resulting pattern (Fig. 3.29) was found to be richer in details of water vapour distribution than conventional observations alone.

This technique has been extended to cover the whole world in association with wind flow analysis which is possible with the water vapour imagery. Such types of analyses benefit data sparse regions particularly over the oceans. More recently, the geosynchronous METEOSAT monitors the earth with a water vapour channel (5.7–7.1 μm) at a more frequent (half-hourly) interval in a higher spatial resolution (5 km) than the polar-orbiting Nimbus 4, thus giving a global view of the moisture pattern of the middle troposphere (300–600 mb) (Fig. 3.30) (Morel, Desbois and Szejwach, 1978). The following large-scale patterns were noted: (1) a long water vapour band stretching from South America to the tip of South Africa, ending in an extra-tropical cyclonic pattern; (2) a well-marked separation between a wet tropical high troposphere and a dry subtropical one following a wavy line from the Carribean Sea to south Tunisia; (3) the presence of extra-tropical perturbations. Clearly, the water vapour channel imagery of a geostationary satellite gives better insights into large-scale mid- and high-tropospheric air motions through studying successive pictures. For a more quantitative approach, an attempt has already been made to relate the radiance and the water vapour mass

Fig. 3.30
Meteosat water vapour image
(5.7−7.1 μm), taken on 1
September 1978 at 1255 GMT
(*Courtesy*: P. Morel)

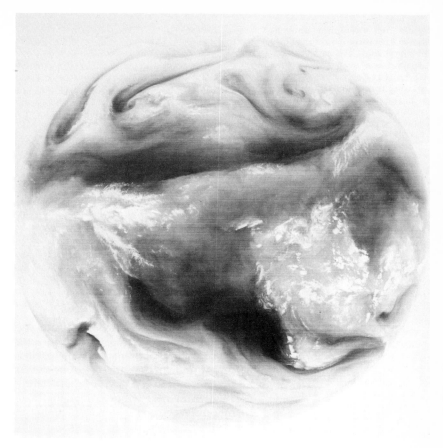

above 600 mb using the Meteosat water vapour channel data with the
aid of conventional radiosonde data (Poc *et al*. 1980).

Wind field analysis

Satellite data have also been employed in determining wind speed and
wind direction. The moisture patterns as detected from the Nimbus 4
THIR 6.7 μm water vapour channel are generally aligned with the
wind field (Steranka *et al*. 1973). This suggests that the water vapour
imagery could be used to infer the directions of horizontal wind flows
(i.e. the so-called *streamline* analysis) shown in Fig. 3.31 and Fig. 3.28.
More commonly, wind flows at different levels can be inferred from
cloud motion as revealed by the animated photography of the spin-
scan cloud cameras onboard some geosynchronous satellites such as
the Applications Technology Satellites (ATS). The sequence of photographs are used to produce a loop movie which is viewed. Cloud
elements (cloud tracers) are selected and their positions at the beginning and the end of the photograph sequence are fixed. The sequence
may require at least 5 photographs covering about 106 minutes. The
displacements of the selected cloud elements can be measured. These

Fig. 3.31
A detailed streamline analysis
derived from the patterns shown
in Fig. 3.28(a) and conventional
observations over the western
USA. The arrowheads were
inferred from the light and dark
patterns in Fig. 3.28(a) that
resulted from upper tropospheric dynamics affecting the
distribution of water vapour
(*Source*: Steranka *et al.* 1973)

will give cloud motion vectors from which the direction and speed of
the wind at the level of the cloud element can be determined after
appropriate corrections for map projection errors (Fujita 1969). The
height of the cloud element can be obtained from infrared data if available. The method can give about the same accuracy as wind speeds
derived from radiosondes. The method has been applied successfully
in mapping the low-level wind fields over the western Indian Ocean
west of 65°E from the METEOSAT images in the visible and thermal
infrared channels, with an error of less than 3 m/sec for speed and 10°
for direction (Cadet and Desbois 1980). The major problem of the
method is the need to ensure precise alignment of the photographs
with each other by using landmarks. It has been possible to show in
a sequence of wind field maps for 10, 12 and 16 May 1978 the abrupt
change of the low-level airflow circulation over the Indian Ocean
known as the burst of the monsoon (Figs 3.32, 3.33 and 3.34). The
establishment of the Somali jet stream in the broad-scale monsoon
system was particularly noted.

Fig. 3.32
Low-level wind vectors determined from cloud displacements
on (a) 10 May 1978–1100 GMT,
(a) 12 May 1978 – 1100 GMT
and (c) 16 May 1978 –
1100 GMT. Each full barb
represents 5 m/s. Refer to
Figs. 3.33 and 3.34 (*Source*:
Cadet and Desbois 1980)

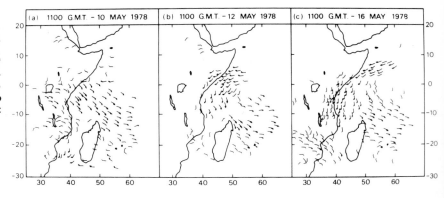

Fig. 3.33
Meteosat image in the visible channel (0.4–1.1 μm), 1 September 1978, 1255 GMT (*Courtesy*: P. Morel)

Fig. 3.34
Meteosat infrared image (10.5–12.5 μm), 1 September 1978, 1255 GMT (*Courtesy*: P. Morel)

The method of obtaining cloud motion has been automated with the computer using the cross-correlation technique which computes the cross-correlation coefficients between two sequential digital data sets after precise geographic matching of these data to a proper map projection (Leese *et al.* 1971). By using the method of fast Fourier Transform in the algorithm, the cross-correlation coefficient can be computed in much less time than that needed in a direct method of computation. This has been applied to single-layer clouds and multiple-layer clouds in low, middle and high levels. It was concluded that the automated method performed particularly well for single-layer clouds, but was handicapped by its inability to discriminate between motions when more than one cloud layer occurred over the same geographical area. It also tended to do better at measuring speed than direction.

Another approach in automatic cloud tracking was to make use of pattern recognition technique (Endlich and Wolf 1981). The two successive pictures were matched with landmarks and then digitally smoothed by averaging the brightness values in a window of pixels. The points that were brighter than the smoothed pictures were selected as cloud tracers and organized into 'touching groups' (or points that touch at least one other member of the group). Extraneous points from large groups were eliminated. After these steps, the two pictures were matched according to the best likeness of the targets in terms of size, brightness and shape at the two moments in time to give the cloud displacements. These cloud displacements were then transformed from photograph co-ordinates to earth co-ordinates. This technique has been tested with the GOES visible and infrared data for the Hurricane Eloise over the Gulf of Mexico on 22 September 1975 and also with the METEOSAT water vapour channel (6.7 μm) for 25 April 1978. It was concluded that the technique performed well from the data for Hurricane Eloise and there was little doubt of the accuracy of the automatically computed cloud motions. The best result was obtained with 4 km resolution data at 10 minute intervals. For pictures separated by intervals of about 30 minutes, data of much coarser resolution were required to track the relatively high speed motions. The method therefore requires the matching of the time and space scales. The automatic tracking of clouds in the water vapour patterns of the METEOSAT was found to be less reliable if the contrast was low. However, in areas where the water vapour fields contained small-scale structures (i.e. in the vicinity of active weather phenomena) the computations were more accurate. On the other hand, the METEOSAT thermal infrared data gave more successful results in tracking clouds. As compared with the cross-correlation technique, the pattern recognition method obtained fewer cloud vectors and the two methods differed by only fractions of a pixel in the computed cloud displacements.

Finally, attention should be given to the possibility of obtaining wind pressure from satellite microwave data such as the Scanning Microwave Spectrometer (SCAMS) onboard Nimbus 6 (Kidder *et al.* 1978). The 55.45 GHz channel data were used to map brightness temperature anomalies of tropical cyclones, which were then related to central pressure by the simple regression technique. From this the surface pressure field could be estimated. This permitted tangenital surface winds to be computed from

$$V_o = C_r^{-x} \qquad [3.18]$$

where V_o is the tangential wind velocity in knots, C is the central pressure in mb, r is the radius of curvature of maximum winds in degrees and x is a constant equal to 0.5. Further research to refine the technique for this area of application is required.

Prediction of severe storms

An important application of satellite remote sensing which has a great impact on weather forecasting is the prediction of such meso-scale weather phenomena as tornadoes, thunderstorms, hailstorms, hurricanes or typhoons. This is particularly facilitated with the availability of the geosynchronous satellite thermal infrared data. Based on his observations on GOES imagery, Purdom (1976) has noted that thunderstorms often form within straight or arc-shaped cloud lines. Adler and Fenn (1979) have shown that using SMS-2 infrared data the occurrence of severe weather such as tornadoes, hail and high wind on ground is closely related to the vertical growth rates and cloud top temperature which indicate the degree of intensity of convection. They discovered that the severe weather elements tended to have cold minimum cloud top temperatures and large rates of growth. This is explained by the model of deep tropical convection proposed by Sikdar and Suomi (1971), that air entering a cumulonimbus cloud from the atmospheric boundary layer rises in the updraught and spreads out beneath the tropopause (Fig. 3.35). The upper part of the cloud is cirrus, composed mainly of ice crystals and is known as the 'anvil'. The colder cloud-top temperature therefore indicates a stronger updraught of the air which may even penetrate the tropopause. By measuring the areal expansions of the cold areas through time, one can therefore detect and monitor the growth rates of the storm. This method has also been applied to the identification of hailstorms in the High Plains in the USA by Reynolds (1980) who made use of enhanced GOES-East (75°W) thermal infrared and visible imagery. Storms penetrating the tropopause were found to have a higher probability of producing damaging hail. A strong relationship was observed to exist between the onset of large hail and cloud-top temperatures getting colder than the environmental tropopause temperature, accompanied by a rapid vertical growth of cloud.

Gentry *et al.* (1980) made use of the infrared (11.5 μm) window channel data of the Medium Resolution Infrared Radiometer (MRIR) onboard the polar-orbiting satellite Nimbus-3 to obtain blackbody temperatures of cloud tops (T_{BB}) to predict the intensity of tropical cyclones. The areal distribution of T_{BB} indicated the amount of latent heat released and the extent to which the clouds of the storm were organized into patterns. The mean T_{BB} parameter, the current

Fig. 3.35
Schematic diagram of the flow in a cumulonimbus cloud

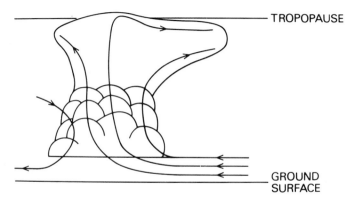

TROPOPAUSE

GROUND SURFACE

Fig. 3.36
Schematic expression of the evolution of the Oklahoma cloud of 29 May 1977 from severe thunderstorm to tornadic cloud: (a) penetrative overshooting above the tropopause with an altitude of 13.5 km, occurred under the intense vertical convention at 0004 GMT, (b) the growing overshooting turret with a height of 2.5 km above the tropopause observed at 0033 GMT. The gravity wave trains being continuously excited during the period from (a) to (c), and (d) rapid collapsing of overshooting cloud tops and the formation of funnel cloud occurred at 0148 GMT (*Source*: Hung and Smith 1982).

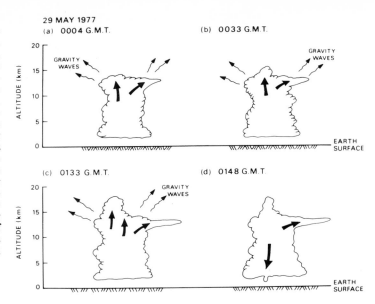

maximum wind and the change in maximum winds during the preceding 24 hours were employed as predictors in regression equations to predict the future intensity of the storm. Hung and Smith (1982) have studied tornadoes using GOES digital infrared data. They found that clouds associated with tornado activity were characterized by both a very low temperature at the cloud top, which suggested a higher penetration above the cirrus canopy, and a very high growth rate of the cold region of the cloud top, an indication of enhanced convection in the cloud. They further observed that clouds penetrating above the tropopause formed overshooting turrets and that the size of the area of the overshooting turret associated with severe storms would be large (of the order of several hundred pixels). But this overshooting turret above the tropopause began to collapse about 15 to 30 minutes before the tornado touchdown. They also found that gravity waves were always present from the clouds associated with severe storms. From these observations, a schematic expression of the evolution of a cloud from severe thundercloud to tornadic cloud can be developed (Fig. 3.36).

Another approach to predict storm intensity is to make use of microwave data in the 19.35 GHz (1.55 cm) region from the Electrically Scanning Microwave Radiometer (ESMR-5) onboard Nimbus 5 to measure the distribution and intensity of rainfall within tropical cyclones which are positively related to storm intensity. Thus, from rainfall measurements one can estimate the latent heat release which is essential to the maintenance and intensification of tropical cyclones at the later stages of development (Hunter, Rodgers and Shenk 1981).

Weather analysis and forecast

Clearly, the availability of polar-orbiting and geosynchronous satellites as platforms for remote sensing of the atmosphere has provided valuable data to improve the accuracy of short-term weather forecasting, particularly for data-sparse areas of the world, such as the tropics. The construction of synoptic weather charts is greatly facilitated with the availability of visible and infrared imagery. Most of the data are also available in digital form which facilitates computer processing. This makes possible near real-image display of the cloud pattern over a place such as has been done in Britain using METEOSAT infrared channel data (Ball *et al.* 1979). Another example is the McIDAS (Man computer Interactive Data Access System) of the University of Wisconsin in Madison, USA which can integrate conventional and satellite data together and permits real-time monitoring of rapidly changing weather situations (Wash and Whittaker 1980). But the use of remote sensing data should be regarded as complementing rather than replacing the more conventional meteorological data. Much useful information can be extracted from an interpretation of the satellite images obtained in the visible and thermal infrared channels of the remote sensors. One should recall here that the grey shades in the visible images are dependent on the reflectivity of the radiating surface whilst those in the thermal infrared images are dependent on the temperature of the radiating surface. The weather images show mostly clouds. Thus, the whiter a cloud appears in the visible image, the thicker is the cloud, but the whiter a cloud appears in the thermal infrared image, the colder and higher is the cloud. This gives a good idea of the spatial distribution and vertical extension of the clouds. One can further examine the configuration and structure of these cloudy areas. Zwatz-Meise (1981) suggested that the distinction of the following four types of cloud configurations, *viz.*, cloud bands, vortex configurations, cellular cloudiness and uniform cloud areas, would be useful to weather analysis and forecast. Indeed, by applying Bjerknes' cyclone model (Petterssen, 1969) together with observations from satellite images of the temperate latitudes, the different types of fronts and wind flows can be identified with certainty as demonstrated by Fotheringham (1979) in his collection of NOAA 5 Very High Resolution Scanning Radiometer (VHRR) acquired images of Europe. On the whole, much achievement has been made in the detection of hazardous weather such as heavy rain, flash flooding, severe thunderstorms, hurricanes (typhoons) or tornadoes which have a great economic impact (Wasserman 1977). The availability of passive microwave data and the possibility of an active spaceborne radar system in future will certainly improve our ability to monitor rainfall, temperature, humidity and winds in the lower atmosphere.

Conclusions

From the foregoing review, a number of significant characteristics of the remote sensing data as applied to study our atmosphere has emerged:

1. Much of the remote sensing data comes from space platforms. Those acquired from the lower altitude polar-orbiting satellites complement those from the high-altitude geostationary satellites. These data are becoming increasingly diverse and multispectral in recent years. Both

imaging and non-imaging data are employed to provide a complete view of the atmosphere in both the horizontal and vertical planes as well as in a time sequence. It should be noted that complicated mathematical reduction is normally required to extract the desired information from the satellite data, especially when remote sounding data are involved.

2. Cloud data provide the focus for most of the remote sensing applications whilst atmospheric absorption of the radiant energy by various gases such as carbon dioxide and water vapour is deliberately allowed to measure a vertical profile of temperatures. This contradicts remote sensing applications to study the earth's surface where cloud cover and atmospheric absorption are regarded as major handicaps.

3. The weather satellite data suffer from poor spatial resolutions and a lack of standardization which present some problems to interpretation. The great variety of meteorological remote sensors in the research and operational satellites is the cause for the incompatibility. In addition, the compatibility between satellite data and conventional ground measured data is poor because of a lack of synchronization of observations and it would be rather difficult to verify the accuracy of the findings where satellite data are used.

4. The major problem of remote sensing applications to the atmospheric environment is data explosion. The large volume of satellite data collected each day requires selection, reduction and automatic handling by the computer. Whilst this accumulation of data will be useful for climatological modelling, meteorologists who face a day-to-day forecasting problem have not only to choose from all possible evaluations of data, those which give the maximum information in the minimum time but also to use satellite images at the right time together with the right parameters.

Despite all these limitations, the scope of remote sensing applications to the atmospheric environment is wide and their outcome has great practical importance. As observed by Barrett and Watson (1977), a new approach in the analysis and application of satellite data is needed so that satellite data can be analysed in terms of their own intrinsic characteristics rather than routinely in terms of the conventional parameters. With the advance in remote sensor technology, one can expect new forms of data which will permit a more penetrative understanding of our atmosphere.

References

Adler, R. F. and Fenn, D. D. (1979) Thunderstorm intensity as determined from satellite data, *Journal of Applied Meteorology* **18**: 502–17.

Ball, A. P., Browning, K. A., Collier, C. G., Hibbett, E. R., Menmuir, P., Owens, R. G., Ponting, J. F., Whyte, K. W. and Wiley, R. L. (1979) Thunderstorms developing over Northwest Europe as seen by Meteosat and replayed in realtime on a fast-replay colour display, *Weather* **34**: 141–7.

Barrett, E. C. (1970) Satellites in geographical research: In *Studies in Geographical Methods, Proceedings of the 3rd Anglo-Polish Geographical Seminar Baranow Sandomierski, September 1–10, 1967*, Polish Scientific Publishers: Warsaw, pp. 243–60.

Barrett, E. C. (1974) *Climatology from Satellites*, Methuen: London.

Barrett, E. C. (1981) Satellite rainfall estimation by cloud indexing methods for desert locust survey and control: In Deutsch, M., Wiesnet, D. R. and Rango, A. (eds), *Satellite Hydrology: Proceedings of the Fifth Annual William T. Pecora Memorial Symposium on Remote Sensing, Sioux Falls, South Dakota, June 10–15, 1979*. American Water Resources Association: Minneapolis, Minnesota, pp. 92–100.

Barrett, E. C., Grant, C. K. and Harris, R. (1976) *Multispectral Characteristics of Clouds Observed by Landsat 2*. ERTS Follow-on Program Study No. 2962A, Fourth quarterly report on mesoscale assessments of cloud and rainfall over the British Isles, NASA-CR-148912: Greenbelt, Maryland.

Barrett, E. C. and Hamilton, M. G. (1982) The use of geostationary satellite data in environmental science, *Progress in Physical Geography* **6**: 159–214.

Barrett, E. C. and Martin, D. W. (1981) *The Use of Satellite Data in Rainfall Monitoring*. Academic Press: London; New York; Toronto; Sydney; San Francisco.

Barrett, E. C. and Watson, I. D. (1977) Problems in analysing and interpretating data from meteorological satellites: In Barrett, E. C. and Curtis, L. F. (eds), *Environmental Remote Sensing 2: Practices and Problems*. Edward Arnold: London, pp. 276–303.

Barry, R. G. (1970) A framework for climatological research with particular reference to scale concepts, *Transactions, Institute of British Geographers*, **49**: 61–70.

Battan, L. J. (1973) *Radar Observation of the Atmosphere*. The University of Chicago Press: Chicago; London.

Brownscombe, J. L. and Nash, J. (1982) Long-term monitoring of stratospheric temperature, *Remote Sensing and the Atmosphere, Proceedings of the Annual Technical Conference held in Liverpool in December 1982*. Remote Sensing Society, pp. 2–11.

Burke, H. H. K., Hardy, K. R. and Tripp, N .K. (1982) Detection of rainfall rates utilizing spaceborne microwave radiometers, *Remote Sensing of Environment* **12**: 169–80.

Cadet, D. and Desbois, M. (1980) The burst of the 1978 Indian summer monsoon as seen from METEOSAT, *Monthly Weather Review* **108**: 1697–1701.

Carlson, T. N., Dodd, J. K., Benjamin, S. G. and Copper, J. N. (1981) Satellite estimation of the surface energy balance, moisture availability and thermal inertia, *Journal of Applied Meteorology* **20**: 67–87.

Conover, J. H. (1963) *Cloud Interpretation from Satellite Altitudes*, Research Note 81, Supplement 1. Air Force Cambridge Research Laboratories: Cambridge, Massachusetts.

Cornillon, P. (1982) *A Guide to Environmental Satellite Data*. University of Rhode Island Marine Technical Report **79**.

Desbois, M., Seze, G. and Szejwach, G. (1982) Automatic classification of clouds on METEOSAT imagery: application to high-level clouds, *Journal of Applied Meteorology* **21**: 401–12.

Endlich, R. M. and Wolf, D. E. (1981) Automatic cloud tracking applied to GOES and METEOSAT observations, *Journal of Applied Meteorology* **20**: 309–19.

Follansbee, W. A. (1973) *Estimation of Average Daily Rainfall from Satellite Cloud Photographs*. NOAA Technical Memorandum NESS **44**: Washington, DC.

Follansbee, W. A. (1976) *Estimation of Daily Precipitation over China and the USSR using Satellite Imagery*. NOAA Technical Memorandum NESS 81: Washington, DC.

Fotheringham, R. R. (1979) *The Earth's Atmosphere Viewed from Space*: University of Dundee, Dundee.

Fujita, T. (1969) Present status of cloud velocity computations from the ATS-1 and ATS-3 satellites, *Space Research IX*: The University of Chicago Press, pp. 551–70.

Gentry, R. C., Rodgers, E., Steranka, J. and Shenk, W. E. (1980) Predicting tropical cyclone intensity using satellite-measured equivalent blackbody temperatures of cloud tops, *Monthly Weather Review* **108**: 445–55.

Greaves, J. R. and Chang, D. T. (1970) *Technique Development to Permit Optimum Use of Satellite Radiation Data*. Final Report on NASA Contract N62306-69-C-0227. Goddard Space Flight Center: Greenbelt, Maryland.

Griffith, C. G. and Woodley, W. L. (1981) The estimation of convective precipitation from GOES imagery with the Griffith/Woodley technique: In Atlas, D. and Thiele, O. W. (eds) *Precipitation Measurements from Space Workshop Report*. National Aeronautics and Space Administration: Goddard Space Flight Center, Greenbelt, Maryland, pp. D154–8.

Griffith, C. G., Woodley, W. L., Grube, P. G., Martin, D. W., Stout, J. and Sikdar, D. N. (1978) Rain estimation from geosynchronous satellite imagery – visible and infrared studies, *Monthly Weather Review* **106**: 1153–71.

Gube, M. (1982) Radiation budget parameters at the top of the earth's atmosphere derived from METEOSAT data, *Journal of Applied Meteorology* **21**: 1907–21.

Harris, R. (1982) Computer assisted nephanalysis: the story so far, in *Remote Sensing and the Atmosphere: Proceedings of the Annual Technical Conference held in Liverpool in December 1982*. Remote Sensing Society, pp. 248–54.

Harris, R. and Barrett, E. C. (1978) Toward an objective nephanalysis, *Journal of Applied Meterology* **17**: 1258–66.

Hasler, A. F. (1981) Stereographic observations from geosynchronous satellites: an important new tool for the atmospheric sciences. *Bulletin of the American Meteorological Society* **62**: 194–212.

Houghton, J. T. (1979) The future role of observations from meteorological satellites, *Quarterly Journal of the Royal Meteorological Society* **105**: 1–23.

Houghton, J. T. and Smith, S. D. (1970) Remote sounding of atmospheric temperature from satellites I. Introduction, *Proceedings of the Royal Society of London* A, **320**: 23–33.

Hung, R. J. and Smith, R. E. (1982) Remote sensing of tornadic storms from geosynchronous satellite infrared digital data, *International Journal of Remote Sensing* **3**: 69–81.

Hunter, H. E., Rodgers, E. B. and Shenk, W. E. (1981) An objective method for forecasting tropical cyclone intensity using Nimbus-5 electrically scanning microwave radiometer measurements, *Journal of Applied Meteorology* **20**: 137–145.

Jacobowitz, H., Smith, W. L., Howell, H. B. and Nagle, F. W. (1979) The first 18 months of planetary radiation budget measurements from the Nimbus 6 ERB experiment, *Journal of the Atmospheric Sciences* **36**: 501–7.

Kidder, S. Q., Gray, W. M. and Vonder Haar, T. H. (1978) Estimating tropical cyclone central pressure and outer winds from satellite microwave data, *Monthly Weather Review* **106**: 1458–64.

Kondrat'ev, K. Ya, Borisenkov, E. P. and Morozkin, A. A. (1970) *Interpretation of Observation Data from Meteorological Satellites*. Israel Program for Scientific Translations: Jerusalem.

Leese, J. A., Novak, C. S. and Clark, B. B. (1971) An automated technique for obtaining cloud motion from geosynchronous satellite data using cross correlation, *Journal of Applied Meteorology* **10**: 118–32.

Minzner, R. A., Shank, W. E., Teagle, R. D. and Steranka, J. (1978) Stereographic cloud heights from imagery of SMS/GOES satellites, *Geophysical Research Letters* **5**: 21–4.

Morel, P., Desbois, M. and Szejwach, G. (1978) A new insight into the troposphere with the water vapor channel of Meteosat, *Bulletin of the American Meteorological Society* **59**: 711–4.

Norton, C. C., Mosher, F. R. and Hinton, B. (1979) An investigation of surface albedo variations during the recent Sahel drought, *Journal of Applied Meteorology* **18**: 1252–62.

Parikh, J. (1977) A comparative study of cloud classification techniques, *Remote Sensing of Environment* **6**: 67–81.

Parikh, J. (1978) Cloud classification from visible and infrared SMS-1 data, *Remote Sensing of Environment* **7**: 85–92.

Parikh, J. A. and Ball, J. T. (1980) Analysis of cloud type and cloud amount during GATE from SMS infrared data, *Remote Sensing of Environment* **9**: 225–45.

Park, S. U., Sikdar, D. N. and Soumi, V. E. (1974) Correlation between cloud thickness and brightness using Nimbus 4 THIR data (11.5 μm channel) and ATS 3 digital data, *Journal of Applied Meteorology* **13**: 402–10.

Petterssen, S. (1969) *Introduction to Meteorology*, McGraw-Hill: New York; St. Louis; San Francisco; Toronto; London; Sydney.

Poc, M. M., Roulleau, M., Scott, N. A. and Chedin, A. (1980) Quantitatative studies of Meteosat water-vapor channel data, *Journal of Applied Meteorology* **19**: 868–76.

Purdom, J. F. W. (1976) Some uses of high-resolution GOES imagery in the mesoscale forecasting of convection and its behavior, *Monthly Weather Review* **104**: 1474–83.

Ray, P. S., Ziegler, C. L., Bumgarner, W. and Serafin, R. J. (1980) Single- and multiple-Doppler radar observations of tornadic storms, *Monthly Weather Review* **108**: 1607–25.

Raschke, E., Vonder Haar, T. H., Bandeen, W. R. and Pasternak, M. (1973) The annual radiation balance of the earth-atmosphere system during 1969–70 from Nimbus 3 measurements, *Journal of the Atmospheric Sciences*, **30**: 341–64.

Reynolds, D. W. (1980) Observations of damaging hailstorms from geosynchronous satellite digital data, *Monthly Weather Review* **108**: 337–48.

Rockwood, A. A. and Cox, S. K. (1978) Satellite inferred surface albedo over northwestern Africa, *Journal of the Atmospheric Sciences* **35**: 513–22.

Rodgers, E. B., Siddalingaiah, H., Chang, A. T. C. and Wilheit, T. (1979) A statistical technique for determining rainfall over land employing Nimbus 6 ESMR measurements, *Journal of Applied Meteorology* **18**: 978–91.

Saunders, R. W. and Hunt, G. E. (1980) METEOSAT observations of diurnal variation of radiation budget parameters, *Nature* **283**: 645–7.

Scofield, R. A. (1981) Visible and infrared techniques for flash flood hydrological, and agricultural applications: In Atlas, D. and Thiele, D. W. (eds) *Precipitation Measurements from Space Workshop Report*. National Aeronautics and Space Administration: Goddard Space Flight Center, Greenbelt, Maryland, pp. D145–53.

Scofield, R. A. and Oliver, V. J. (1981) A satellite derived technique for estimating rainfall from thunderstorms and hurricanes: In Deutsch, M., Wiesnet, D. R. and Rango, A. (eds) *Satellite Hydrology*. American Water Resources Association: Minneapolis, Minnesota, pp. 70–76.

Shenk, W. E., Holub, R. J. and Neff, R. A. (1976) A multispectral cloud type identification method developed for tropical ocean areas with Nimbus-3 MRIR measurements, *Monthly Weather Review* **104**: 284–91.

Sikdar, D. N. and Suomi, V. E. (1971) Time variation of tropical energetics as viewed from a geostationary altitude, *Journal of the Atmospheric Sciences* **28**: 170–80.

Smith, W. L., Hickey, J., Howell, H. B., Jacobowitz, H., Hilleary, D. T. and Drummond, A. J. (1977) Nimbus-6 earth radiation budget experiment, *Applied Optics* **16**: 306–18.

Staelin, D. H., Rosenkranz, P. W., Barath, F. T., Johnston, E. J., and Waters, J. W. (1977) Microwave spectroscopic imagery of the earth, *Science* **197**: 991–3.

Steranka, J., Allison, L. J. and Salomonson, V. V. (1973) Application of Nimbus 4 THIR 6.7 μm observations to regional and global moisture and wind field analyses, *Journal of Applied Meteorology* **12**: 386–95.

Stout, J. E., Martin, D. W. and Sikdar, D. N. (1979) Estimating GATE rainfall with geosynchronous satellite imagery, *Monthly Weather Review* **107**: 585–98.

Sutton, O. G. (1967) *Understanding Weather*, Penguin Books: Harmondsworth; Baltimore; Victoria.

Vonder Haar, T. H. and Suomi, V. E. (1971) Measurements of the earth's radiation budget from satellites during five-year-period. part I: Extended time and space means, *Journal of the Atmospheric Sciences* **28**: 305–314.

Wark, D. Q. and Hilleary, D. T. (1969) Atmospheric temperature: successful test of remote probing, *Science* **165**: 1256–8.

Wash, C. H. and Whittaker, T. M. (1980) Subsynoptic analysis and forecasting with an interactive computer system, *Bulletin of the American Meteorological Society* **61**: 1584–91.

Wasserman, S. E. (1977) The availability and use of satellite pictures in recognizing hazardous weather: In Clough, D. J. and Morley, L. W. (eds) *Earth Observation systems for Resource Management and Environmental Control*. Plenum Press: New York and London, pp. 419–35.

Wilheit, T., Chang, A. T. C., Rao, M. S. V., Rodgers, E. B. and Theon, J. S. (1977) A satellite technique for quantitatively mapping rainfall rates over the oceans, *Journal of Applied Meteorology* **16**: 551–60.

Zwatz-Meise, V. (1981) Use of satellite images and derived meteorological parameters for weather analysis and forecast, in Cracknell, A. P. (ed) *Remote Sensing in Meteorology, Oceanography and Hydrology*. Ellis Horwood Ltd: Chichester, pp. 412–51.

Chapter 4 The lithosphere: geology, geomorphology and hydrology

Characteristics of the lithospheric environment

In contrast to the atmosphere, the lithosphere is the *solid* realm called the earth where the human population lives. It is the source of chemical elements and compounds vital to life. These substances are released during interactions between the earth's surface and the atmosphere, in the presence of heat and water which are essential components of the hydrological cycle. As a result, physical disintegration and chemical decomposition of the solid rock occur, leading to the formation of soils on the surface of the earth. In addition, the lithosphere also supplies the mineral resources (hydrocarbon fuels, metals and non-metals) that are most essential to man's economic well-being.

The earth itself is made up of four concentric zones: the solid inner iron-nickel core (radius 1,255 km), the liquid outer iron-nickel core (radius 2,221 km), a mantle of ultramafic rock (radius 2,895 km), and the crust which is the thin outermost layer of the earth ranging in thickness from 16–40 km (Harris 1973). Despite its insignificant depth in comparison with the mantle and core, the crust of the earth is the most important from the human population point of view because all relief features are found there. In a strict sense, the term lithology refers to the crust and the uppermost mantle which combine to form the earth's rigid shell. The crust is much less dense than the mantle and is largely composed of oxygen (93.8 per cent by volume and 46.6 per cent by weight), silicon (0.9 per cent by volume and 27.7 per cent by weight), aluminium (0.5 per cent by volume and 8.1 per cent by weight) and iron (0.4 per cent by volume and 5.0 per cent by weight). The other elements include calcium, sodium, potassium and magnesium which are essential for soil fertility and plant growth. The crust is much thicker under the continents than beneath the ocean basins. The upper layer of the continental crust is granitic (sial) and the lower layer is basaltic (sima). It is separated from the mantle by a surface called the Mohorovicic discontinuity.

The application of remote sensing to the lithospheric environment has naturally focused on the crust with a view to extract geologic, geomorphologic and hydrologic information. Such information is in turn utilized to help in mineral exploration. In order to meet these requirements, the highest spatial and spectral resolution capability of the remote sensing systems is called for and a great variety of imaging platforms is employed to provide complementary data at varying scales for these tasks.

Objectives and approaches

The major objectives of the remote sensing applications in geology, geomorphology and hydrology are to detect, identify and map earth features on the surface and subsurface and to infer the processes at work through the synoptic vantage point afforded by the imagery acquired from aerospace platforms. Specifically, the remotely sensed imagery has to be interpreted for elusive lithological and structural information as well as for the more directly observable landform, land cover and water availability characteristics. Remote sensing is particularly suited for use in studying the dynamic aspect of the terrain features, particularly the genetic origin of landforms in geomorphologic applications. This dynamic aspect emphasizes processes and temporal changes such as in the study of coastal changes or river channel changes or glacier retreats which are also closely related to hydrology. In all applications, both the qualitative and quantitative aspects are adopted (Lo 1976; Lewis 1974). The qualitative approach provides identification and description of the terrain features. This involves the evaluation of the image characteristics, in particular, tone, pattern, texture, and form or relief. The quantitative approach, on the other hand, involves metric measurement of the terrain characteristics, notably linear, area, volume, height and direction data. The dip-slope measurements are especially important to geologists, geomorphologists and hydrologists. Another form of quantitative approach involves the measurement of image tone or density with the aid of a microdensitometer. There is a strong correlation between image density data and the corresponding terrain features (Verstappen 1977, pp. 79–81). These density data can be easily handled by the computer to provide enhancements, such as contrast stretch or density slicing, to aid the image interpretation. The more general availability of the computer and multispectral image data in digital form have also tremendously improved the accuracy of interpretation.

Applications using aerial photography

Conventional vertical aerial photography assumes a dominant role among all forms of remote sensing in the study of the lithospheric environment through its excellent spatial resolution (in the order of 20–30 line pairs/mm for low contrast targets at 1.6:1 ratio) and the stereoscopic capability which facilitates the extraction of detailed qualitative and quantitative information of the surface features. The usual approach in photo-interpretation is adopted, which by proceeding from the general to the specific and from the known to the unknown (Lo 1976), allows one to obtain an overall view of the terrain before focussing down to the recognition of landforms, drainage patterns, soils, land use and land cover, all of which serve as clues to the geology, soils and hydrology of the area.

In geology, the use of aerial photography is so well established that the terms *photogeology* or *aerogeology* have been coined to refer to this process of photo-interpretation. The aims is to unravel the lithological and structural characteristics of an area (Miller 1961; Von Bandt 1962; Allum 1966; Ray 1960 and Mekel 1978). In *lithological interpret-*

ation, or the recognition of rock types from aerial photographs, Allum (1966) recommended an approach which combined geomorphological and structural analysis together in a regional setting so that one should start by recognizing the climatic and erosional environments. A so-called generalized photogeological legend or key was developed based on photographic evidence alone to suggest the *type* rather than the specific *name* of the rock. This is a typical qualitative approach. On the other hand, Mekel (1978) emphasized the importance of photographic characteristics of sedimentary, igneous and metamorphic rock types as an aid to interpretation. It was observed that areas of sedimentary rock might yield more information on lithology and structure than those formed by igneous and metamorphic rocks. Differential erosion due to the occurrence of alternating resistant and less resistant beds as well as banding indicated by vegetation cover have helped the interpretation. On the other hand, intrusive igneous rocks tend to be homogeneous over large areas, frequently exhibiting joints or fractures. Extrusive igneous rocks are usually associated with special landforms such as volcanoes and are more readily identifiable. Metamorphic rocks may be revealed by characteristically more rounded and subdued landforms, the occurrence of folding in a sedimentary area plus the presence of cleavage, schistosity or foliation, banding and fracturing. Mekel (1978) has been able to develop keys for the major sedimentary and igneous rock types based on their characteristics, observable on aerial photographs.

In *structural interpretation*, the purpose is to detect the attitude of beds, folds, faults and joints, from which the structural relationships of the rocks in the area can be deduced. Because of the close correlation between structure and relief, a good knowledge of geomorphology is required. Drainage, relief, lineaments and vegetation, tone (or colour) and texture can all throw light on the bedrock distribution and structural relations of the area. For this purpose, low sun-angle photography, i.e. aerial photography in the early morning and late afternoon, taken with the sun at an angle of 10° or less from the horizon, is especially suitable because it shows subtle differences in relief and texture whilst the shadows also tend to emphasize linear features. The detection of photo-lineament in terms of its type, orientation and length is an important step to the understanding of the geological structure of an area.

The stereoscopic capability of aerial photography facilitates the study of relief in the qualitative approach, but one should beware of the vertical exaggeration which makes the slopes appear steeper than they actually are. This impression is caused by the different base-height ratios. Aerial photography taken with a narrow-angle camera from a high altitude will have a more unfavourable base-height ratio which reduces the apparent relief of the stereomodel than that taken with a wide-angle camera at a lower altitude. It is possible to estimate the relief exaggeration by applying the following formula cited by Verstappen (1977, p. 44):

$$E_r = B.f/b_a.f_s \qquad [4.1]$$

where E_r is the relief exaggeration, B is the air base, f is the focal length of the camera, b_a is the eyebase and f_s is the focal length of the stereoscope lenses used to observe the stereomodel. In studying the attitude of the beds, the strike and dip of the bedding planes can also be determined from aerial photographs (Verstappen 1977, p. 44). The simplest method is to measure the height difference between two points by a parallax bar and to determine the horizontal distance between the two points by a radial triangulation method. The angle of slope can then be obtained from the equation:

$$\cot \alpha = \frac{D}{\Delta h} \qquad [4.2]$$

where α is the slope angle, D is the horizontal distance and Δh is the difference in height. There are instrumental aids such as the ITC–Slope Templet or the Mekel slope comparator which will allow the determination of dip and slope with ease (Mekel, Savage and Zorn 1967).

Drainage pattern plays an especially important role in structural interpretation. The form and density of the drainage network provide clues to the underlying geology and hydrology. As early as 1932 Zernitz classified drainage patterns into six basic types and twenty-four modifications (Zernitz 1932). A more logical grouping is shown in Fig. 4.1. It is clear that the drainage pattern can be structurally controlled and the drainage density depends on the infiltration-runoff ratio controlled by the resistance and permeability of the surface material. It was generally observed that the highest density

Fig. 4.1
Major Types of drainage patterns

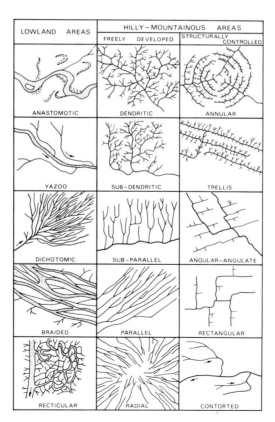

occurred in shales, medium density in phyllites, granites, sandstones and siltstones, and the lowest density in limestones and gravels (Fezer 1971). The drainage density (D) can be quantitatively determined by the equation:

$$D = \frac{\Sigma L}{A} \qquad [4.3]$$

where L is the length of each channel segment in a basin and A is the drainage area of the basin as defined by the watersheds. The streams can also be classified into different orders by using the method developed by Strahler (1957). This designates the smallest tributaries as order 1. Where two first-order channels join, a channel segment of order 2 is formed; where two second-order channels join, a channel segment of order 3 is formed, and so forth. The main stream where all discharge passes is the stream segment of the highest order. A quantity called *bifurcation ratio* (R) can be computed by:

$$R_b = \frac{N_u}{N_u + 1} \qquad [4.4]$$

where N_u is the number of channel segments of a given order, and N_u+1 refers to the number of segments in the next higher order. It was observed that bifurcation ratios range normally between 3.0 and 5.0 for watersheds in which geological conditions do not distort the drainage pattern. In other words, any abnormal bifurcation ratios will indicate the occurrence of dominant geological control (Shreve 1966). Also, abnormal bifurcation ratios may be related to flood discharges so that basins with a high bifurcation ratio would yield a low but extended peak flow whereas basins with a low bifurcation ratio would produce sharp peaks. An understanding of the watershed geometry is important in hydrology. Hence, one sees the close relationships among geological structures, landforms and hydrological characteristics. These relationships are in turn useful in predicting the potential environments for mineral resources.

In geomorphology, the interpretation of aerial photographs is more direct and descriptive because landforms, our objects of interest, are highly visible. Hence, a lower degree of *deduction* is involved. In the *genetic* approach, the *processes* responsible for the formation of the landforms are to be inferred. Deduction is certainly required (Fezer 1971). Verstappen (1977, p. 6) has pointed to the importance of aerial photography in forming a link with the disciplines of geology, soil science and hydrology, thus contributing substantially to the transformation of geomorphology from a subject of academic interest into a modern science with numerous applications. Of particular importance is the study of the large-scale processes such as the temporal change in landforms or sediment movement patterns in estuaries, resulting in the emergence of *applied geomorphology* with the aim of tackling practical problems (Mather 1979).

In hydrology, the main application areas of aerial photography lie in recording and monitoring the amount of water in the surface and subsurface environments. The visible portion of the electromagnetic spectrum (0.4–0.7 μm) is particularly suited to delineate flood plains, detect water turbidity or colour, monitor changes in depth of water bodies and identify snow and ice on the surface of the lithosphere.

In the foregoing discussions, the black-and-white panchromatic aerial photography has been assumed. But it is advantageous to make use of colour and colour infrared photography for geological, geomorphological and hydrological interpretation, especially when a direct qualitative approach is adopted. This is because the human eyes are capable of distinguishing many more combinations of hue, value and chroma in colour than shades of grey (Strandberg 1968), thus permitting a greater number of surface features to be identified by their colours. However, at high altitudes the colour attenuation caused by atmospheric scattering and absorption has resulted in reduced contrast in true colour photographs whereas the colour infrared, because of its longer wavelength (0.7 μm), continues to show good colour contrast and has been found to be best for geological mapping (Pressman 1968).

In the following sections, some case studies are presented to illustrate the various approaches explained above.

Case Study 1:

Geological reconnaissance of concealed terrains in harsh environments

Aerial photography has contributed most significantly to reconnaissance surveys of the geology of such difficult environments as the Amazon Basin and Antarctica where basic geological information is lacking. A good example is Howard's (1965) study of the Middle Amazon Basin using 238 photographs acquired with a $f = 152$ mm aerial camera from a height of 4,560 m, giving a nominal scale of about 1:30,000. The total area covered amounted to 10,619 km² (Fig. 4.2). The basin is flat and covered with dense forests which make the area highly inaccessible. The area, however, presented a challenge to the photogeologist because the photographs recorded only the tree tops without revealing any distinct tonal variations associated with geological structures (Fig. 4.3). Much of the photogeological interpret-

Fig. 4.2
Location map of study area, Middle Amazon Basin (*Source:* Howard 1965)

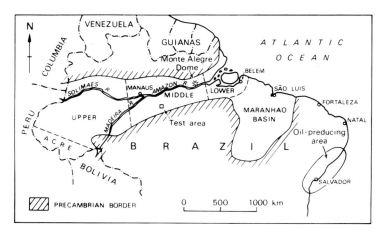

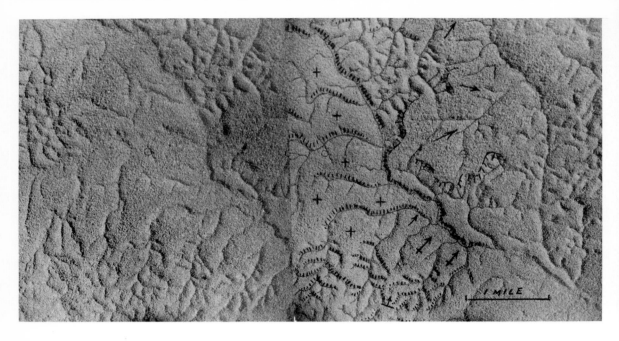

Fig. 4.3
A representative stereopair of the study area (Middle Amazon Basin). Simple hachures indicate the steeper of two valley sides; hachures connected by broken line mark edge of undissected upland; plus signs indicate horizontal surfaces; arrows, sloping surfaces; arrows with cross bar, smooth descents from horizontal surfaces; dip symbol, the probable direction of dip of sediments exposed by dissection of mantle formation. See also the legend in Fig. 4.4 for further explanation of some of the symbols (*Courtesy*: A. D. Howard; *Source*: Howard 1965)

ation therefore had to be based on deduction. To compound the problem, the geology of the area was largely unknown. The approach adopted by Howard was to examine the drainage pattern, from which a detailed map was made to provide the base for plotting structural data (Fig. 4.4). The forest cover in fact tended to emphasize some drainage lines. Two distinct crown heights of the forest were observed to be closely associated with the nature of the terrain: (1) the low crown trees (determined photogrammetrically to be 21–23 m high) found in the periodically flooded areas of the river plain; (2) the high crown trees (27–30 m high) in the dry valley sides. The dendritic drainage was regarded as the 'norm' based on the postulated northward dip of the surface formations, the generally northward slope of the surface and the low relief. Any deviations from this 'norm' were then noted and regarded as caused by the control of local structures.

Another guide to aid the interpretation of structures was topography. The gentle slopes were all exaggerated when viewed under the stereoscope so that minute slope contrasts could be easily detected. It was found that a greater dissection of the terrain by rivers resulted in greater local relief in the eastern and southeastern part of the area west of Rio Maues. Variations in vegetation type might also reflect differences in soils which in turn often reflected differences in underlying rocks. Because of the mixtures of hundreds of species of trees in a single square kilometre in the rainforest, the analysis of vegetation was difficult, but it did reveal a definite correlation between the topographic setting and the dominant crown height of the forest: the high crown trees tended to dominate on the crest of a slope and decreased in abundance downslope. There were only limited natural clearings observed in the rainforest, where grass, shrubs and clumps of trees were present. Thus, by making use of drainage, topography and vegetation as guides, a structural interpretation of the area was made. It had been possible to reveal from this study that a thin

Fig. 4.4
A part of the photogeologic map of the study area shown in Fig. 4.2 (*Courtesy*: A. D. Howard; *Source*: Howard 1965)

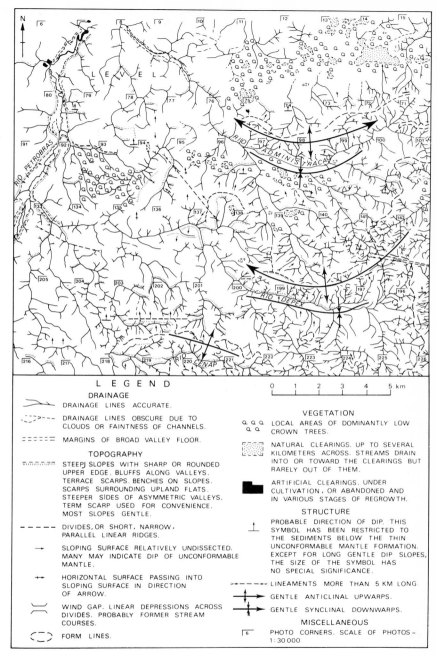

formation on the surface appeared to rest with angular unconformity on a thick northward dipping sedimentary sequence and structural warps were detected as undulations in the mantle formation, probably caused by active deformation or differential compaction of sediments.

From the Amazon one moves to a climatically contrasting environment: the Antarctica. Smith's (1967) study has also demonstrated the useful role played by aerial photography as a geological reconnaissance tool. Besides the nature of the terrain problems faced by geologists were the extensive snow and ice cover. It was found that in Antarctica structural regularity was better developed in the younger rocks than in the older rocks. In most regions, the bedrock surface was usually concealed from view by a cover of soil and surface deposits accumulated by the action of weathering, gravity, water, wind and glacial ice in addition to the ice cover. Therefore, bedrock exposures were extremely limited. The photography used was obtained by a tri-metrogon camera, i.e. a three-camera system designed to acquire wide-angle coverage, which was intended primarily for reconnaissance topographic mapping. The photographic scale was too small (about 1:40,000) for detailed geological interpretations. Despite these limitations, it was possible to distinguish between layered and non-layered rocks by examining the small and scattered outcrop area, the ice-free valley system and the associated mountains and nunataks (or rock peaks) in the McMurdo region from the oblique tri-metrogon photographs (Fig. 4.5). The most distinctively layered rocks were the younger ones, notably sandstones and shales of Devonian to Jurassic age, intercalated with thick sills of dolerite. The sills were revealed by the darker tone and more prominent relief. Minor deformation caused by tilting and faulting was also observed. Stratigraphical interpretation was hampered by the absence of distinctive horizon markers, and the fact that the tonal appearance of the rock sequence was too similar in the black-and-white panchromatic photography despite the colour differences among these rock strata. They would probably be more easily distinguishable on colour photography. As for the non-layered rocks, plutonic intrusives and metamorphic rocks

Fig. 4.5
Tri-metrogon oblique photograph of upper Wright Valley, showing the rock units in the McMurdo region of Antarctica. The rock within the valley belongs to the basement complex of pre-Devonian age and is intruded by a dyke complex represented by the striped pattern to the left of the frozen lake. The thick dark band of rock rimming the valley is a dolerite sill. The light banded rocks in the right foreground are sediments of the Beacon Group, and towards the background these are seen to be capped or intercalated with a higher sill. Morainal and mass movement detritus along the valley bottoms and sides are the more prominent geomorphic features (*From* Smith 1967)

were found. The metamorphic rocks showed banding which could be easily seen on the aerial photographs. Intrusive dykes revealed themselves by the colour contrasts and protruding relief caused by differential erosion. Finally, recent volcanics were also readily recognized from their topographic form, but the details of form were obscured by the low contrast and the very dark colour of the rock. Other features of geomorphological interest such as glaciers, cirques and sand dunes were all easily recognizable. Of special interest was the ice deformation revealed by the crevasse patterns. From the aerial photographs one could obtain data on the spatial distribution and change of these crevasses. This study concluded by recommending the use of vertical and oblique aerial photography in black-and-white and colour to supplement the tri-metrogon photography for photogeologic interpretation in Antarctica.

In contrast to the above two examples, the hot deserts provide the ideal environment for photogeology because of the clarity of the atmosphere and the scarcity of vegetation cover which permits much of the rock outcrop to be exposed.

Case Study 2:

Detection of landform changes by time-lapse or sequential photography

Aerial photography is commonly employed to monitor changes in landforms with a view to unravel the processes at work, i.e. the genetic approach in geomorphology (Verstappen 1977, pp. 112–137). Three examples drawn from the volcanic, fluvial, and coastal environments are presented below to illustrate such an approach.

The volcanic environment has been regularly monitored by remote sensing methods, notably with the thermal infrared scanner. However, the thermal infrared scanner has been found to be more suited to monitor secondary volcanic activity, such as, fumaroles, gas vents, etc. than the primary volcanic activity of the open-chimney volcanoes in paroxismal phase (Cassinis *et al.* 1977). Conventional aerial photography is normally used to record volcanic eruptions because of the ease with which a camera can be used. The photography can record the different components of the explosion cloud during eruption and the landform change afterwards. In recording the eruption of Mount St Helens in southwestern Washington, USA on 18 May 1980, the Stoffels (1980) recorded three components of the explosion cloud, as a black projectile-laden ash cloud originating in the detachment plane, a vertical white steam cloud originating in the depression located north of the summit crater and a vertical grey ash-laden cloud originating in the summit crater itself. The eruption had devastated an area of 500 km². The top 400 m of the summit of the volcano had been blasted off from its original height of 2,950 m and the crater was enlarged to about 2 km in diameter and 1.5 km in depth. The original symmetrical crater was transformed into a huge steep walled amphitheatre facing northwards, from which steams could be seen rising up. All these changes can be revealed by examining the stereopairs of aerial photographs of Mount St Helens acquired before and after the

Fig. 4.6
Stereogram of Mount St Helens taken on 21 July 1978 (before the eruption) North is to the left (*Courtesy*: Mark Hurd Aerial Surveys)

eruption (Figs. 4.6, 4.7), from which topographic maps or three-dimensional digital terrain models (Fig. 4.8) can be constructed.

In the fluvial environment, one common application of time-lapse photography is to study the river channel change which is related to the rates of bank erosion. Channel changes in the form of meander development or of braiding sequences occur during equilibrium conditions. On the other hand, rapid channel changes may be the result of climate or land use changes. Evidence of the rate and nature of channel change can be obtained by direct measurement in the field, from historical maps and from vegetation patterns. All these three aspects can be adequately satisfied with the use of aerial photography. On the whole, as long a time span as possible is required to study the channel change, but it is still handicapped by the fact that considerable changes may have occurred outside the time range for which the photography is available. Collateral information such as that acquired from historical maps is therefore necessary to give a better insight into the process. With the availability of stereomodels and ground control points, the channel changes can be accurately depicted and quantified photogrammetrically. Lewin and Weir (1977) presented a case study of the evolution of the meander loop on River Rheidol in Wales, UK using photography taken in 1969 and 1970 (Fig. 4.9) and the loop was photogrammetrically mapped and contoured (Fig. 4.10). By using in addition various historical maps and old aerial photography taken by the Royal Air Force in the UK, the channel changes in both the short and the longer terms were detected (Figs. 4.11 and 4.12). In this way, the processes of bank erosion and sedimentation over nearly a decade and the evolution of one loop over a period of about a century could be revealed.

In the coastal environment, studies of coastal processes and landforms using aerial photography are particularly popular as witnessed

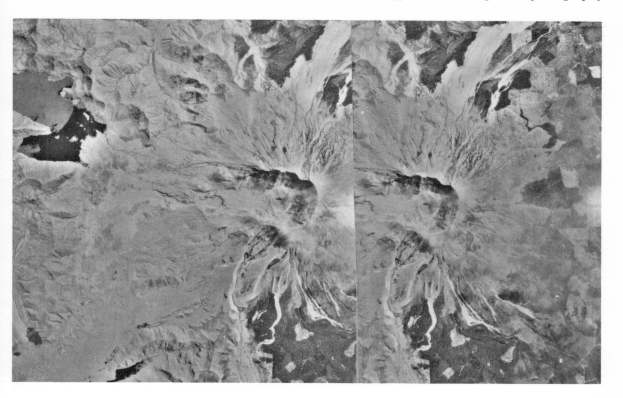

Fig. 4.7
Stereogram of Mount St Helens taken on 26 July 1980 after the eruption on 18 May 1980. Lava and ash from the eruption obliterated the surface features on the Northern (left) side of the cone within the area shown on Fig. 4.6. To the South, features can be identified on both Figs. 4.6 and 4.7. The explosion reduced the peak by 396 m from its original elevation of 2,249.5 m. The resulting crater is 1,609 m wide and 762 m deep (*Courtesy*: Mark Hurd Aerial Surveys)

by the volume of papers collected by El-Ashry (1977). Through its synoptic vantage point, aerial photography reveals details of macroscopic coastal features and circulation patterns of the sea water, such as the size and spacing of waves, direction of wave fronts and the distribution of shallow-water sediments to depths of about 7 m. By using sequential aerial photography, changes in the coastal and near-shore erosional and depositional features, both naturally and artificially formed, can be easily observed. In this way, a record of the long-term trends of the coastline can be obtained for coastal protection purposes. However, the major difficulty with sequential coastal aerial photography is that the interpretation of the change is restricted by the time interval of photography. One cannot determine the exact moment when the changes have occurred. Classic examples of this type of approach are the works by Cameron (1965), and El-Ashry and Wanless (1967). Cameron studied the coastal changes in Rose Bay and Advocate Harbour, Nova Scotia, Canada using sequential aerial photographs obtained at intervals ranging from 2–20 years. It was possible by comparing these photographs to determine the directions of longshore drift and growth of bars and spits. El-Ashry and Wanless examined the effects of coastal morphology with examples drawn from North Carolina, South Carolina, Florida and Texas, in the USA. It was found that a violent storm could cause rapid changes in coastal form as a result of the rise in water level which brought the zone of wave activity further inland, but after the major storm had passed, the storm-induced changes were gradually smoothed and obliterated so that the coast appeared to have resumed the form it had before the

Fig. 4.8
Digital Terrain Model of Mount St Helens produced from contours compiled on a Wild A-10 photogrammetric plotter based on aerial photography taken on 26 July 1980 from an altitude of 50,000 ft (15,240 m) (*Courtesy*: Mark Hurd Aerial Surveys, Inc.)

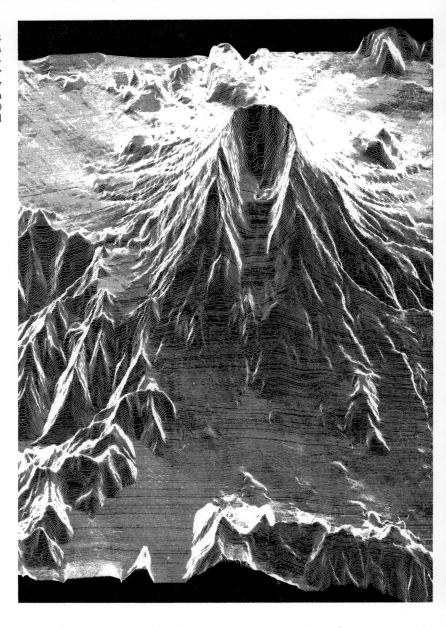

storm. However, quantitative measurements using photogrammetric methods should be employed before one could conclude whether any changes had really taken place.

In recent years, Welsted (1979) followed a similar approach in studying the Bay of Fundy coast in Canada. He drew attention to the difference in appearance of the coast at low- and high-tide conditions so that one should avoid using just one photograph or one set of photographs to study the coast (Fig. 4.13). At low tide, one could see much of the foreshore and the contact between the land and water was approximately 300 m further seaward. His study illustrated four approaches in the use of aerial photographs: (1) as a detailed descrip-

Fig. 4.9
An aerial view of a meander
loop on the River Rheidol,
Wales, UK, taken on 14 June
1970 (*Courtesy*: Department of
Geography, the University
College of Wales, Aberystwyth)

Fig. 4.10
Contour plot of the meander
loop on the River Rheidol,
Wales, UK (*Source*: Lewin and
Weir 1977)

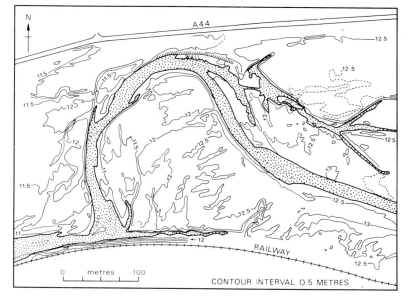

tion of the coastal morphology; (2) as a means to study coastal
processes at work; (3) as a technique to determine coastal changes;
(4) as evidence in determining the long-term trends of the coast. For
descriptive morphology, large-scale aerial photographs (scale 1:16,260)
taken near the time of low tide were employed to obtain a full view
of the foreshore features. It has been possible to interpret the geology

Fig. 4.11
Recent channel changes on the
meander loop on the River
Rheidol, Wales, UK (*Source*:
Lewin and Weir 1977)

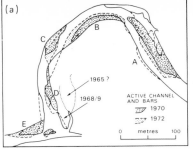
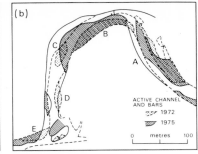

Fig. 4.12
Historical development of the
meander loop on the River
Rheidol in Wales, UK,
1845–1948 (*Source*: Lewin and
Weir 1977)

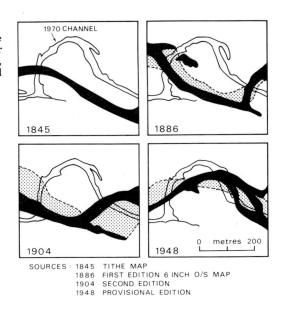

(lithology and structure) of the coast because the igneous and sedimentary rocks exposed on a shore platform appeared differently through the contrast between the irregular jointing of the igneous rocks and the regular bedding of the sedimentaries. Other geomorphological features detectable were (1) sea cliffs, their presence and absence, heights and gradients; (2) the associated rock falls and areas of slumping; (3) the shore platform, the beach material and small-scale features such as beach cusps, beach ridges and runnels; (4) depositional features such as spits, bars, tombolos and cuspate forelands; the appearance of which depended on the time of photography; (5) tidal mudflats, natural and reclaimed marshes. A detailed description of the coastal morphology provides a basis for subsequent studies of the coastal processes, coastal changes and long-term evolution of the coastline.

In understanding coastal processes, additional information on water depth, waves, currents and the drifts of sediments is required, which can be obtained from aerial photographs. A qualitative approach is just as useful as the quantitative approach. The wave refraction

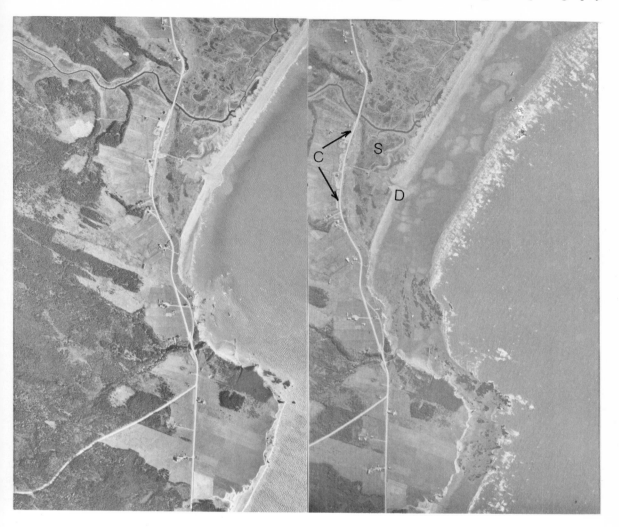

Fig. 4.13
Stereogram of Red Head and Anderson Hollow, New Brunswick, Canada. Left photo: high tide (No. 510–165, taken on 19 October 1962); right photo: low tide (No. 516–51, taken on 20 October 1962). C = raised beach and cliff; D = tidal delta; S = salt marsh (*Courtesy*: J. Welsted; *Source*: Welsted 1979)

pattern from aerial photographs provides clues to offshore water depths, sheltered or unsheltered areas and direction of the approaching waves whilst the photographic tonal variations reveal the degree of turbidity of the water. The use of sequential aerial photographs permits changes in the depositional landforms to be detected more easily than those in the erosional landforms. Cameron's (1965) study of the Advocate Harbour, Nova Scotia bears this out clearly (Figs 4.14, 4.15). Finally, to understand the long-term trends of the coastline, one should make inference based on the detailed description and knowledge of the geomorphological history of the area. Thus, Welsted (1979) has been able to infer that considerable retreat of the cliffs has occurred along the Bay of Fundy coastline, since the formation of the river terraces, but at the recent time the presence of spits across the seaward side seemed to suggest that erosion is no longer the dominant process.

Fig. 4.14
An aerial photograph of the spits in Advocate Harbour, Nova Scotia, Canada 1964 (Photo No: A18353-57, National Air Photo Library, Ottawa). The photograph was taken during the ebb and a strong tidal current can be seen at the harbour entrance. This prevents longshore drifting from closing the entrance and creating a baymouth bar. B = bars formed between 1960 and 1964; C = tidal current; H = hooks or recurved spits; L = direction of dominant longshore drift (*Courtesy*: J. Welsted; *Source*: Welsted 1979)

Case Study 3:

Floodplain delineation and ground-water flow system detection

Applications of aerial photography to hydrology are characterized by a deductive approach which infers the occurrence of the features through an interpretation of the geological, geomorphological and biological facts. In floodplain delineation, the purpose is primarily to detect the flood and flood prone areas. One indicator of these areas is the vigour or health of the vegetation which serves as a surrogate to the underlying soil and surface morphology. This in turn is also related to the ground water situation. As the water table is high in the floodplain, one can expect the occurrence of more healthy plants and a high foliage density even during the dry season. This explains why colour infrared photography is normally preferred in this type of study. Conway and Holz (1973) employed Kodak Ektachrome infrared

Fig. 4.15
The changing outline of the spits as mapped from sequential aerial photographs, 1939–1960 (*Source*: Welsted 1979)

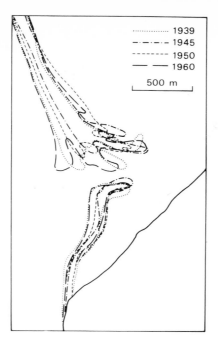

```
·············· 1939
–·–·–· 1945
–––––– 1950
–– –– 1960
```

500 m

aerofilm type SO-117 air photographs taken with Wild RC-8 (23 × 23 cm format) cameras from a height of 3,048 m (nominal scale 1:20,000) for the delineation of the alluvial floodplain of the Rio Lavayen, northeast of the city of Salta in Northwestern Argentina. The photographs were taken during the dry season. The volume of water in the river was therefore low. The vegetation type in the study area was predominantly a semi-deciduous plant association which was either in flower or in the leaf-dropping stage, thus reducing foliage density in the area and revealing much of the undergrowth. However, the proximity of the water table to the surface in the floodplain encouraged the growth of tropical vegetation. As the near infrared is highly reflected by the spongy mesophyll of the leaves of healthy plants, the plant vigour can be revealed by the varying degrees of red or magenta colour according to the strength of infrared reflectance response. In this way, a distinct vegetation change at the limit of river flooding was easily observed, thus facilitating the precise delineation of the floodplain. However, this estimate was regarded as conservative, and multiple-time photography should be acquired to improve the estimate because the flooding situation may well change.

In addition to this delineation task, the sequence of development of the river's meandering channel could be reconstructed through an interpretation of the stages of colonization of plants in the abandoned channel segments. This was based on the accepted theory of plant succession which postulated that the early colonizers would foster an environment for other plants to follow (Braun 1956). Thus, different plants would sequentially colonize the abandoned course until they reached maturity. The different stages of growth of the plants were revealed by their heights, number of species and density of coverage. In this way, abandoned channels could be identified and classified by age, as revealed by the development stage and biomass of the vegetation found on them. Similarly, the varying vegetative response also illustrated minor changes in the topography of the floodplains indicating the positions of terraces and early terrace levels marked by a lack of healthy response of the vegetation to the infrared. In another application by Currey (1977) the origin of floodwaters was traced with the use of true colour aerial photographs and satellite imagery. The colour aerial photographs were in 70 mm format and were printed to a scale of 1:20,500. The flood water edges were readily identified and the directions of the flood water flow paths as well as the origin of the flood water were determined by the colour of the sediments in the water which was clearly visible on the colour aerial photographs. The different colours of the sediments reflected the geological features and soils from which they originated.

In the detection of ground-water flow systems, even more sophisticated inference from geological, geomorphological and biological interpretation of the aerial photographs is required. A good example is Sauer's (1981) study of the hydrogeology of glacial deposits in Saskatchewan, Canada. The ground-water flow system can be represented by the following equation:

$$\text{change in storage} = \text{recharge} - \text{discharge}$$

On a long-term basis, discharge (or outflow) will equal recharge (or inflow) so that the storage will be constant. Clearly, storage is dependent on porosity whereas flow rate is dependent on permeability and hydraulic gradient. Through an interpretation of landforms, one can identify the recharge, discharge and storage areas. Thus, moraines and glacio-fluvial outwash plains are good recharge areas whilst ice-contact valleys may act as a recharge, discharge or storage area in any part according to the presence of surface ponding, springs along slopes or valley bases and indications of soil salinity caused by ground-water seepage and evaporation. Flow phenomena of ground-water are revealed in aerial photographs by springs and natural piping, variations in the grey tones, vegetation and plant communities, soil salinity and salt accumulation, slope instability, stream flow characteristics and floodplain features as well as the land use activities of man.

Natural piping is a form of erosion caused by ground–water seepage pressure as the ground-water flows out to the surface. The grey tone of the soil reflects its water content. As the water content increases, the spectral reflectance is reduced. In clay deposits the reduction extends over a range of wavelengths from 0.4–1.0 μm whereas for sandy deposits this reduction occurs in the range of red light (0.6–0.7 μm) and near-infrared (0.7–1.0 μm). Plants are good indicators of ground–water occurrence as already discussed in connection with floodplain delineation. A strong correlation was found between ground–water and plant species in arid and semi-arid environments. In the prairie environment of Saskatchewan, Canada, this correlation also holds: water-loving plant species indicate the occurrence of ground-water storage. In arid or cold environments fewer species of plants can survive; hence a sudden lush growth of vegetation suggests the occurrence of saturated permeable strata. Salinity is related to ground–water discharge if the water discharging from springs is saturated with sulphates. Evaporation produces the large salt flats or alkali flats commonly found in the prairie region. The high spectral reflectance of salt permits these salt flats to be detected easily from aerial photographs. Slope instability is closely related to the geology of the deposits, their water content and ground-water flow rates. Slumping is a good indication of either low strength clays or high rates of seepage or both. Stream flow and floodplain morphology are often affected by ground-water recharge or discharge as already discussed. Similarly, land use activities of man such as mining, irrigation and forest cutting all tend to affect the recharge, discharge and to some extent storage of ground-water. Thus, direct and indirect photographic evidence can be extracted to reveal the ground-water flow system in an area.

Applications using thermal infrared scanner data

Thermal infrared scanner data normally tap the reflected and emitted radiant energy in the two transmission windows of 3–5 μm and 8–14 μm. There are two major approaches in applying these scanner data to the geological, geomorphological and hydrological studies of the lithosphere: (1) the direct visual correlation of the thermal image data with the ground features and (2) the quantitative analysis of repetitively acquired images using a theoretical thermal model. The following case studies examine these two different approaches in greater depth.

Case Study 1:

Detection and mapping of geothermal phenomena and other associated features

Thermal infrared images can display surface temperature variations of the ground and are most effective in picking out high temperature targets such as active volcanic craters, hot springs, hot gases and forest fires. The temperature contrast of these targets to the background is high and their temperatures are relatively constant with respect to time. According to Wien's displacement law, the wavelength of peak radiant emittance shifts to the shorter wavelengths with increasing blackbody temperature. Thus, the most suitable wavelengths to use to detect these high temperature targets lie in the 1.0–6.0 μm region (Shilin *et al.* 1969). The major problem, however, is to ensure that solar heating is eliminated so as not to confuse it with the geothermal heat source to be detected. The surveillance of volcanic activities using thermal infrared scanners has been successfully carried out by Fischer *et al.* (1964) over Hawaiian volcanoes, Moxham and Alcara (1966) over the Philippine volcanoes, Shilin *et al.* (1969) over Kamchatka volcanoes in the USSR and Cassinis *et al.* (1971, 1977) over Italian volcanoes. The work by Shilin *et al.* (1969) made use of the spectral interval of 3.2–5.3 μm and a spectral filter to eliminate reflected solar radiation to survey volcanoes in Kamchatka. These volcanoes varied in their nature, ranging from those characterized by weak fumarole activities to those with active eruptions. In all cases the thermal air survey was carried out in the daytime and panchromatic photography was also taken simultaneously. By comparing the thermal infrared image and the aerial photograph of the volcano together, a thermal activity map could be produced for the volcano. The thermal activity map exhibited areas of high thermal anomalies. By repeated aerial surveys at regular intervals, the changes in thermal activity of the volcano could be used to predict the periods of activization of the volcano. However, the interpretation has been hampered by the fact that the thermal infrared air survey was conducted in the day time when solar heating was strong. It was therefore recommended that the best time for the thermal infrared survey should be the evening or night-time when the interference from solar heating is absent.

The work of Cassinis *et al.* (1977) reported further progress in the surveillance of the volcanic environment through an on-going programme in Italy. They found that the use of thermal infrared in conjunction with the multispectral method would be best to survey volcanic activity in an area. While thermal infrared images detect directly the energy radiated from the surface, which is produced by internal volcanic activity, the multispectral method permits reflectivity characteristics of the vegetation to be sensed. It was observed that the small but continuous amounts of magmatic gases filtering through the soil could influence the spectral behaviour of the vegetative canopy. It appeared that the decrease of the reflectivity of the vegetation in the near infrared region could be accompanied by a simultaneous increase of thermal infrared radiance. By studying an active volcano, Mount Etna in 1974 using thermal infrared in the spectral regions of $9–11 \, \mu m$ and $1.5–2.0 \, \mu m$ in addition to colour infrared photography from the aerial survey, it was found that the ratio between the near infrared reflected, and the thermal infrared emitted, could emphasize slight fluctuations in the reflectivity of the vegetation, picking out areas affected by gas vents. Aerial surveys carried out before and during the eruption of Mount Etna revealed the reduced possibility of forecasting volcanic eruptions using only the thermal infrared due to the small amount of heat transfer in the volcanic area and that the surface heat distribution is strongly related to the thermal conductivity of the layer beneath the earth's surface and to the velocity of the magma movements inside the volcanic structure. On the other hand, thermal infrared would be more suitable for monitoring secondary volcanic activities such as fumaroles, gas vents, etc. typified by the Vulcano volcano. By detecting the total emitted power, the shifts in position of the thermal barycentres and the trends of radiance levels as well as thermal gradients through time (Fig. 4.16), the areas of thermal interaction together with the various heating sources can be monitored.

The use of thermal infrared imagery to detect ground-water discharge has also been found to be most effective. Wood (1972) demonstrated the visual detection of water discharges along the banks of the Lehigh River in Pennsylvania with thermal infrared image in the $8–14 \, \mu m$ region. He observed that much ground water had a nearly constant temperature of about $12° \, C$ while the river temperature varied seasonally, being much warmer in the summer and cooler in the winter than the nearly constant ground-water temperature.

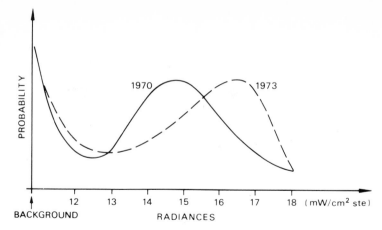

Fig. 4.16
Comparison between the thermal physiognomy curves of 1970 and 1973 for Vulcano Island, Italy (*Source*: Cassinis *et al.* 1977)

Because the river was shallow and turbulent, the thermal anomalies stood out well, thus permitting the location of ground-water discharges. It was recommended that the best time to acquire the thermal infrared images for this type of detection was during the pre-dawn hours and the best season was winter when the river temperature was about the same as the land temperature (about 4° C) and ground-water discharges were relatively warm (10° C to 15° C). In another application by Lee (1969), using thermal infrared imagery in the 3–3.5 μm and 8–14 μm windows, a more quantitative approach was adopted to determine the volume of the ground-water discharge from shoreline springs at Mono Lake in California. The thermal anomaly was measured from the imagery by determining the areas between successive temperature contours and the temperature difference between the fresh water and open salt water (Fig. 4.17). The values so obtained were substituted in the following equation:

$$\text{Thermal Anomaly} = \sum_{i=1}^{n} A_i \left(T_n - \frac{T_i + T_{i-1}}{2} \right) \qquad [4.5]$$

where A is area in square metres and T is temperature in °C. The total thermal anomaly was expressed in deg.m². It was also discovered that the temperature difference between lake surface and spring water at its point of discharge and the temperature difference between the lake surface and air had a great impact on the development of the anomaly. For a given spring temperature, air temperature and lake temperature, the magnitude of the anomaly was directly proportional to the spring's discharge. By combining these temperature differences with the thermal anomaly, it was possible to establish a simple linear regression equation with the actually measured ground-water discharge volume. By applying this equation for the appropriate temperature conditions, the volume of ground-water discharge could be determined. The average error was about 12 per cent, and the best result was obtained with a large volume of ground-water discharge (i.e. greater than 5.0 litres/sec). It was concluded that the optimum time for a thermal anomaly to develop was when the air was colder than the spring water which in turn was colder than the open salt

Fig. 4.17
Plan view of temperature contours between the spring and open water (*Source*: Lee 1969)

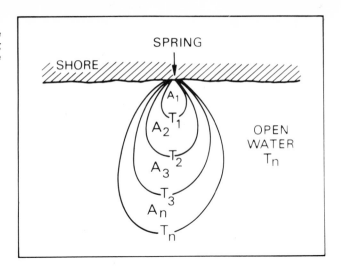

water. These conditions occurred during the pre-dawn hours in autumn and late spring.

Before leaving this category of application, one should note that the temperature anomaly between the land and ground-water gives rise to a tonal contrast in the pre-dawn thermal infrared imagery which helps to emphasize structural lineaments, thus making it resemble low sun-angle photographs (Wolfe 1971; Lattman 1963).

Case Study 2:

Thermal inertia mapping

Under this category the thermal infrared imagery is used to detect near-surface conductive heat transfer which shows only low-level temperature anomalies. These low-level temperature anomalies are easily obscured by (a) the *cloud cover* which cuts down the amount of solar radiation reaching the ground; (b) the *sky temperature* which determines the amount of re-radiation downwards from the atmosphere to the ground; (c) the *air temperature* near the surface which influences the rate of ground cooling and (d) the *wind* which creates convective and evaporative cooling of the surface. The *topographic* variations and *vegetation* cover can also produce 'noise' effects on the temperature anomalies. Thermal inertia refers to the resistance of a material to a change of temperature and is normally defined mathematically as:

$$P = \sqrt{k\rho c} \qquad [4.6]$$

where P is the thermal inertia, k is the thermal conductivity, ρ is the density and c is the specific heat of a material. Watson (1975) has shown that for ground surfaces with different thermal inertias (Fig. 4.18) but the same albedo, the maximum thermal contrast occurs just prior to sunrise and about 1.5 hours after mid-day. Little or no

Fig. 4.18
Diurnal temperature curves for varying (a) thermal inertia (in cal/cm²/s⁴) and (b) albedo (in fractions < 1) (*Source*: Watson 1975)

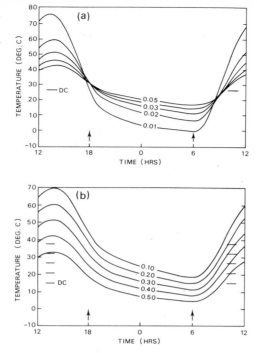

contrast is seen between 7.00 a.m. and 9.00 a.m. while the rates of temperature change are high throughout most of the day apart from 12 noon to 3.00 p.m. As for ground surfaces with different albedos but the same thermal inertia (Fig. 4.11b), the maximum thermal contrast occurs near noon and the minimum contrast at dawn. It is worthy to note that the ground surface with low albedo (0.10) has higher temperatures at all times and a greater temperature range than surfaces with high albedo (0.50). The thermal inertia of a dry material is found to be linearly correlated with its density (Fig. 4.19). Thus, one

Fig. 4.19
Thermal inertia (cal/cm²/s⁴) versus density (g/cm³) for a variety of rock-forming minerals, rocks and soils. The linear correlation line, bracketed by lines showing the standard deviation, shows the increase of thermal inertia with increasing density. Rocks and rock-forming minerals which are high in silica (e.g. quartz) tend to fall above the line. Those which are low in silica (e.g. olivine) tend to fall below the line (*Source*: Watson 1975)

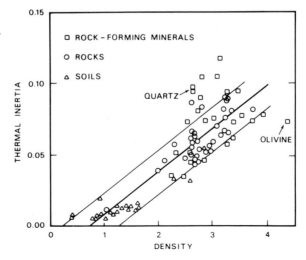

sees in Figure 4.19 that minerals and rocks high in silica (e.g. quartz, granite, rhyolite) have high thermal inertias while those low in silica and high in iron (e.g. olivine, gabbro, basalt) have low thermal inertias. The addition of moisture to dry soils will significantly increase their thermal inertias as a result of the changes in specific heat and conductivity. Thus, by mapping the thermal inertias one can discriminate rock types in areas where vegetation cover is thin. This makes the technique useful for geological mapping in arid and semi-arid regions.

The thermal inertia of the ground surface is estimated from two primary variables: the diurnal temperature range (ΔT) and the albedo (A). By acquiring the thermal infrared imagery at two separate times during one diurnal cycle, i.e. in the pre-dawn time to record the minimum temperature and in the mid-day (12.00–3.00 p.m.) to record the maximum temperature, the temperature difference ΔT can be obtained. By taking panchromatic aerial photography during the midday flight, the albedo, A, can be estimated after having the photography converted to a digital format for computer processing. By making good spatial registration of these three types of images, thermal inertia images can be produced (Pratt *et al.* 1978; Kahle *et al.* 1976).

Kahle *et al.* (1976) reported a good example of the application of this approach to the creation of a computer-derived image of the thermal inertia of Pisgah Crater-Lavic Lake in the Mojave Desert, California (Fig. 4.20). An eleven-channel multispectral scanner was flown over the site at an altitude of 1,200 m. Ten spectral channels of visible and near infrared spectral reflectance data in the 0.5–1.1 μm range and one channel of the thermal infrared in the 8–14 μm band were acquired at 2.00 p.m. near the time of maximum temperature and the thermal data were obtained again at 5.00 a.m. near the time of minimum temperature. Ground measurements were also made of soil moisture, temperature, surface spectral reflectance and meteorological conditions during the 48 hours prior to and during the aircraft flights. An albedo image was constructed from the 10 channels of visible and near infrared data in combination with ground measurements. Day time and night-time temperature images were created separately from the thermal infrared data calibrated with onboard blackbody sources. These day and night thermal images were separately registered together after proper rectification and geometric corrections to a 7.5-minute topographic map. Finally, the two corrected temperature images were matched together to produce a day and night temperature difference image. A thermal model of heat flow was solved for the top 50 cm of the earth's surface which took into account the effects of solar heating, albedo, slope, sky radiation, ground radiation, heat conductivity of the ground and sensible heating. Solution of the equations gave the temperature in a model as a function of depth and time (similar to Fig. 4.18a). In this way, the diurnal temperature range (ΔT) was related to albedo, slope, slope direction and thermal inertia for the latitude, elevation, time of the year and meteorological conditions in the study area. Thus, given ΔT, albedo and topography, the thermal inertia for each pixel of the image could be worked out. By assigning the appropriate grey level to the thermal inertia value for each pixel, the thermal inertia image (Fig. 4.20) was produced. This image brought out the dominant feature – the Pisgah basaltic lava

Fig. 4.20
Thermal inertia image of Pisgah Lava Flow and Lavic Lake, Mojave Desert, California. Note that high thermal inertia values appear bright. A = cinder cone (Pisgah Crater), B = dry playa (Lavic Lake), C = an older lava flow, D = areas of cinders, E and F indicate the boundary between the lava flow and the dry lake and G = a thin layer of sand overlying the basalt outwash (*Source*: Kahle *et al.* 1976)

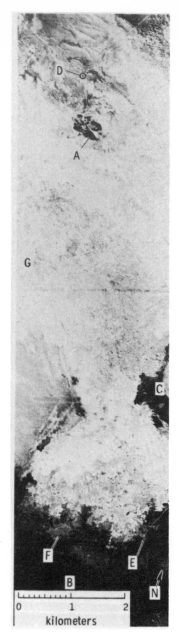

flow which appeared bright, indicating a relatively high thermal inertia but dark in the albedo image because of the low reflectivity of the basalt. Many other geomorphological features such as a cinder cone, a dry playa, older lava flow and alluvial fans could be detected. Also some ideas concerning the nature of the materials (whether hard rocks or unconsolidated sediments) composing these features could be obtained.

Another application of thermal inertia for geological mapping was reported by Pratt and Ellyett (1979) in Australia where the method was employed to map residual soils in the vast inland arid regions as a guide to subsurface geology. This made use of the fact that thermal inertia changed with different soil moisture contents as mentioned before. Therefore, provided the residual soils derived from different rock types had contrasting porosities, geological mapping could be carried out. In the arid zones, the top 10 cm of the soil was unlikely to be completely dry, especially for the fine-textured soils such as clays. It was also unlikely that different soils contained the same amount of moisture, except after complete saturation by rain. Thus, for a clay soil with a moisture content of $0.1 \, \text{m}^3 \, \text{m}^{-3}$ and a dry sand, both with a porosity of 40 per cent, the thermal contrast was found to be about $0.017 \, \text{cal cm}^{-2} \, \text{s}^{-1/2} \, \text{K}^{-1}$, corresponding to a ΔT contrast range of about 8.0 K according to the local meteorological condition. This permitted a large enough thermal inertia to be detected. The porosity of the residual soils is related to mineralogical variations and thus is a useful property for geological mapping. From this example of application, one sees that the thermal inertia method can also be applied to determine soil moisture content. Pratt and Ellyett (1979) also described such an application to determine soil moisture, based on the porosity variable. Based on the method developed by De Vries (1963) to calculate the dielectric constant of granular materials, they obtained a simple equation to determine the heat capacity per unit volume (C) of a soil:

$$C = 4.186 \times 10^6 [0.46(X_s + X_c) + X_w] \qquad [4.7]$$

where C is heat capacity per unit volume in $\text{Jm}^{-3}\text{K}^{-1}$, X_s is the sand volume fraction, X_c is the clay volume fraction and X_w is the water volume fraction. A simulation study suggested that thermal inertia exhibited a strong dependence on soil moisture and only a small dependence on soil type. Idso *et al* (1975) discovered from field work that the volumetric water contents of surface soil layers between 2–4 cm thick were linearly correlated to the amplitude of the diurnal surface soil temperature wave for clear day and night periods (Fig. 4.21). If the air temperature was also incorporated with the soil temperature, the regression correlation coefficients for 0–2 cm and 0–4 cm depth increment relations were improved by about 1/2 of 1 per cent. This implied that the extra effort involved to get additional air temperature data was not worthwhile in view of such small improvement, and the relation between soil water content and diurnal surface soil temperature difference alone could satisfactorily be used to assess the soil water status in the uppermost few centimetres of bare soil.

However, the diurnal surface soil temperature wave was not a unique descriptor of soil water content. It was necessary to know the soil type before good estimates of water content could be obtained from the amplitude of the diurnal surface soil temperature wave. If soil type was unknown, one could not determine water content, but one could estimate *pressure potential* (i.e. the tension with which water is held by the soil particles) for the soil. From the pressure potential one could then estimate soil water content based on the relations demonstrated in Fig. 4.22. According to the findings of Pratt and Ellyett (1979), the accuracy of determination of soil moisture was found to be about 18 per cent having taken into account the uncertainty in determining

Fig. 4.21
The amplitude of the diurnal surface soil temperature wave on clear day-night periods versus the mean daylight volumetric soil water content of two different depth intervals (0–2 cm and 0–4 cm). The points enclosed by diamonds represent volumetric water contents obtained by direct gravimetric sampling of the soil, while all remaining points represent volumetric water contents obtained by the albedo measurement technique (*Source*: Idso *et al.* 1975)

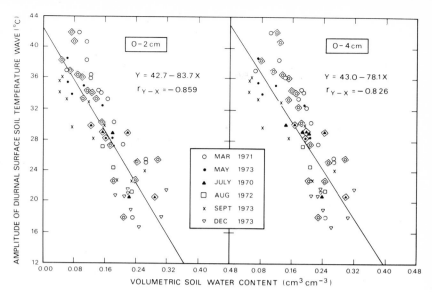

Fig. 4.22
(a) The amplitude of the diurnal surface soil temperature wave versus the mean daylight soil water pressure potential of the 0–2 cm depth increment for four different soils and (b) volumetric soil water content versus soil water pressure potential for the four different soils tested (*Source*: Idso *et al.* 1975)

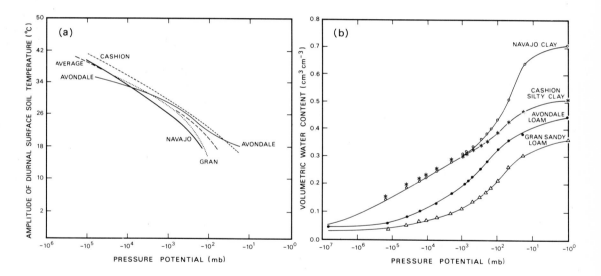

thermal inertia (±4 per cent), soil composition (±3 per cent) and porosity (±11 per cent).

Case Study 3:

Multispectral thermal infrared sensing for the detection of silicate rocks

This specific application of thermal infrared imagery is to make use of a special property of the silicate rocks and minerals (SiO_4); they display spectral emittance minima in the 8–12 μm wavelength region

(Vincent and Thomson 1971). This phenomena is believed to be caused by interatomic vibrations. This occurrence of the major molecular absorption bands (reststrahlen bands) in the thermal infrared can be made use of in the discrimination of the silicate (felsic) rocks from the non-silicate (mafic and ultramafic) rocks which exhibit quite different emittance minima. From Fig. 4.23, one sees that the position of the silicate emittance minimum shifts to longer wavelengths as the content of SiO_2 decreases. However, in order to map these emittance minima, more than one thermal infrared channel is required and very few scanners can split the 8.0–14.0 μm into more than one channel (Vincent 1975). Vincent (1975) reported an application using a two-element Hg:Cd:Te thermal detector constructed by Minneapolis-Honeywell which was capable of detecting radiances in two channels simultaneously: (1) 8.2–10.9 μm and (2) 9.4–12.1 μm. The application was carried out with this detector over a sand quarry near Mill Creek, Oklahoma, USA on 25 June 1970 from an aircraft at an altitude of 1,000 m above mean terrain. The images (Figs 4.24a, 4.24b) after proper calibration revealed that cold objects were dark and warm objects were light. In order to distinguish the silicate rocks from the non-silicate ones, a ratio image was produced by dividing the radiance of each pixel in channel 1 by the corresponding pixel in channel 2 (Fig. 4.24c). The quartz and quartz-sandstone with approximately 90–100 per cent silica appeared dark while the non-silicates such as vegetation, water, carbonate-rich soil and buildings appeared bright in the ratio image. This is because the silica reststrahlen band was present mainly in the first channel (i.e. 8.2–10.9 μm). Thus, objects with a high percentage of silica appeared dark if the temperature differences over the whole image were not too large (Vincent, Thomson and Watson 1972).

Fig. 4.23
Emissivity spectra of silicate rocks (*Source*: Vincent and Thomson 1971)

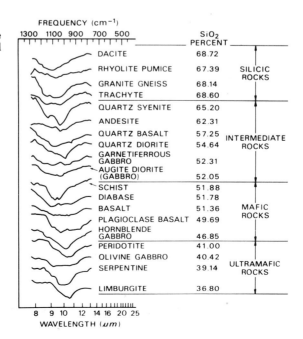

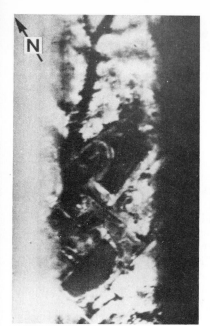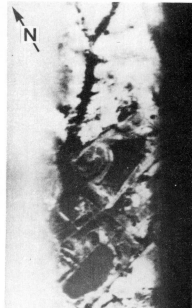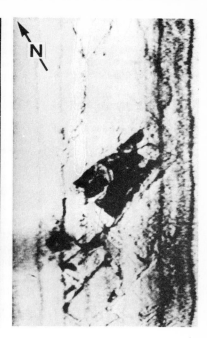

Fig. 4.24
Two-channel analogue infrared images of a sand quarry at Mill Creek, Oklahoma, USA: (a) channel 1: 8.2–10.9 μm and (b) channel 2: 9.4–12.1 μm. Cold objects are dark and warm objects are light. Hot and cold reference plates are at the left and right margins of each image, respectively. Image dimensions are roughly 1 × 2 km. (c) Ratio image (radiance in channel 1 divided by the radiance in channel 2). Dark areas occur where the ratio is less than the average silica ratio. Because the silica reststrahlen band is present mainly in Channel 1, objects with a high percentage of silica will appear dark if the temperature differences over the scene are not *too* large (*Source*: Vincent *et al.* 1972)

Applications using active and passive microwave imagery

Under this section, most of the applications discussed will be using active microwaves, i.e. radar, although some mention of passive microwaves will be made towards the end.

The side-looking airborne radar (SLAR) imaging system is the most important active remote sensing system which is capable of acquiring imagery over areas constrained by adverse weather conditions. The radar imagery acquired is quite similar in appearance to that of aerial photography and the techniques developed for photogeologic interpretation are to a great extent also applicable. Indeed, because of the sideways-looking aspect of the antenna in the radar imaging system, the SLAR image exhibits an abundance of shadows so that the SLAR image may be compared to a low sun angle aerial photograph. This is an advantage which permits subtle terrain features and lineaments to be enhanced. However, one should note that in the case of radar, the sun angle is replaced by the depression angle of the antenna, which decreases from the near range to the far range across the whole radar image. The occurrence of radar shadows therefore depends on whether the back slopes of the features are greater than the depression angles at which these features are imaged. Thus, even the same feature imaged at different locations will give distinctly different shadow lengths. Another important difference between radar imagery and aerial photography is that the tonal variations on a radar image largely reflect the different degrees of surface roughness of the terrain and to some extent the dielectric properties of the surface material which are highly dependent on the porosity of the soil and its water content. Therefore, the radar imagery is particularly suited to land-

147

form analysis from which geological and hydrological inferences can be made. The commonly used radar wavelengths for geological applications are 0.8 cm (K band), 3 cm (X band) and 25 cm (L band). It is possible therefore to experiment with multi-channel radar imaging in addition to the use of dual polarization (HH and HV) and different look-directions. Both the real aperture and synthetic aperture radars have been employed.

Case Study 1:

Quantitative determination of terrain slope

It has been mentioned above that radar shadows are the result of the interaction between the depression angle (β) of the radar beam and the slope of the terrain (α) facing away from the beam so that three cases result: (1) if $\alpha < \beta$, the back slope is fully illuminated, no shadow results, (2) if $\alpha = \beta$, the back slope is partly illuminated, commonly referred to as grazing; and (3) if $\alpha > \beta$, the back slope is obscured from the radar beam, giving rise to radar shadows (Fig. 4.25; Lewis and Waite 1973). Thus, radar shadowing will be more intensive in the far range, i.e. at the lower depression angles. However, the above relationships assume that the flight line of the aircraft is parallel to the strike of the terrain crestline. As the angle θ between the two increases, the angle at which α will shadow at a given depression angle also increases. The effect of θ on the angle at which α will shadow at a given β is shown graphically in Fig. 4.26. This can be used to determine by how much a terrain slope (α) is greater given β and θ. From Fig. 4.26 it can be seen that if β and θ are 40°, the terrain back slope must be greater than 47.5°.

For determining the mean regional slope, it will be quicker to adopt a different approach: to determine the depression angle (β) where grazing occurs provided that the landforms on both the near-range and far-range sides of grazing are homogeneous. However, by increasing

Fig. 4.25
Relationship of radar shadow (slant range length) with depression angle β (*Source*: Lewis and Waite 1973)

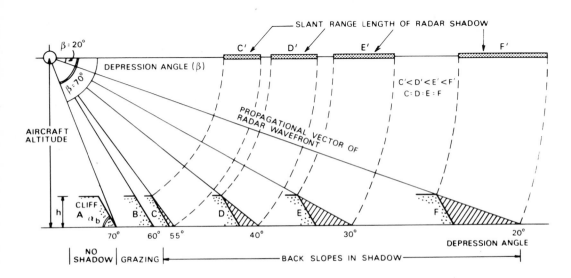

148

Fig. 4.26
Relationship of θ, the angle described by the flight direction and strike of the crestline, with α_b, the true backslope angle of terrain in shadow. This is also a nomogram for correcting the apparent backslope angle to true backslope angle. For example, if a slope is at a 40° depression angle β and the angle between the flight path and strike of the crestline θ is 40°, then the true blackslope must be greater than 47.5° (*Source*: Lewis and Waite 1973)

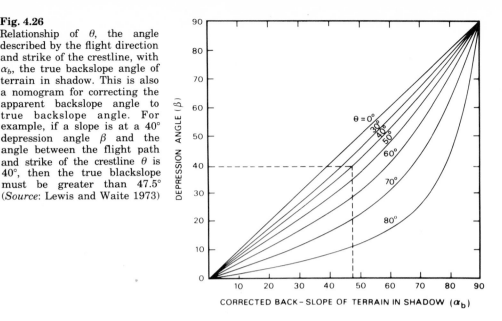

the number of look-directions of the radar beam, even this assumption can be relaxed. The disadvantages of this approach are: (1) the grazing areas are difficult to delimit; (2) it is highly probable that the grazing areas are hidden in radar shadows of the terrain features in front. A new approach, therefore, is to determine simply whether there are shadows or no shadows along a crest line. This decision will allow one to know whether the terrain slope is bigger or smaller than the depression angle of the radar beam which is known. Lewis and Waite (1973) adopted such an approach to produce a cumulative frequency curve of terrain slope over an area. Obviously, this approach is most suitable for mountainous regions with high terrain slopes (greater than 15°). Their result suggested good correspondence between cumu-

Fig. 4.27
Cumulative frequency curves showing good correspondence between map- and radar-derived terrain slope (α) data (*Source*: Lewis and Waite 1973)

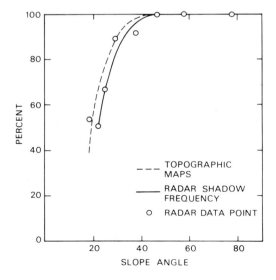

lative frequency curves of radar-derived and topographic map-derived terrain slope (Fig. 4.27). The method was found to be able to provide a more realistic cumulative frequency curve of slopes for areas of moderate to high relief than topographic maps of 1:62,500 scale. Strictly speaking, this method of terrain slope determination is semi-quantitative in nature because it does not produce actual values for individual slopes. For this, one should make use of parallax measurement on stereo SLAR images for height determination to be discussed in Ch. 9. From the known height and the horizontal distance, terrain slope angle can be easily determined (Verstappen 1977: 53–4).

The interaction between the depression angle of the radar beam and terrain slope permits a qualitative approach to classify landforms into mountains, high hills, low hills and plains based on radar shadow lengths (McCoy and Lewis 1976). For mountainous areas, radar shadowing occurs for over half of the range; for hills, radar shadowing occurs for less than half of the range. In particular, for the low hills, the tonal change from front to back is gradual; for high hills, the tonal change is abrupt. For plains, a uniform tonal signature is observed at the macroscale but a salt and pepper appearance is obvious at the microscale. The different vegetation covers may cause darker or lighter tones. In all cases, no radar shadows should be found.

Case Study 2:

Detection of structural linear features

The shadow patterns of the radar imagery help to emphasize relief and structure. With suitable tonal contrasts and a proper orientation of the flight line, lineaments on the earth's surface are easily detected. The detection of these lineaments can lead to the discovery of geological structures which are indicators of the occurrence of mineral resources (Martin-Kaye, Norman, Skidmore 1980). In the cold environment of Pilgrim Springs on Seaward Peninsula, Alaska, the synthetic aperture X-band (2.4–3.8 cm) radar imagery with like-polarization (HH) acquired from an altitude of about 18,000 m was successfully utilized to detect structural lineaments (i.e. faults, fractures, contacts, etc.; Dean *et al.* 1982). Three classes were mapped as a result: (1) the prominent E-W trending lineaments; (2) the shorter and less distinct N-S trending lineaments; (3) the subsidiary NNE–WNW trending lineaments (Fig. 4.28). These were confirmed in the field and from other sources, to be faults. The analysis suggested closely spaced block faulting with highest fracture density in the eastern half of the study area. Some of these faults might have provided conduits for the emergence of hot water at the springs.

In the Los Andes region of northwestern Venezuela, a case was described by Vincent (1980) who applied synthetic aperture X-band (3.12 cm wavelength) like-polarization (HH) radar imaging (with a spatial resolution of 10 m) from an altitude of 12,000 m for mineral and petroleum exploration. These were north-south and south-north flights but the look-direction was always westward. The radar images were interpreted for lineaments which were suspected of being faults or joints. These were mapped. Landsat false colour composites at the

Fig. 4.28
Linear trends interpreted on the radar imagery (*Source*: Dean *et al.* 1982)

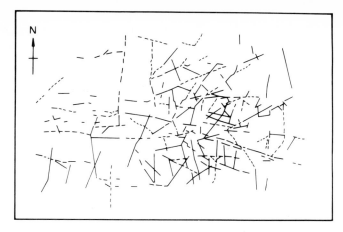

scale of 1:250,000 were also employed to detect lineaments. It was found that the radar data added 27 per cent more lineaments to those mapped from Landsat images. These lineaments were then analyzed to find the area of concentration and the directional trends. From such an analysis, one could locate areas where numerous fault intersections occurred, leading to the discovery of an anticline which could be oil-bearing.

In this type of geological interpretation, care must be exercised to avoid interpreting 'spurious lineaments'. Also, one should note the general reduction of linear data in the look-direction. This directional bias can only be overcome by obtaining radar images in two look-directions at an angle of 135° between them.

Case Study 3:

Extraction of drainage network parameters

The employment of radar imagery to analyse drainage basins has attracted much attention especially because of the varied opinions on its usefulness. The classic study undertaken by McCoy (1969) confirmed the potential of K-band SLAR (wavelength about 1 cm) for the identification, mapping and measurement of such hydrologic parameters as drainage area, stream length, stream order, basin perimeter and ratios of bifurcation, length and circularity. McCoy found that the radar imagery at a scale of 1:200,000 could produce drainage information comparable to that derived from a 1:62,500 topographic map. He has been careful in pointing out the need to enlarge the scale of the radar image to three times and the importance of adopting the correct method for area measurement (i.e. counting grids rather than using a planimeter in view of the different scale variation of the radar image in the range and azimuth directions) as well as length measurement (i.e. using transparent scales rather than a map measuring wheel to measure short stream segments). The other steps to follow were (1) to orientate the radar image so that the shadows of objects would fall towards the observer to avoid a false impression of inverted topography, (2) to trace all obvious trunk streams and major tributaries first and (3) to draw in every visible low-order stream around each major tributary by using highlights, shadows, slight change in film density, occurrence of ponds, etc. as clues. The radar shadows in steep mountainous areas tend to enhance the first-order streams on the one hand but obscure large areas on the other. In areas of low relief, no shadows occur and small streams can only be detected by vegetation changes which cause variations in film density. In areas of rough topography, multiple looks of the terrain by the imaging radar can produce much more information of first-order streams (Figs 4.29, 4.30). McCoy's analysis of 28 drainage basins of varying sizes (from less than 26 km² to larger than 52 km² in areal extent) with different terrain characteristics gave results in drainage area, bifurcation ratio, average-length ratio and basin perimeter which were comparable to those obtained from topographic maps.

However, the main criticism of McCoy's work is the loss of detail of first- and second-order streams in radar imagery. Mekel (1972) also noted the considerably less detail of drainage patterns in the radar

Fig. 4.29
K-band radar image of Trinchera Creek, Colorado (*Courtesy*: R. M. McCoy; *Source*: McCoy 1969)

single
look direction

Fig. 4.30
Interpreted drainage orders of a basin from radar imagery (K-band). Top: using composite look-direction and bottom: using single-look direction. See Fig. 4.29 (*Source*: McCoy 1969)

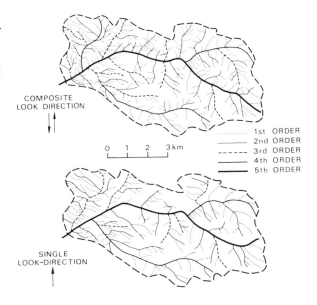

COMPOSITE
LOOK DIRECTION

0   1   2   3 km

1st ORDER
2nd ORDER
3rd ORDER
4th ORDER
5th ORDER

SINGLE
LOOK-DIRECTION

Fig. 4.31
SAR 580 imagery (sample 'frame' 126) of the middle section of the Rouge drainage basin, Quebec. Hydrological features interpreted are: *A* – strong stippled return = agnatic vegetation; *K* – weak return with subdued texture = muskeg; *M* – strong edging return – aspect angle-orientation effect = meander scar; *Q* – strong curvilinear return = vegetation on point bars; *W* – dark return – specular effect = water; *Y* – strong sinuous return with dark areas = partially infilled meander cut-off; *Z* – surrogate return from riparian vegetation = exposed section of point-bar. Other features of interest interpreted are: *E* = highway; *H* = hydro transmission towers; *L* = bright linear returns from power lines; *P* = power line cut swath; *R* = rail line and embankment; *T* = trails and logging cuts; *V* = radar shadow. (*Courtesy*: J. T. Parry; *Source*: Parry *et al.* 1980)

image than in the high altitude aerial photography. It appears that the detectability is controlled by terrain characteristics as well as operational factors. It is generally agreed that low relief and forest cover tend to suppress details of drainage. The orientation of the drainage channels in relation to the radar beam obviously affects their detectability, with considerable loss of detail in drainage systems orientated orthogonally to the flight line. It was further noted that the character of the drainage system would have a direct effect on the amount of network detail imaged. The mature dendritic systems employed in McCoy's examples are more likely to display a near-complete channel network on radar imagery than other drainage systems.

An investigation using a dual-frequency and dual-polarised radar system for drainage analysis was carried out by Parry, Wright and Thomson (1980) in Canada. A high-resolution, synthetic-aperture system (SAR 580) developed by the Environmental Research Institute of Michigan was used. The system operates in two frequencies: X-band (3 cm or 9.45 GHz) and L-band (25 cm or 1.35 GHz) transmitting horizontally polarized energy and receiving both like (HH) and cross (HV) polarized backscattered energy. The output gives two image films, one with X-band HH and HV imagery and the other with L-band HH and HV imagery. The nominal range and azimuth resolutions for the X-band imagery are 1.5 m and 2.1 m respectively, whereas for the L-band they are 2.3 m and 2.1 m respectively. The study area was the drainage basins of the North and Rouge Rivers in the Canadian Shield which is characterized by an irregular, forested topography. The flight was designed in such a way that an area with two look-directions was obtainable. By manually mapping the hydrol-

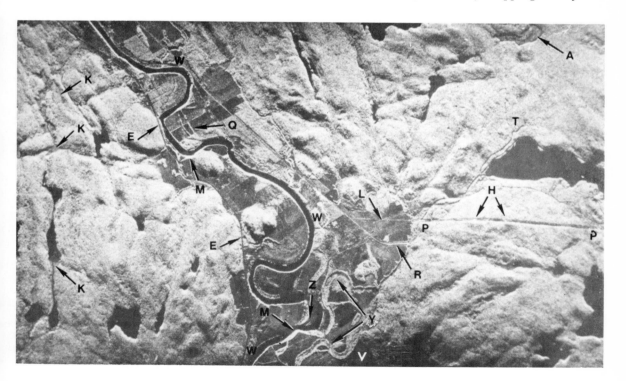

Fig. 4.32
Interpreted hydrological detail from SAR 580 imagery in the middle section of the Rouge drainage basin, Quebec (*Source*: Parry *et al.* 1980)

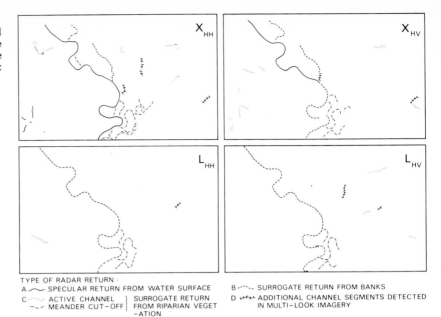

TYPE OF RADAR RETURN :
A ⌒ SPECULAR RETURN FROM WATER SURFACE   B ⌒ SURROGATE RETURN FROM BANKS
C ⌒ ACTIVE CHANNEL ⎤ SURROGATE RETURN   D ⌒ ADDITIONAL CHANNEL SEGMENTS DETECTED
  ⌒ MEANDER CUT–OFF ⎦ FROM RIPARIAN VEGET   IN MULTI–LOOK IMAGERY
                       –ATION

ogical details according to the types of radar return from the four types of radar imagery (2–4 × enlargement) for comparison with the corresponding topographic map (1:50,000 scale) details (Figs 4.31, 4.32), it was concluded that the radar performance in detecting drainage channels was relatively poor especially in detecting first- and second-order streams. Only 37 per cent of the drainage network appearing on the 1:50,000 topographic map was detected. The best performance was achieved with the X-band like-polarized (HH) imagery which provided 64 per cent of the actual drainage network. The same type of radar imagery was also most effective in recording channel detail and in identifying rapids and depositional features such as point-bars. There was a marked increase in information when two look-directions were employed resulting in a 19 per cent increase in the amount of drainage detail. Lakes were readily identifiable on all X-band like-polarized (HH) radar imagery. Muskeg and aquatic vegetation were also easily differentiated because of their unique signatures. Clearly, the SLAR imagery is important as a reconnaissance tool in hydrological studies.

Case Study 4:

Geological mapping in heavily forested terrain

A more comprehensive geological mapping application over the Amazon jungle in Brazil using SLAR was presented by Correa (1980). The synthetic aperture radar operating in the X-band (3 cm wavelength) in like-polarisation (HH) from an altitude of 11,000 m was also used, which provided imagery with a spatial resolution of about 20 m, supplemented by vertical aerial photographs at 1:130,000 and 1:73,000 scales in selected areas (cloud cover conditions permitting). The resulting radar image scale was 1:400,000. The depression angle of the radar beam varied from 45° near range to 13° far range. Because of the thick jungle cover of the study area, characterized by a predominance of broadleaf trees with an average height of 20–30 m, wide canopies and straight stems and trunks, geological mapping has to make use of the correlation of the terrain with the vegetation types (cf. case study 1, pages 123–5, previously discussed under aerial photography).

Based on the concept that the presence of a given vegetation group is related to an ecological system where biological conditions are in balance with the physical environment, a classification of the Amazon jungle into five major types was developed: (1) tropical rain forest, (2) woodland forest, (3) semi-deciduous woodland forest, (4) savanna vegetation and (5) floodplain vegetation (Table 4.1 and Fig. 4.33(a,b)).

Table 4.1
Classification scheme for the Amazon jungle
(*Source*: Correa 1980)

Vegetation Types	Vegetation Characteristics	Environmental Characteristics
1. Tropical Rain Forest	Dense, multistoreyed, uniform tree height	diverse, from alluvial plains to mountains (above 1,000 m) and to deeply dissected igneous and metamorphic terrain
2. Woodland forest	Less dense than (1)	diverse
(a) woodland forest with palm trees		present in valleys and gently rolling terrain with fertile soil
(b) liana forest		present in mountain slopes and hill tops; deeply dissected terrain
3. Semi-deciduous woodland forest	Low average tree height; tree trunks and branches not straight; many species losing leaves in dry season	in low mountains and deeply eroded igneous, metamorphic and sedimentary terrains
4. Savanna vegetation	Small trees, large evergreen leaves and contorted trunks with rough bark, distributed sparsely over grass	in areas with well-defined dry and rainy seasons, undulating relief, crystalline basement outcrops; thin soil, sandy to clayey
(a) Woodland savanna		
(b) Parkland savanna		
5. Floodplain vegetation		in alluvial plains periodically flooded by rivers
(a) Alluvial forest		occupying areas with well-developed soils
(b) Shrubland		dominating river margins where the soil is sandy

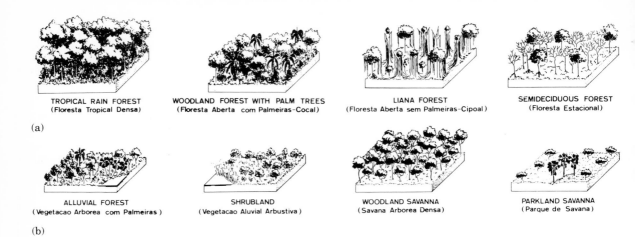

(a)

TROPICAL RAIN FOREST
(Floresta Tropical Densa)

WOODLAND FOREST WITH PALM TREES
(Floresta Aberta com Palmeiras-Cocal)

LIANA FOREST
(Floresta Aberta sem Palmeiras-Cipoal)

SEMIDECIDUOUS FOREST
(Floresta Estacional)

(b)

ALLUVIAL FOREST
(Vegetacao Arborea com Palmeiras)

SHRUBLAND
(Vegetacao Aluvial Arbustiva)

WOODLAND SAVANNA
(Savana Arborea Densa)

PARKLAND SAVANNA
(Parque de Savana)

Fig. 4.33
Amazon vegetation types (*Courtesy* A. C. Correa; *Source*: Correa 1980)

Geological mapping requires studying the radar return from the various vegetation types, i.e. the degree of surface roughness affecting the image texture and tone. The application of the technique was illustrated with three examples of interpretation in the floodplain, sedimentary terrain, and igneous and metamorphic terrain, all with dense forest covers. In the floodplain, the radar return from the alluvial deposits such as sand beaches and levees which are usually associated with grassland vegetation was controlled by ground surface roughness and moisture content. For the forested areas in the swamp and plateau, the radar return was scattered by tree canopies, hence a function of vegetation surface roughness. In the sedimentary terrain the changes of landforms were revealed by textural changes in the radar image, which corresponded to lithological contacts. The diabase sills were easily mapped because of the relatively uniform fine texture and grey tone on the radar images, indicating some edaphic control

Fig. 4.34
Radar mosaic of a study area in the Amazon Basin. The town of Alenquer, Brazil is shown in the lower left (*Courtesy*: A. C. Correa; *Source*: Correa 1980)

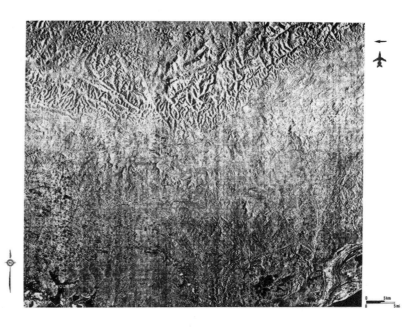

Fig. 4.35
Generalized geological map of the study area in the Amazon Basin corresponding to the radar mosaic in Fig. 4.34 (*Source*: Correa 1980)

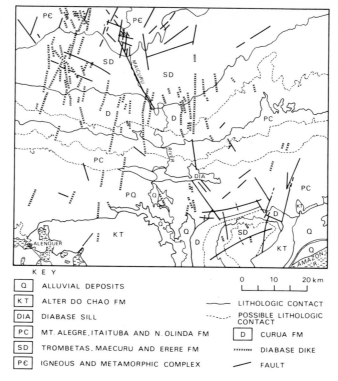

KEY

Q	ALLUVIAL DEPOSITS
KT	ALTER DO CHAO FM
DIA	DIABASE SILL
PC	MT. ALEGRE, ITAITUBA AND N. OLINDA FM
SD	TROMBETAS, MAECURU AND ERERE FM
PꞒ	IGNEOUS AND METAMORPHIC COMPLEX

——	LITHOLOGIC CONTACT
------	POSSIBLE LITHOLOGIC CONTACT
D	CURUA FM
....	DIABASE DIKE
——	FAULT

0   10   20 km

of a unique vegetation assemblage of tree canopy to topographic relief. Faults could also be delineated even under the thick tropical rainforest (Figs 4.34, 4.35). In the igneous and metamorphic terrain, the different rock types were detected on radar images by characteristic landforms, drainage and structural patterns as well as the distribution of vegetation. The igneous and metamorphic rock complex was covered largely by thick tropical rain forests. The hilly topography and V-shaped valleys were the result of deep erosion partly controlled by fracture patterns. The coarse textured areas of the radar image were found to be associated with granites and granodiorities after field checks (Figs 4.36, 4.37). Three different landforms were identifiable, viz., the elongated inselberg-type hills with flat tops and savanna-type vegetation; a flatter terrain with a smooth texture; and a hilly area with dense vegetation cover. A qualitative approach was adopted in geological mapping in the Amazon jungle, which attempted to correlate tonal information with the vegetation type and its associated landforms. It demonstrated the usefulness of SLAR as a reconnaissance geological tool.

Case Study 5:

Mapping of surface deposits in desert regions

In contrast to the previous case study, radar's role in the sparsely vegetated desert area is much more direct, as exemplified by the work

157

Fig. 4.36
Radar mosaic for an area in the Precambrian Basement Complex of northern Brazil. The Paru do Oeste River, one of the tributaries of the Amazon, lies along the left side of this mosaic (*Courtesy*: A. C. Correa; *Source*: Correa 1980)

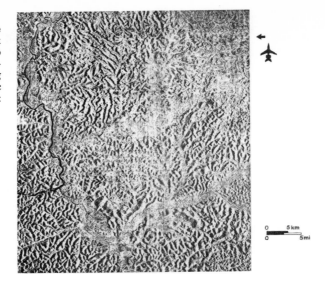

Fig. 4.37
Generalized geological map of the study area in the Amazon Basin corresponding to the radar mosaic in Fig. 4.36 (*Source*: Correa 1980)

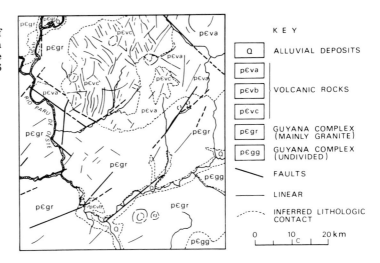

of Sugiura and Sabins (1980) in the Mojave Desert, California. X-band (3 cm wavelength) synthetic aperture SLAR system using like-polarization flown at an altitude of 10,200 m along a flight path oriented N14°E was employed. The look-direction was westwards with a depression angle of 30° for the near range and 10° for the far range. The original imagery was recorded in a ground range format at a scale of 1:400,000 with a spatial resolution of 12 m. This was subsequently enlarged to form an uncontrolled radar mosaic at the scale of 1:250,000 which provided the base for interpretations of surface deposits (Fig. 4.38). According to the brightness of the radar return, physiological setting as well as drainage pattern and texture, six radar-rock units of the surface deposits in the Bristol Lake/Granite Mountain area were discriminated, viz.; (1) cobble-boulder fan deposits; (2) gravel-cobble fan deposits; (3) sand-gravel fan deposits;

Fig. 4.38
X-band radar mosaic of the Bristol Lake-Granite Mountain area, California. Letters refer to typical localities of surface deposits detected from the image. For full descriptions, see Table 4.2 (*Source*: Sugiura and Sabins 1980)

(4) desert pavement; (5) aeolian sand deposits; (6) playa deposits (Fig. 4.39 and Table 4.2). Notice the very great contrast in tone between the very bright 'cobble-boulder fan deposits' and the very dark playa deposits, a result of the difference in the size of these deposits and their moisture-retention capability.

It is interesting to note the further differentiation of the 'aeolian sand deposits' into four types of sand dunes according to radar signatures in a part of the study area known as Kelso Dunes (marked as 'E' on Fig. 4.38). The radar image for this area was specially enlarged

Fig. 4.39
Interpretation of surface deposits from the radar mosaic in Fig. 4.38 (*Source*: Sugiura and Sabins 1980)

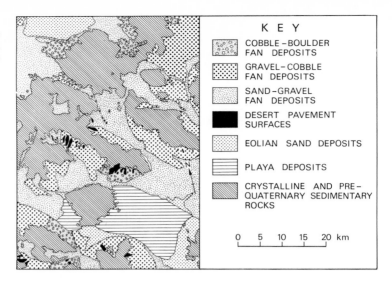

KEY

	COBBLE – BOULDER FAN DEPOSITS
	GRAVEL – COBBLE FAN DEPOSITS
	SAND – GRAVEL FAN DEPOSITS
	DESERT PAVEMENT SURFACES
	EOLIAN SAND DEPOSITS
	PLAYA DEPOSITS
	CRYSTALLINE AND PRE-QUATERNARY SEDIMENTARY ROCKS

0 5 10 15 20 km

(1:80,000) (Fig. 4.40). The dark signature of the dune areas with little or no vegetation was particularly noteworthy. As the vegetation cover increased, an intermediate to dark signature was apparent. Four types of sand dunes were identified on this basis: (1) longitudinal dunes; (2) irregular to rounded dunes; (3) transverse dunes; (4) sand ridges (Fig. 4.41). The characteristic appearance of these dunes in the image and in the field were summarized in Table 4.3.

Table 4.2
Characteristics of the surface deposits of the Bristol Lake/Granite Mountain area, Mojave desert, California on the radar imagery
(*Source*: Sugiura and Sabins 1980)

Case Study 6:

Snowfield mapping

An area of application in hydrology where both active and passive microwave remote sensing can contribute is snowfield or snow cover

Radar-rock unit	Measured Surface roughness, h (average) in cm	Radar signature Observed	Physiographic setting	Typical locality (as shown on Fig. 4.38)
(1) Cobble-boulder fan deposit	30	Bright	Adjacent to bedrock outcrops and upper fan areas; generally dissected by drainage channels	A
(2) Gravel-cobble fan deposits	10	Bright to intermediate	Adjacent to bedrock outcrops and mid-fan areas	B
(3) Sand-gravel fan deposits	0.5	Dark to intermediate	Upper to lower fan areas; and valleys between mountain ranges	C
(4) Desert pavement	1	Dark	Adjacent to bedrock outcrops and upper to mid-fan areas	D
(5) Aolian sand deposits	<.05	Dark to intermediate	Lower fan areas and along margins of the playas	E
(6) Playa deposits	<.01	Dark	Enclosed basin area	F

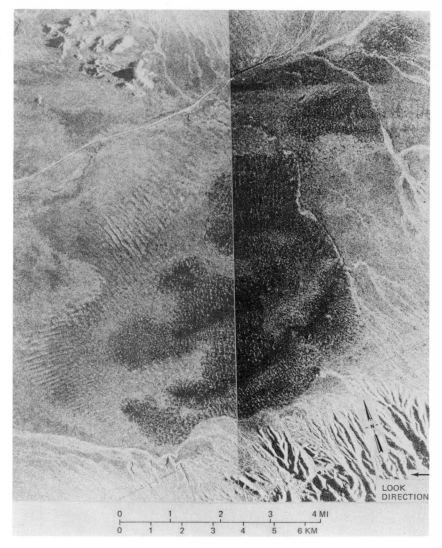

Fig. 4.40
Radar image of the Kelso dunes
area (locality 'E' on Fig. 4.38)
(*Source*: Sugiura and Sabins
1980)

LOOK
DIRECTION

| 0 | | 1 | | 2 | | 3 | | 4 MI |
| 0 | 1 | 2 | 3 | 4 | 5 | 6 KM |

mapping. The extent of snow cover, the stored amount of water and
the state of the snow metamorphism, in particular the date of the
snow-melt and the intensity of runoff, are of prime importance to
hydrology and water management in large regions. The active microwave in the form of radar using Ka-band frequencies can be used to
map accurately the areal extent of the snowfields and glaciers (Waite
and MacDonald 1970). In addition, the freshly fallen, dry snow can be
differentiated from the old snow (firn and névé) based on the difference
in their brightness of radar return. The former permits radar signals
at almost any wavelength to penetrate to great depths, thus giving a
low radar return. The latter, on the other hand, provides a very high
return on imaging radars because of the higher degree of compaction
in the old snow. For quantitative estimation of the runoff from the
snow, the passive microwave remote sensor has to be used. The intensity of microwave radiation received is a function of the temperature,

Fig. 4.41
Interpretation of sand dune types of the Kelso Dunes area from the radar mosaic in Fig. 4.40.

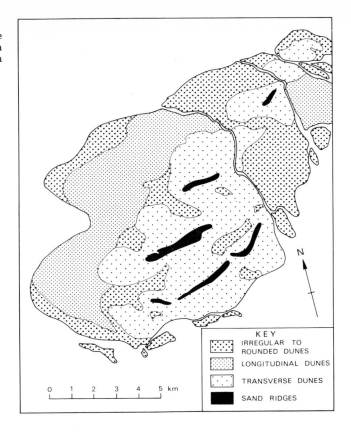

KEY

- ▨ IRREGULAR TO ROUNDED DUNES
- ▨ LONGITUDINAL DUNES
- ⠂ TRANSVERSE DUNES
- ■ SAND RIDGES

0 1 2 3 4 5 km

N

Table 4.3
Radar and field characteristics of different types of sand dunes in Kelso Dunes
(*Source*: Sugiura and Sabins 1980)

Dune type	Radar Signature	Field Description
(1) Longitudinal dunes	Intermediate	Subdued longitudinal dunes; vegetation sparse to moderate
(2) Irregular to rounded dunes	Intermediate to dark	Irregular to rounded subdued dunes; dunes often separated by nearly flat areas; vegetation sparse to moderate
(3) Transverse dunes	Dark with discrete bright tones	Transverse dunes with gentle windward slopes and steep lee slopes, dune heights up to 10 m; vegetation absent to sparse
(4) Sand ridges	Dark	Large linear sand ridges with steep slopes, heights range from 45 m to 167 m; vegetation absent to sparse

density crystal size and free water content of the snow layer and is commonly referred to as the brightness temperature (T_B) (see Ch. 3). By making use of these microwave emission properties of snow, three snow conditions in a high altitude Alpine area can be discriminated according to the average spectral values in horizontal and vertical polarization (Schanda *et al.* 1983): (1) high winter conditions, i.e. snow without any melting metamorphism of the snow cover; (2) a thick wet firn layer (at least several cm) of wet quasi-spherical ice crystals (1–3 mm diameter); (3) a thick crust (several cm) of refrozen firn

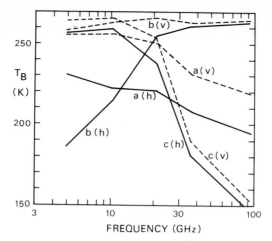

Fig. 4.42
Average spectra of brightness temperatures observed at horizontal and vertical polarisation and at 50° nadir angle. (a) During high winter condition, snow depths of all measurements normalized to 48 cm water equivalent, (b) wet firn layer of at least several centimetres on the surface of the snow cover, and (c) thick (at least several centimetres) crust of refrozen firn (*Source*: Schanda *et al.* 1983)

(Fig. 4.42). It was evident that wet snow exhibited a flat spectrum for vertical polarization whereas dry snow showed a decreasing spectrum for both polarizations. The refrozen firn exhibited a steeper spectrum. The winter dry snow displayed a stronger polarizing effect than other types of snow conditions.

These variations in spectural behaviour can be explained by the volume scattering of the snow crystals of different snow conditions. The winter snow has small snow crystals as compared with the larger but more isotropically arranged crystals of refrozen firn. The wet snow tends to exhibit the same spectrum as water. By using this fact, one can relate snow conditions with the equivalent amount of water stored in each type of snow cover. The penetration depth of microwaves through dry winter snow or refrozen spring snow is larger than the usual thickness of snow cover over land (one or a few metres). For wet snow, the penetration depth of microwaves is drastically reduced to a few centimetres or less. Hence, only the uppermost surface layer actually contributes to the observed radiation. Therefore, only when the snow is dry can the water equivalent of the whole snow cover be determined. Figure 4.43 is an example of using a passive microwave radiometer at 36 GHz frequency with 50° nadir angle to detect the brightness temperature of the snow cover in both horizontal and vertical polarizations. One can see the pronounced effect of the water equivalent below 20 cm, corresponding to a snow depth of 60–80 cm. At longer wavelengths, the effect of water equivalent on the brightness temperature is much less pronounced. Thus, by using the difference of brightness temperatures at 36 and 18 GHz (i.e. a multifrequency approach) one can estimate the water equivalent. It was concluded that a radiometer of the highest possible frequency should be used to map dry snow, preferably with dual-polarization and a second radiometer at a different frequency to differentiate the type of snow. As for wet snow the mapping should be done with a scatterometer or an imaging radar looking off-nadir with an incidence angle more than 30° at frequencies higher than 10 GHz. The combination of a passive microwave radiometer with a synthetic aperture radar is ideal in mapping both the areal extent and the brightness temperature of the snow cover.

Fig. 4.43
Brightness temperature at 36 GHz horizontal and vertical polarization function of the water equivalent of dry snow from the beginning of the snow season to late winter conditions at the end of March. Data taken between March 1977 and December 1980 (*Source*: Schanda *et al.* 1983)

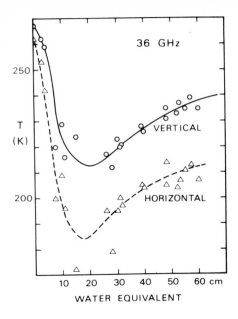

Applications using satellite remote sensing

The introduction of satellite remote sensing has revolutionized the study of our lithospheric environment. The general availability of the land resources satellite remote sensing data such as Landsat to the public has resulted in an increasing number of applications taking advantage of the regional perspective afforded by these data and the wide capabilities of the multispectral sensors. The availability of these data in digital form means that sophisticated manipulations of the data by computer are possible and necessary in order to fully exploit the spectral, spatial and temporal content of the data. Much emphasis has been placed on the development of the softwares for general-purpose and specific applications. The impact of this development has led to such new disciplines as 'geologic remote sensing' (Goetz and Rowan 1981), 'applied geomorphology' (Verstappen 1983) and 'satellite hydrology' (Deutsch, Wisnet and Rango 1981).

Case Study 1:

Mapping regional structures and lithology

The synoptic view of the earth from space provides the best vantage point to study the relationships between landforms and major structural features. The early manned and unmanned space flight programmes, such as the Mercury, Gemini and Apollo series in the USA, have provided excellent black-and-white and colour photography of the earth for regional structure and landform analysis, as exemplified by the works of Lowman (1968, 1972) and Bodechtel and Gierloff-Emden (1969). As an example, the Gemini XI photograph of the Gulf of Aqaba and Gulf of Suez taken in September 1966 (Fig. 4.44) provides information about different rock types (discrimination based on colour differences) and fault alignments (Fig. 4.45). The Gulf of Aqaba-Dead Sea rift dominated the photograph and was regarded by some geologists as a graben which would imply movement perpendicular to its trend. With the availability of Landsat MSS and RBV imagery acquired from a more permanent platform orbiting around the earth at an altitude of 900 km, a spate of geological applications emerged focussing on structural and lithologic interpretations as well as mineral exploration. The MSS imagery has attracted the most attention, and the visual interpretation of structures is normally carried out with the use of a false colour composite formed by combining bands 4, 5 and 7 through blue, green and red colours respectively. However, for lineament analysis, the band 6 or band 7 image (near infrared bands) which emphasized the topography and the drainage pattern is best employed. The sun angle is also an important consideration. In order to bring out the topographic detail, a suitable sun angle is 2° to 3° less than the slopes on the ground (Slaney 1981). In an interpretation of the Landsat image of the Los Angeles-Ventura area of southern California (Fig. 4.46), a system of active faults is clearly visible. On the whole, linear features varying in length from a few kilometres to hundreds of kilometres which can be equated with structural elements such as faults, joints or fractures are recognizable from Landsat images.

Based on an interpretation of Landsat images the existence of major

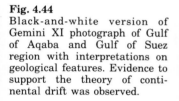

Fig. 4.44
Black-and-white version of Gemini XI photograph of Gulf of Aqaba and Gulf of Suez region with interpretations on geological features. Evidence to support the theory of continental drift was observed.

Fig. 4.45
Interpretation of the geological structures of the Red Sea region from Gemini XI photograph (see Fig. 4.44) (*Source*: Bodechtel and Gierloff-Emden 1974)

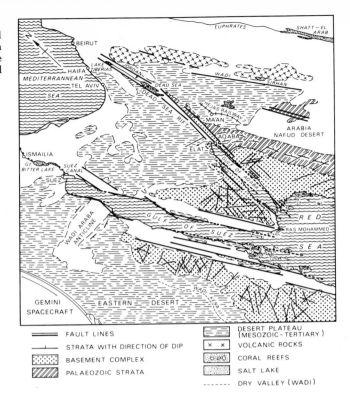

thrust faults on the north and south sides of Tien Shan in China was noted. This was particularly clear on the south side where folded sedimentary formations were visible. By combining the interpreted structural information with earthquake data, Molnar and Tapponnier (1975) were able to estimate the amount of crustal shortening across the eastern portion of Tien Shan to be about 200–300 km. Likewise other strike-slip faults were recognized from Landsat images in Altyn Tagh and Kunlun mountains. From this, the relative motions of the Indian subcontinent and Eurasia could be deduced and it was concluded that crustal deformation extended over a large area with as much as 1,000 km of lateral movement along subparallel strike-slip faults, similar to the San Andreas fault system.

It has already been shown in previous case studies that airborne active microwave imaging systems can contribute additional structural information to that of aerial photography. Similarly, spaceborne radar imaging systems, such as Seasat SAR and Shuttle Imaging Radar-A (SIR-A), have also complemented the Landsat multispectral sensors in this respect. Seasat was a short-lived satellite designed to observe the earth's oceans with the use of microwave sensors launched on 28 June 1978 by the USAs National Aeronautics and Space Administration. It carried three microwave radars, one microwave radiometer, and one visible/infrared radiometer in a nearly circular orbit with an inclination angle of 108° (Fu and Holt 1982). On 10 October 1978 it failed in orbit due to a massive short circuit in its electrical system, but it collected an extensive set of observations of the earth's oceans during its 105 days in orbit. The Seasat SAR (synthetic aperture radar) was

Fig. 4.46
Delineation of major faults and other lineaments of the greater Los Angeles area of Southern California (Path 44 Row 36) based on an interpretation of the Lansat image (band 7 – 0.8–1.1 μm), taken on 4 November 1978.

one of the microwave radars which has provided high-resolution images of the earth's oceans and land from an altitude of 795 km. The wavelength employed was 23.5 cm or a frequency of 1.275 GHz (L-band) and a system bandwidth of 19 GHz with like polarization (HH). The depression angle of the radar beam was about 67° and the incidence angle of the surface was 23°. The theoretical resolution of the surface was 25 × 25 m. The SIR-A was a similar synthetic aperture radar system carried onboard the second Space Shuttle Columbia launched on 12 November 1981 on its second orbital mission. The experiment was aimed at acquiring radar images of a wide variety of different geologic regions around the earth. Although the SIR-A radar system shared essentially similar characteristics of those of Seasat SAR system, it exhibited differences in its orbital altitude (259 km), inclination angle (38°), depression angle (43°), incidence angle (50°), image resolution (40 × 40 m) and system bandwidth (6 MHz) (Ford, Cimino and Elachi 1983; Elachi 1982; Elachi *et al.* 1982).

In an evaluation of the Seasat SAR imagery for geologic mapping in Tennessee-Kentucky-Virginia area (Valley and Ridge Province) in the USA, Ford (1980) found that because of the Seasat SAR's single right-side-looking system orbiting in a plane inclined to the earth's axis of rotation, two different look-directions of the same terrain were possible. The study area was covered by radar imagery acquired looking N67.5°E (Fig. 4.47) and looking N67.5°W. This fact compen-

Fig. 4.47
Mosaic of optically correlated Seasat SAR imagery from revolutions 407 and 163 of the Cumberland Plateau area between Tennessee and Kentucky, USA. Look-direction is N67.5°E (*Courtesy*: J. P. Ford; *Source*: Ford 1980)

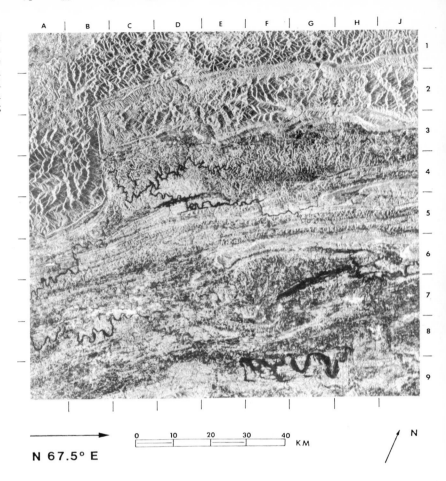

N 67.5° E

0   10   20   30   40   K M

N

sated for the directional bias of the radar imagery in suppressing lineaments that struck more or less parallel to the radar look-direction. As a result, the number of minor geomorphic lineaments (less than 10 km long) mapped from the SAR images was greater than that mapped from the corresponding Landsat MSS image by a factor of about 2 (Fig. 4.48). The SAR images also exhibited high contrast as a result of the local change in terrain slope in the study area. This was attributable to the low inclination (20°) of the Seasat SAR radar beam which produced layover of all slopes that exceeded about 20°. This layover caused steep slopes to appear to be excessively compressed towards the imaging direction and the loss of geologic information. On the other hand, Landsat MSS images did not show any significant geometric distortion (Fig. 4.49). However, the maximum possible enhancement of good Landsat imagery did not provide the topographic detail of the orbital Seasat SAR imagery. The SIR-A system which acquired radar imagery with a larger inclination angle (38°) than that of Seasat SAR was also applied to the same Ridge-and-Valley type topography of the Appalachian Plateau (Ford 1982). It was found that the SIR-A radar imagery could locate most lineaments that were less than 15 km in length, which the Seasat SAR failed to locate. This

Fig. 4.48
Interpreted geomorphic lineaments from (a) Seasat SAR image looking N67.5°E (Fig. 4.47) and (b) enhanced Landsat MSS band 6 image (Fig. 4.49) (*Source*: Ford 1980)

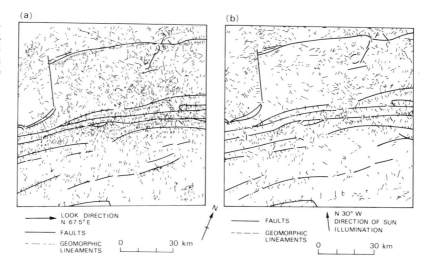

(a)

(b)

LOOK DIRECTION
N 67.5° E

—— FAULTS

– · – GEOMORPHIC
LINEAMENTS

N

0   30 km

—— FAULTS

– – – GEOMORPHIC
LINEAMENTS

N 30° W
DIRECTION OF SUN
ILLUMINATION

0   30 km

Fig. 4.49
Landsat MSS (band 6) of the area covered by Seasat SAR looking N67.5°E (Fig. 4.47). The image has been filtered and stretched to enhance contrast. Sun illumination direction is N30°W (Azimuth = 150°), elevation angle is 26° (Scene ID: 1858–15303, Sevierville, Tennessee, 28 November 1974 (*Courtesy*: J. P. Ford; *Source*: Ford 1980)

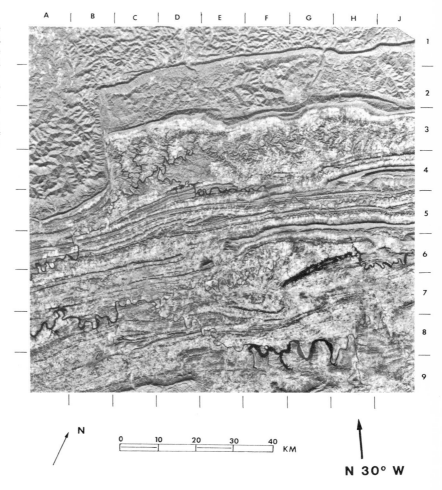

A   B   C   D   E   F   G   H   J
1
2
3
4
5
6
7
8
9

N

0   10   20   30   40
KM

N 30° W

169

amply demonstrated the importance of the relationship between the radar look-angle and the angle of surface slope in the direction of radar illumination to the perception of linear topographic features. The radar look-angle should always exceed the surface slope by a certain amount to avoid excessive layover.

The different types of satellite imagery have also been used for the extraction of lithologic information. In general, the various clues explained under aerial photography are applicable here to discriminate rock types. Thus, as Slaney (1981) has shown with reference to Landsat imagery in Canada, sedimentary rocks could be visually recognized because of distinctive differences in their drainage and fracture patterns or in reflectance especially in bands 4 and 5. The highly reflective outcrop of limestone in the Landsat image of Frobisher Bay, Canada (Fig. 4.50) makes it stand out from the darker granite surrounding it. However, when the vegetation cover exceeds 10 per cent of the surface area, recognition of rock types by reflectance alone is not possible. As for igneous rocks, they can be recognized by their irregular shape and image texture. Generally, granite exhibits a darker tone than the surrounding rocks but texture provides the more useful clue. In Fig. 4.50, the granite is characterized by the large number of lakes scattered about the surface. Metamorphic rocks are even more difficult to recognize if they do not show foliation planes or planes of schistosity. The gneiss in Fig. 4.50 shows foliation and is

Fig. 4.50
Landsat image (band 7) of Frobisher Bay, South Baffin Island, Canada. Note the large number of lakes in the area, being underlain by a highly fractured, massive, reddish-biotite granite. The light coloured patches extending east from the lake in NW are thin Palaeozoic limestone outcrops. However, the similar light toned patch to the NE is glacial drift which marks the southeastern margin of a moraine system. The linear features seen here indicate ice movement from NW to SE (*Source*: Slaney 1981)

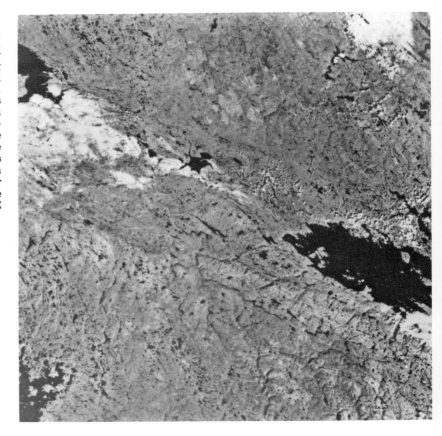

easily detected. In another evaluation of the Landsat imagery for lithologic mapping based on visual interpretation of false colour composites in the semi-arid environment. Viljoen *et al.* (1975) drew attention
to the importance of seasonal variations, colour tones, patterns and
textures in mapping the stratigraphic sequence of sediments in South
Africa. The wet summer Landsat image was found to be superior to
the dry winter image in mapping the stratigraphic components within
the Dolomite series because of stronger tonal variations caused by
changes in soil moisture content, atmospheric haze and the association
of characteristic vegetation with underlying geology. Seasat SAR
imagery has also been employed for lithologic interpretation as
demonstrated by Ford (1980) over the Ridge and Valley Province study.
area of the Appalachians mentioned on pages 167–9. Image textures on
the SAR images provided the basis for lithologic interpretation. The
textural features included granularity (from coarse to fine), definition
(from high to low) and contrast (from high to low). Five image-textures
could be discriminated with characteristics shown in Table 4.4 and
mapped in Fig. 4.51. They provided a measure of the Seasat SAR backscatter from the vegetation cover on the level to sloping terrain in the
study area. A good correlation between the image textures and topography was also noted. The type of topography was found to correlate
reasonably well with bedrock associations. In this way, the five image
textures were found to correspond to five lithologic associations as
indicated in Fig. 4.51.

So far, only the visual interpretations of the satellite imagery have
been discussed. As Landsat imagery is available in computer compatible tape (CCT) format, computer processing can easily be employed
to facilitate evaluation of the multispectral characteristics of the data.
Taranik (1978) listed out the digital image processing procedures as
data preprocessing, image enhancement and image classification.
Under image enhancement which is aimed to optimize display of the
data to the analyst, the following types of data manipulation are possible: contrast enhancement, edge enhancement, ratio enhancements,
temporal ratioing and simulated natural colour enhancements. For the
extraction of lithologic information, ratio enhancements have been
found to be most fruitful. One of the commonly used methods is to
produce a colour ratio composite image by ratioing Landsat MSS spectral bands (4/5, 5/6 and 6/7 ratios) pixel by pixel which are then
contrast-stretched to enhance the spectral differences. The three
ratioed images are then copied into cyan, yellow and magenta diazo
films respectively and combined. Thus, continuous variation in colour

Table 4.4
Characteristics of image-texture
units mapped in Fig. 4.51
(*Source*: Ford 1980)

Texture	Granularity	Definition	Contrast
1	Coarse	High	High
2	Fine	Low	High
3	Fine	High; closely spaced, subparallel, linear orientations; or finely granular	Medium
4	Fine	High	Medium
5	Medium	Medium; linear to curvilinear orientations	High

Fig. 4.51
Image texture map drawn from Seasat SAR images (Fig. 4.47). Texture units 1–5 correspond with lithologic associations as follows: (1) sandstone-shale-silt-stone (north); quartzite-shale-conglomearate (south); (2) clay-shale-limestone; (3) conglom-erate-sandstone-shale; (4) sand-stone-shale-siltstone; (5) dolomite-chert-crystalline lime-stone. Image characteristics of texture units are given in Table 4.4 (*Source*: Ford 1980)

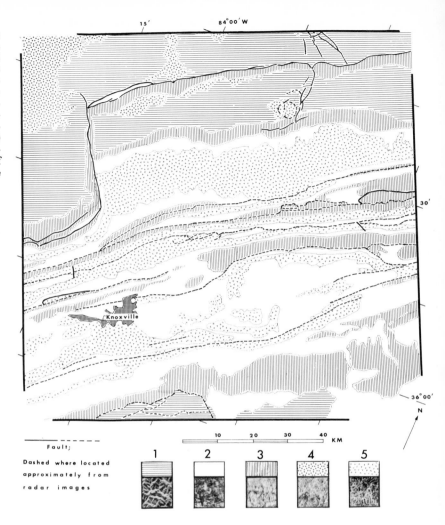

is produced with cyan as the lowest ratio (highest vegetation density) and magenta as the highest ratio (lowest vegetation density). This method enhances subtle spectral reflectivity differences and minimizes radiance variations due to albedo and topography. This method has been employed to detect hydrothermally altered areas where mineral resources may occur and to discriminate major rock types in south-central Nevada, USA (Rowan *et al*. 1976). In such a colour ratio composite, the hydrothermally altered areas appear as green to dark green and brown to red-brown. The green and dark green colours are indicative of the occurrence of limonitic rock (which may be hydro-thermally altered) because both ferric absorption bands near 0.4 μm and 0.9 μm are well developed. Materials with these spectral properties are green on the colour ratio composite and red to the human eye in the field. The brown area represents light coloured hydrothermally altered volcanic rocks and the red-brown pattern reveals limonite-free, silica-rich, light-colour volcanic rocks. The vegetation is orange with the darker hue representing denser vegetation (Pl. 4, page 36). In the south-central Nevada example (Pl. 4), the striking correspondence between the green patterns and the Goldfield mining district suggests that the mapping of hydrothermally altered limonitic rock distribution can lead to the discovery of mineral resources. This technique, however, is limited to areas with no more than 35–40 per cent vegetation cover. However, because of the acid soil conditions that characterize many hydrothermally altered areas, vegetation cover is usually sparse.

It is possible to make use of geobotanical information in combination with the colour ratio composite images to locate mineral resources. Geobotanical associations are found to be closely related to regional lithologic variations whereas barren areas and certain shrubs and herbs are specifically related to the concentrations of metals. An excellent example of application is provided by Raines, Offield and Santos (1978) who conducted a search for uranium in the Powder River Basin, Wyoming using Landsat MSS imagery. They made use of the MSS colour ratio composite images to locate the exposed limonitic rocks by substituting the 4/6 ratio for the 5/6 ratio which gave better discrimination of the hydrothermally altered areas, because the 4/6 ratio was less affected by vegetation than the 5/6 ratio. However, the problem of obscuration of many mineralized areas by vegetation chiefly high prairie sage, grass, and herbs covering 50–75 per cent of the ground remained. They therefore decided to use a colour coded MSS 5/6 ratio image to show the variation in the vegetation cover. This displayed areas of high vegetation density as cyan (low ratio) and areas of low vegetation density as magenta (high ratio). This image also revealed a basin-axis lineament as indicated by a marked change in vegetation density and the character of several stream channels crossing it. Comparison with the colour ratio composite image revealed a spatial correlation between moderate vegetation cover and rocks having an intermediate ratio of sandstone to mudstone with which the uranium deposits were associated. In addition to structural information derived from lineament analysis, a model of occurrence of uranium in the Powder River Basin could be formulated. This approach was followed by mapping ultramafic rocks in a heavily vegetated terrain near Crescent City, California (Raines and Wynn 1982). Because the ultramafic rocks could only support lesser amounts of vegetation than the surrounding rocks and exhibited a rounded topography, an accurate map of ultramafic rocks could be obtained with the Landsat imagery.

Case Study 2:

Landform analysis

There are numerous applications of satellite imagery for landform analysis *per se*. Of special interest are the investigations on the desert and volcanic landforms on the global scale using Landsat imagery, Skylab photography, Seasat SAR imagery and SIR-A data. The investigation of the global desert was initiated by McKee and Breed (1974) using Landsat-1 imagery. This involved identification, description and measurement of characteristic forms and structures of the sand bodies, with a view to determine the distribution pattern of various dune types and their classification as well as to infer the processes responsible for each dune type. The synoptic coverage of Landsat imagery at a uniform scale facilitates direct comparisons of widely separated areas and understanding of the relations of the sand areas to their surrounding features. Landsat mosaics can be easily made to provide a map base for superimposition of data. Altogether 15 desert or semidesert areas were studied. Visual interpretation of the false colour composite images was found to be most effective because the sand features could be easily discerned with their characteristic yellow colour (McKee and Breed 1976). From this investigation, a preliminary classification of sand dunes into five types based on forms was possible: (1) parallel straight or linear; (2) parallel wavy or crescentic; (3) star or radial; (4) parabolic or U-shaped; (5) sheets or stringers. It was found that linear dune complexes occurred in the Simpson Desert of Australia, the Kalahari Desert of South Africa, the Empty Quarter of Saudi Arabia and the Sahara of North Africa. Parallel wavy or crescentic dune complexes were represented in the Kara Kum Desert of the USSR and in the Nebraska Sand Hills of the western USA. Variants of this basic type were found to occur in the Great Eastern Erg of Algeria (fish-scale type), Saudi Arabia (giant crescent type), Gobi Desert in China (bulbous type) and Takla Makan desert (basketweave type) in China. Star or radial dune complexes were found to scatter at random in parts of the Great Eastern Erg of Algeria, in the Gran Desierto de Sonora in Mexico and the Empty Quarter of Saudi Arabia. Parabolic dunes were found in Rajasthan Desert of India and at White Sands, New Mexico, USA. Finally, in some areas such as near Lima, Peru, the sand dunes failed to develop definite geomorphic forms and the sand accumulated in flat sheets or in stringers as in South West Africa.

This study of the desert sand seas was continued with colour handheld photography obtained by the crew members of Skylab-4 in 1973 from an altitude of 435 km. This high-resolution photography permitted detailed identification of the components of each sand dune pattern and gave a better understanding of the depositional environments of aeolian sand bodies by means of sand-distribution mapping. The relationship between the sand distribution and the surface wind regimes could be established to provide insights into the processes of sand dune formation (McKee, Breed and Fryberger 1977). With the availability of Seasat SAR, it was found that the radar backscatter from the sand dunes was primarily specular because of their smooth surfaces, causing a dark response, unless the sand dunes had steep slopes oriented normal to the incident radar beam (Blom and Elachi 1981). Provided that the orientation was favourable in relation to the radar illumination and that the interdune spacing was adequately wide, star, dome, linear and crescentic dunes could all be recognized. However, sand sheets and stringers did not give a bright radar return unless they had vegetation cover. With the use of SIR-A data applied to large dunes in Africa and Asia, the conclusion given above was confirmed (Breed *et al.* 1982). The SIR-A imaging experiment had succeeded in recording major dunes in parts of several major deserts, including the Takla Makan, Badan Jaran, Ulan Bah and Mu Us in China; the Kara Kum Desert in Turkmen SSR, the An Nufud of the northern Arabian Peninsula and the Sahara of Northern Africa, in addition to lesser dune fields of the USAs deserts. The major dunes are usually 2 km or more in width or diameter and as much as 200–300 m in height. They are readily identifiable on the SIR-A images. One should be warned, however, that the dune-slope orientations as revealed on SIR-A imagery could be ambiguous and misleading if only the properly orientated parts of the slip faces gave a bright return.

In connection with the application of SIR-A imagery to desert study, one should note an interesting discovery of subsurface valleys in the Eastern Sahara as a result of the penetrating ability of the radar beam (24 cm wavelength) through the dry sand surface into the bed rock underneath. According to the nature of the surface the radar beam penetration can be at least 1 metre in sand sheet and drift sand and 2 or more metres in sand dunes (McCauley *et al.* 1982).

Satellite imagery has been employed for volcanic activity monitoring and landform study. When Landsat MSS imagery was first available, it was applied immediately to study volcanic geomorphology and to detect geothermal areas in Iceland which is one of the most active areas of volcanism in the world (Williams *et al.* 1973; Williams 1976). The outline of two primary craters on the volcanic island of Surtsey could be clearly seen. The lava flows from the eruptions were easily delineated. Because of the high latitude location of Iceland, it has variable solar elevation angles during the time of satellite passage throughout the year from about 2° to 53°. By using low sun angle imagery combined with a snow cover, geologic structure and terrain morphology are greatly enhanced. The extreme linearity of fissures, crater rows and grabens were particularly evident on the Landsat MSS imagery. Geothermal areas could also be detected. Under optimum conditions, a small geothermal area (2.5 km²) with a heat output of $25-125 \times 10^6$ cal/sec could be detected by its snow-melt pattern.

One of the most useful applications of Landsat imagery is to detect landform changes before and after a volcanic eruption. This was done during the eruption of Heimaey, Vestmann Islands, Iceland beginning on 23 January 1973. Comparison of band 7 imagery acquired on 21 November 1972 and that on 3 July 1973 revealed the areal increase of the island caused by lava flows to the east and northeast as well as the major changes in coastline configuration. The airborne eruptive products called tephra covered up areas and were responsible for lowering the reflectance of objects on MSS bands 6 and 7 of the post-eruption imagery. A more recent example of such a type of comparison was done also for the eruption of Mount St Helens on May 18 1980. A comparison between the false colour composite image taken on 11 September 1979 and that on 19 August 1980 clearly revealed the crater having been changed to a horse-shoe shaped outline and a wide area (479 km²) of devastated forest land being covered with ash. All the rivers and lakes leading from the volcano appeared light blue, indicating water made turbid by the sediments from the eruption (Pl. 5a and 5b, page 36).

The handheld colour photography which was obtained by Skylab-4 crew members, specifically for observation of volcanoes, provided details on volcanic ring infrastructures and caldera landforms through the display of subtle differences in coastal landforms, topographic textures, relief, drainage pattern and albedo. In this way, previously unobserved relationships between geomorphology and structure could be shown (Friedman and Heiken 1977).

Environmental satellite systems can also be employed to monitor volcanic eruptions and study volcanic landform changes. Cochran and Pyle (1978) reported of the use of US Department of Defense Block 5 (DMSP) meteorological satellite's infrared sensor (spatial resolution 0.5 km) to detect the eruption site of Kilauea Volcano, Hawaii at night-time on 13 September 1977. The day-time pictures of the geostationary SMS-2 satellite's infrared sensor (spatial resolution 8 km) could detect the hazy cloud or volcanic smog formed by the ejected volcanic ash, smoke from burning vegetation, condensed water vapour and dust.

SIR-A data have also been applied to study volcanic landforms (Masursky 1982). The coarse surface texture of the lava flows tended to give bright radar returns, thus distinctly outlining each lava flow or eruptive centre.

Case Study 3:

Mapping of surface and ground waters

Satellite remote sensing permits hydrologic problems involving large areas and trends to be solved, thus complementing the conventional data (Anderson 1981). It is fortunate that water is spectrally unique. Clear water transmits radiant energy in the blue-green band and absorbs energy in the near infrared band. Turbid water with heavy concentrations of sediment, however, also reflects some of the radiation in visible wavelengths but absorbs nearly all the near infrared energy. Thus, on the band 5 (0.6–0.7 μm) image of Landsat MSS image, the sediments in the rivers and along the coast can be easily distinguished (Fig. 1.26). On the band 7 (0.8–1.1 μm) image, all hydrologic features are registered black (Fig. 1.27). In a false colour composite made up of all 4 MSS bands, clear water appears as dark blue whilst turbid water appears as light blue (Pl. 1). This characteristic provides the basis for mapping river flooding as exemplified by the work of Deutsch and Ruggles (1978). The purpose was to map the areal extent of the flood which occurred in the Indus River Valley of Pakistan during August and September 1973 using Landsat imagery. The whole Indus River system is covered by 25 Landsat scenes. Of these only 13 flood scenes were used for the mapping. In order to enhance the contrast between water and wet surfaces as well as surrounding dry areas on the scene, the Landsat 70 mm negative transparencies were 'contrast stretched' by photo-optical methods. From the reprocessed positive transparencies of the four spectral bands, a false colour composite image was formed with the aid of the colour additive viewer by projecting the transparencies through appropriate colour filters. To depict details in the flooded area, it was recommended that band 4 be projected in green light, band 5 in blue light and band 7 in red light. The flooded area appeared as cyan and vegetation as magenta. In order to depict the extent of the flood, a temporal composite was formed by projecting MSS band 7 imaged on 26

September 1972 through a red filter and that imaged on 3 September 1973 through a green filter. The flooded areas appeared in red in such a temporal composite and a mosaic of temporal composites (Pl. 6, page 36) provided a synoptic view not only of the total flooded area but also the normal water distribution over the lower Indus River Valley (400,000 km^2 in area). The flooded area was found to exceed 20,000 km^2 from the mosaic. Apart from depicting the distribution of flooding, Landsat false colour composites can also reveal streamlines of light and dark toned water caused by different sediment concentrations and particle size (Green, Whitehouse and Outhet 1983). When these stream-lines were observed at different flood stages, they defined the spatial pattern of flooding by showing the location of areas of active flow. This information is useful to assist in planning flood mitigation measures, especially the building of levees.

Another useful characteristic of water is that it has the highest specific heat of any substance. It is possible to detect water or moisture by sensing the emitted thermal energy as cool areas during warm days or as warm areas on cool nights. The Heat Capacity Mapping Mission (HCMM) satellite launched in April 1978, which carried a two-channel radiometer (0.55–1.1 μm and 10.5–12.5μm) in a sun-synchronous orbit (altitude 620 km) collected data at approximately 0230 and 1330 local standard time with a repeat cycle of 5 or 16 days depending on latitude. The day-time and night-time thermal infrared images (10.5–12.5 μm) reveal clearly hydrological features, such as lakes, rivers and seas through temperature anomalies compared with surrounding features. Some studies by Heilman and Moore (1981, 1982) suggested the possibility of detecting ground water using HCMM thermal infrared imagery acquired at 1330 hours Local Standard Time in conjunction with Landsat photographic and ground data. A relationship between radiometric temperatures from HCMM scenes and water table depths of up to 5 m was established. This relationship is most significant at night-time when the effect of the vegetation cover is least. In a study carried out by Heilman and Moore (1982) in the Big Sioux River Basin, Brookings County, South Dakota which has a predominantly agricultural land use of small grains, row crops and pasture (Fig. 4.52), they estimated surface soil temperatures T_s in °C from the following equation:

$$T_s = 0.79\ T_c\ e^{(-0.80\ PC)} + 20.35 \qquad [4.8]$$

where T_c is a composite radiometric temperature in °C comprising radiance contributions from the soil and crop and PC is the percentage of land cover expressed as a fraction. The T_c was obtained from HCMM temperatures corrected for atmospheric effects and the land cover percentage was estimated from a Landsat false colour composite superimposed on HCMM computer grey map within a pixel with the aid of a Bausch and Lomb Zoom Transferscope. Four dates of HCMM 1330 local standard time thermal infrared data (in June, July, August and September 1978) were used. The predicted soil surface tempera-tures (T_s) revealed a linear relationship with depths of water table (data acquired from United States Geological Survey observation wells in the study area). This correlation increased in strength as the season progressed and the temperature anomalies became maximized (Fig. 4.53). These predicted soil temperatures also showed a strong

Fig. 4.52
Photographic enlargement of HCMM thermal infrared image (scene ID AA0125-08340) taken at night on 29 August 1978, showing the Big Sioux River Basin in Brookings County in southeastern South Dakota. Note that dark tone is cool and bright tone is warm (*Courtesy*: J. L. Heilman; *Source*: Heilman and Moore 1981)

correlation with soil moisture (Fig. 4.54). (See the previous section on thermal inertia mapping, pages 140–5). Multiple regression analysis of the September data yielded the equation:

$$Ts = 26.60 - 0.05 \ SWC + 2.50 \ D \qquad [4.9]$$

with a coefficient of determination (r^2) of 0.87, where *SWC* is the volumetric soil water content in the 0–4 cm layer in percentage and *D* is water table depth in metres. Research is needed to separate the effects of soil moisture from those of the water table in ground water studies.

Snow, a useful source of water, was probably the first 'earth resource' observed from space. When the first US satellite (TIROS-1) was launched in April 1960, snow could be detected in eastern Canada from the initial low-resolution images acquired (Bowley and Barnes 1981). The problem of recognition was to separate snow from cloud because of their similar reflectance. With the availability of more advanced remote sensing systems carried by the satellites, the monitoring of snow cover on a regular basis became possible and the importance of snow cover as a parameter in the prediction of snowmelt derived runoff was realized. Snow mapping can now be carried out with the NOAA VHRR (Very High Resolution Radiometer), GOES

Fig. 4.53
Relationship between predicted soil surface temperature and water table depth on 4 September 1978 (*Source*: Heilman and Moore 1982)

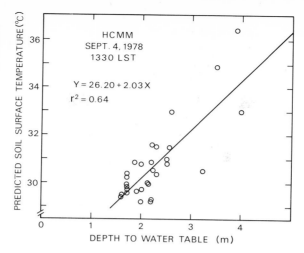

Fig. 4.54
Relationship between predicted soil surface temperature and volumetric soil water content in the 0–4 cm layer on 4 September 1978 (*Source*: Heilman and Moore 1982)

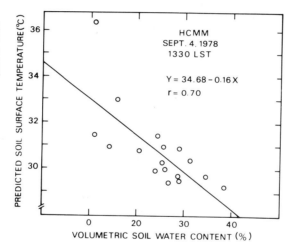

(Geostationary Operational Environmental Satellite), VISSR (Visible and Infrared Spin-Scan Radiometer) and Landsat MSS sensors. Despite the rather poor spatial resolution of the NOAA and GOES images, as compared with that from Landsat, NOAA and GOES provided a larger coverage of the area at more frequent intervals. NOAA gave repeat coverage at half hourly intervals both day and night. The Landsat MSS and RBV data have been particularly accurate in snow mapping because of the high spatial resolution (80–100 m) which permitted more correct delimitation of the snow cover area. In the low spatial resolution NOAA VHRR images, for example, the snowline tended to be integrated and small snow patches undetected. In an evaluation conducted by Barnes *et al.* (1974), it was found that the extent of snowpacks in Arizona and southern Sierra Nevada could be mapped from Landsat in more detail than was depicted in aerial survey snow charts. The Landsat band 5 was particularly useful for this purpose, and with the aid of a Bausch and Lomb Zoom Transferscope, the Landsat image could be superimposed optically onto a base map. Through adjusting the image projection size and stretching of the image along either axis, the original Landsat image was rectified to fit the exact scale of the base map. The snow extent was then directly transferred from the Landsat image to the map. These Landsat snow cover observations, when separated on the basis of watershed elevation, have been found to be strongly correlated to the runoff rates for lower elevation watersheds (Fig. 4.55) as demonstrated by Rango *et al.* (1975). With the new Landsat 4 Thematic Mapper which is a second generation high resolution remote sensing system, a spectral band 5 (1.55–1.75 μm) is designed to detect vegetation and soil moisture content as well as to differentiate snow from clouds.

Another approach to map snow is to use the passive microwave sensors – the Electrically Scanning Microwave Radiometer (ESMR) carried onboard Nimbus 5 and Nimbus 6. As explained in previous sections, microwaves are mostly unaffected by clouds and can penetrate various snow depths depending on their wavelengths. Short wavelength radiation is scattered by small snow crystals whilst longer wavelength radiation is affected by very large crystals. The microwave radiometers detect the brightness temperature of snow. On the whole, deeper and denser snow allows greater scattering, thus lowering the brightness temperature. As the liquid water in the snow increases, the brightness temperature increases. Foster *et al.* (1980) have employed both Nimbus 5 (1.55 cm) and Nimbus 6 (0.81 cm) ESMR horizontally and vertically polarized data to monitor snow depths in areas of the Canadian high plains, the Montana and North Dakota high plains and the steppe of central Russia. In all cases, correlations between brightness temperature and snow depth were found to be significant at the 0.001 level, and the Nimbus 6 (0.81 cm) ESMR data consistently produced higher correlations than Nimbus 5 (1.55 cm) ESMR data, probably because the shorter wavelength microwave radiometer could not sense emission from as deep in the snowpack as the longer wave-

Fig. 4.55
Landsat derived snow cover estimates versus measured runoff (1973 and 1974) for four watersheds less than 3.050 m mean elevation in the Wind River Mountains, Wyoming (*Source*: Rango *et al.* 1975)

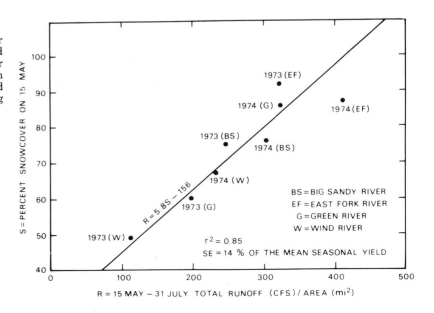

length one. As a result, the Nimbus 6 data were less affected by the variable nature of the underlying ground conditions of the snowpack. In view of the difference in snowpack condition, land cover, underlying soil condition and surface temperature of the different geographic areas under study, difficulty was experienced in extrapolating relationships between microwave brightness temperature and snow depth from one area to another.

Conclusions

The survey in this chapter has brought out some significant aspects of applied remote sensing to the study of the lithosphere – the solid realm of our earth. There is a great variety of remote sensors being used, which complement each other in the extraction of geological, geomorphological and hydrological information at different levels of required accuracies. At least, three directions of development are obvious. First, there is a gradual shift from aerial cameras towards the use of non-photographic sensors, notably in the thermal infrared and microwave channels, as witnessed by the interest shown in thermal inertia mapping, active microwave sensing (synthetic aperture radar) in the detection of structural lineaments, and passive microwave sensing in monitoring water resources. Secondly, the availability of satellite remote sensing data in digital form, such as the Landsat MSS CCT, permits sophisticated computer preprocessing, enhancement and classification, notably, the employment of ratioing techniques in detecting hydrothermally altered rocks in relation to mineral exploration. Thirdly, the increasing interest in satellite remote sensing has led to the adoption of a regional approach in studying the lithosphere. Landsat imagery has been widely used for regional structure interpretation (Cochrane and Wan 1983) and for environmental geomorphology (Verstappen 1977). This further leads to integrating the geological, geomorphological and hydrological aspects with the land use and land cover of a region (Woldai 1979). The success of applied remote sensing in this field depends on understanding the value of the various techniques and integrating them into a problem-solving methodology. Development of new insights into geology, geomorphology and hydrology is necessary before any new tools can be fully exploited.

References

Allum, J. A. E. (1966) *Photogeology and Regional Mapping.* Pergamon: Oxford; London; Edinburgh; New York; Toronto; Sydney; Paris; Braunschweig.

Anderson, D. G. (1981) Roles of satellites in hydrology: In Deutsch, M., Wiesnet, D. R. and Rango, A. (eds) *Satellite Hydrology.* American Water Resources Association: Minneapolis, Minnesota, pp. 144–55.

Barnes, J. C., Bowley, C. J. and Simmes, D. A. (1974) *The Application of ERTS Imagery to Mapping Snow Cover in Western United States.* Final Report under Contract NAS5-21803, Environmental Research and Technology Inc: Concord, Massachusetts.

Blom, R. and Elachi, C. (1981) Spaceborne and airborne imaging radar observations of sand dunes, *Journal of Geophysical Research* **86**: 3061–73.

Bodechtel, J. and Gierloff-Emden, H. G. (1974) *The Earth from Space*. David and Charles: Newton Abbot.

Bowley, C. J. and Barnes, J. C. (1981) Satellite snow mapping techniques with emphasis on the use of Landsat: In Deutsch, M., Wiesnet, D. R. and Rango, A. (eds) *Satellite Hydrology*. American Water Resources Association: Minneapolis, Minnesota, pp. 158–64.

Braun, E. L. (1956) The development of association and climax concepts: their use in interpretation of the deciduous forest, *American Journal of Botany* **43**: 906–11.

Breed, C. S., McCauley, J. F., Schaber, G. G., Walker, A. S. and Berlin, G. L. (1982) Dunes on SIR-A images: In Cimino, J. B. and Elachi, C. (eds) *Shuttle Imaging Radar-A (SIR-A) Experiment*. Jet Propulsion Laboratory: Pasadena, California, pp. 4.52–87.

Cameron, H. L. (1965) Coastal studies by sequential air photography, *The Canadian Surveyor* **19**: 372–81.

Cassinis, R., Lechi, G. M., Marino, C. M. and Tonelli, A. M. (1977) Monitoring volcanic environments – results of a remote sensing programme in Italy: In Van Genderen, J. L. and Collins, W. G. (eds) *Monitoring Environmental Change by Remote Sensing*, Remote Sensing Society.

Cassinis, R., Marino, C. M. and Tonelli, A. M. (1971) Evaluation of thermal I.R. Imagery on Italian volcanic areas – ground and airborne surveys, *Proceedings of the Seventh International Symposium on Remote Sensing of Environment*, Vol. 1, University of Michigan: Ann Arbor, Michigan, pp. 81–9.

Cochran, D. R. and Pyle, R. L. (1978) Volcanology via satellite, *Monthly Weather Review* **106**: 1373–5.

Cochrane, G. R. and T. Wan (1983) Interpretation of Structural characteristics of the Taupo volcanic zone, New Zealand, from Landsat imagery, *International Journal of Remote Sensing* **4**: 111–28.

Conway, D. and Holz, R. K. (1973) The use of near-infrared photography in the analysis of surface morphology of an Argentine alluvial floodplain, *Remote Sensing of Environment* **2**: 235–42.

Correa, A. C. (1980) Geological mapping in the Amazon jungle – a challenge to side-looking radar, in *Radar Geology: An Assessment*. Jet Propulsion Laboratory: Pasadena, California, pp. 385–416.

Currey, D. T. (1977) Identifying flood water movement, *Remote Sensing of Environment* **6**: 51–61.

De Vries, D. A. (1963) Thermal properties of soils: In Van Wyk, W. R. (ed.) *Physics of the Plant Environment*, John Wiley: New York, pp. 210–35.

Dean, K. G., Forbes, R. B., Turner, D. L., Eaton, F. D. and Sullivan, K. D. (1982) Radar and infrared remote sensing of geothermal features at Pilgrim Springs, Alaska, *Remote Sensing of Environment* **12**: 391–405.

Deutsch, M. and Ruggles, Jr., F. H. (1978) Hydrological applications of Landsat imagery used in the study of the 1973 Indus River flood, Pakistan, *Water Resources Bulletin* **14**: 261–74.

Deutsch, M., Wiesnet, D. R. and Rango, A. (eds) (1981) *Satellite Hydrology*, American Water Resources Association: Minneapolis, Minnesota.

El-Ashry, M. T. (ed.) (1977) *Air Photography and Coastal Problems*. Dowden; Hutchinson and Ross: Stroudsburg; Pennsylvania.

El-Ashry, M. T. and Wanless, H. R. (1967) Shoreline features and their changes, *Photogrammetric Engineering* **32**: 184–89.

Elachi, C. (1982) Radar images of the earth from space, *Scientific American* **247**: 46–53.

Elachi, C. et al. (1982) Shuttle imaging radar experiment, *Science* **218**: 996–1003.

Fezer, F. (1971) Photo interpretation applied to geomorphology – a review, *Photogrammetria* **27**: 7–53.

Fischer, W. A., Moxham, R. M., Polcyn, F. and Landis, G. H. (1964) Infrared surveys of Hawaiian volcanoes, *Science* **146**: 733–42.

Ford, J. P. (1980) Seasat orbital radar imagery for geologic mapping: Tennessee–Kentucky–Virginia, *The American Association of Petroleum Geologists Bulletin* **64**: 2064–94.

Ford, J. P. (1982) Analysis of SIR-A and Seasat SAR images of Kentucky–Virginia, in Cimino, J. B. and Elachi, C. (eds) *Shuttle Imaging Radar-A (SIR-A) Experiment*. Jet Propulsion Laboratory: Pasadena, California, pp. 4.10–4.19.

Ford, J. P., Cimino, J. B. and Elachi, C. (1983) *Space Shuttle Columbia Views the World with Imaging Radar: The SIR-A Experiment*. Jet Propulsion Laboratory: Pasadena, California.

Foster, J. L., Rango, A., Hall, D. K., Chang, A. T. C., Allison, L. J. and Diesen, III, B. C. (1980) Snowpack monitoring in North America and Eurasia using passive microwave satellite data, *Remote Sensing of Environment* 10: 285–98.

Friedman, J. D. and Heiken, G. (1977) Volcanoes and volcanic landforms, in *Skylab Explores the Earth*, Lyndon B. Johnson Space Center. National Aeronautics and Space Administration: Washington, DC, pp. 137–71.

Fu, L. L. and Holt, B. (1982) *Seasat Views Oceans and Sea Ice with Synthetic-Aperture Radar*. Jet Propulsion Laboratory: Pasadena, California.

Goetz, A. F. H. and Rowan, L. C. (1981) Geologic remote sensing. *Science* 211: 781–91.

Green, A. A., Whitehouse, G. and Outhet, D. (1983) Causes of flood streamlines observed on Landsat images and their use as indicators of floodways, *International Journal of Remote Sensing* 4: 5–16.

Harris, P. (1973) The composition of the earth: In Gass, I. G., Smith, P. J. and Wilson, R. C. (eds) *Understanding the Earth: A Reader in the Earth Sciences*. The Artemis Press: Sussex, pp. 53–69.

Heilman, J. L. and Moore, D. G. (1981) Ground water applications of the Heat Capacity Mapping Mission. In Deutsch, M., Wiesnet, D. R. and Rango, A. (eds) *Satellite Hydrology*, American Water Resources Association: Minneapolis, Minnesota, pp. 446–9.

Heilman, J. L. and Moore, D. G. (1982) Evaluating depth to shallow groundwater using Heat Capacity Mapping Mission (HCMM) data, *Photogrammetric Engineering and Remote Sensing* 48: 1903–6.

Howard, A. D. (1965) Photogeologic interpretation of structure in the Amazon Basin: a test study, *Geological Society of America Bulletin* 76: 395–406.

Idso, S. B., Schmugge, T. J., Jackson, R. D. and Reginato, R. J. (1975) The utility of surface temperature measurements for the remote sensing of surface soil water status, *Journal of Geophysical Research* 80: 3044–47.

Kahle, A. B., Gillespie, A. R. and Goetz, A. F. H. (1976) Thermal inertia imaging: a new geologic mapping tool, *Geophysical Research Letters* 3: 26–8.

Lattman, L. H. (1963) Geologic interpretation of airborne infrared imagery, *Photogrammetric Engineering* 29: 83–87.

Lee, K. (1969) Infrared exploration for shoreline springs at Mono Lake, California, test site, *Proceedings of the Sixth International Symposium on Remote Sensing of Environment*, The University of Michigan: Ann Arbor, Michigan, pp. 1075–1100.

Lewin, J. and Weir, M. J. C. (1977) Monitoring river channel change: In Van Genderen, J. L. and Collins, W. G. (eds) Monitoring Environmental Change by Remote Sensing, Remote Sensing Society, pp. 23–27.

Lewis, A. (1974) Geomorphic-geologic mapping from remote sensors: In Estes, J. E. and Senger, L. W. (eds) *Remote Sensing: Techniques for Environmental Analysis*. Hamilton Publishing Company: Santa Barbara, California, pp. 105–26.

Lewis, A. J. and Waite, W. P. (1973) Radar shadow frequency, *Photogrammetric Engineering* 39: 189–96.

Lo, C. P. (1976) *Geographical Applications of Aerial Photography*. Crane, Russak: New York; David and Charles: Newton Abbot; London; Vancouver.

Lowman, Jr., P. D. (1968) *Space Panorama*. Weltflugbild: Zurich, Switzerland.

Lowman, Jr., P. D. (1972) *The Third Planet*, Weltflugbild: Zurich, Switzerland.

McCauley, J. F., Schaber, G. G., Breed, C. S., Grolier, M. J., Haynes, C. V., Issawi, B., Elachi, C. and Blom, R. (1982) Subsurface valleys and geoarchaeology of the Eastern Sahara revealed by shuttle radar, *Science* 218: 1004–20.

McCoy, R. M. (1969) Drainage network analysis with K-band radar imagery, *Geographical Review* 59: 493–512.

McCoy, R. M. and Lewis, A. J. (1976) Use of radar in hydrology and geomorphology, *Remote Sensing of the Electro Magnetic Spectrum* 3: 105–22.

McKee, E. D. and Breed, C. S. (1974) An investigation of major sand seas in desert areas throughout the world, *Proceedings of Third Earth Resources Technology Satellite-1 Symposium*, Vol. I. National Aeronautics and Space Administration: Washington DC, pp. 665–679.

McKee, E. D. and Breed, C. S. (1976) Sand seas of the world, in Williams, Jr., R. S. and Carter, W. D. (eds) *ERTS-1: A New Window on our Planet*, US Government Printing Office: Washington, DC, pp. 81–8.

McKee, E. D., Breed, C. S. and Fryberger, S. G. (1977) Desert sand seas, in *Skylab Explores the Earth*. National Aeronautics and Space Administration: Washington, DC, pp. 5–47.

Martin-Kaye, P. H. A., Norman, J. W. and Skidmore, M. J. (1980) Fracture trace expression and analysis in radar imagery of rain forest terrain (Peru) in *Radar Geology: an Assessment*, Report of the Radar Geology Workshop: Snowmass, Colorado, July 16–20, 1979. Jet Propulsion Laboratory: Pasadena, California, pp. 502–7.

Masursky, H. (1982) Volcanic field east of Raton, New Mexico, in Cimino, J. B. and Elachi, C. (eds) *Shuttle Imaging Radar-A (SIR-A) Experiment.* Jet Propulsion Laboratory: Pasadena, pp. 4.19–21.

Mather, P. M. (1979) Theory and quantitative methods in geomorphology, *Progress in Physical Geography*, **3**: 471–87.

Mekel, J. F. M. (1972) *The Geological Interpretation of Radar Images*, ITC Textbook of Photo-Interpretation. Enschede, Vol. VIII-2.

Mekel, J. F. M. (1978) *The Use of Aerial Photographs and other Images in Geological Mapping*, International Institute for Aerial Survey and Earth Sciences (ITC). Enschede: the Netherlands.

Mekel, J. F., Savage, J. F. and Zorn, H. C. (1967) *Slope measurements and Estimates from Aerial Photographs*, International Institute for Aerial Survey and Earth Sciences (ITC), Delft, The Netherlands.

Miller, V. C. (1961) *Photogeology.* McGraw-Hill: New York.

Molnar, P. and Trapponnier, P. (1975) Cenozoic tectonics of Asia: effects of a continental collision, *Science* **189**: 419–26.

Moxham, R. M. and Alcara, A. (1966) Infrared surveys at Taal volcano, Philippines, *Proceedings of the Fourth Symposium on Remote Sensing of Environment*, University of Michigan: Ann Arbor, Michigan, pp. 827–44.

Parry, J. T., Wright, R. K. and Thomson, K. P. B. (1980) Drainage on multiband radar imagery in the Laurentian area, Quebec, Canada, *Photogrammetria* **35**: 179–98.

Pratt, D. A. and Ellyett, C. D. (1979) The thermal inertia approach to mapping of soil moisture and geology, *Remote Sensing of Environment* **8**: 151–68.

Pratt, D. A., Ellyett, C. D., McLauchlan, E. C. and McNabb, P. (1978) Recent advances in the application of thermal infrared scanning to geological and hydrological studies, *Remote Sensing of Environment* **7**: 177–84.

Pressman, A. E. (1968) Geologic comparison of Ektachrome and infrared Ektachrome photography, in Smith, Jr., J. T. (ed.) *Manual of Color Aerial Photography*, American Society of Photogrammetry: Falls Church, Virginia, pp. 396–397.

Raines, G. L., Offield, T. W. and Santos, E. S. (1978) Remote-sensing and subsurface definition of facies and structure related to uranium deposits, Powder River basin, Wyoming, *Economic Geology* **73**: 1706–23.

Raines, G. L. and Wynn, J. C. (1982) Mapping ultramafic rocks in a heavily vegetated terrain using Landsat data, *Economic Geology* **77**: 1755–69.

Rango, A., Salomonson, V. V. and Foster, J. L. (1975) Seasonal stream-flow estimation employing satellite snowcover observations. Preprint X-913-75-26, 34 pp.

Ray, R. G. (1960) *Aerial Photographs in Geologic Interpretation and Mapping.* United States Government Printing Office: Washington, DC.

Rowan, L., Wetlaufer, P. H. and Goetz, A. F. H. (1976) Discrimination of rock types and detection of hydrothermally altered areas in south-central Nevada: In Williams, Jr., R. S. and Carter, W. D. (eds) *ERTS-1: A New Window on our Planet.* Geological Survey Professional Paper 929, US Government Printing Office: Washington, DC pp. 102–105.

Sauer, E. K. (1981) Hydrogeology of glacial deposits from aerial photographs, *Photogrammetric Engineering and Remote Sensing* **47**: 811–22.

Schanda, E., Matzler, C. and Kunzi, K. (1983) Microwave remote sensing of snow cover, *International Journal of Remote Sensing* **4**: 149–58.

Shilin, B. V., Gusev, N. A., Miroshnikov, M. M. and Karizhenski, Y. Y. (1969) Infrared aerial survey of the volcanoes of Kamchatka, *Proceedings of the Sixth International Symposium on Remote Sensing of Environment*, Vol. 1. The University of Michigan: Ann Arbor, Michigan, pp. 175–187.

Shreve, R. L. (1966) Statistical law of stream numbers, *Journal of Geology* **74**: 17–38.

Slaney, V. R. (1981) *Landsat Images of Canada: A Geological Appraisal*, Geological Survey of Canada. Ottawa, Canada.

Smith, H. T. U. (1967) Photogeologic interpretation in Antarctica, *Photogrammetric Engineering*, **33**: 297–304.

Stoffel, D. B. and Stoffel, K. L. (1980) Mt. St. Helens seen close up on May 18, *Geotimes* **25**(10): 16–17.

Strahler, A. N. (1957) Quantitative analysis of watershed geomorphology, *Transactions of American Geophysical Union* **38**: 913–20.

Strandberg, C. H. (1968) The sensation of color: In Smith, Jr., J. T. (ed.) *Manual of Color Aerial Photography*, American Society of Photogrammetry: Falls Church, Virginia, pp. 3–11.

Sugiura, R. and Sabins, F. (1980) The evaluation of 3-cm-wavelength radar for mapping surface deposits in the Bristol Lake/Granite Mountain area, Mohave Desert, California, in *Radar Geology: An Assessment*. Jet Propulsion Laboratory: Pasadena, California, pp. 439–56.

Taranik, J. V. (1978) Principles of computer processing of Landsat data for geologic applications. *Open-File Report 78–117, US Geological Survey*, Sioux Falls, South Dakota.

Verstappen, H. T. (1977) *Remote Sensing in Geomorphology*. Elsevier: Amsterdam; Oxford; New York.

Verstappen, H. T. (1983) *Applied Geomorphology: Geomorphological Surveys for Environmental Development*. Elsevier: Amsterdam; Oxford; New York.

Viljoen, R. P., Viljoen, M. T., Grootenboer, J. and Longshaw, T. G. (1975) ERTS-1 imagery: an appraisal of applications in geology and mineral exploration, *Minerals Science and Engineering* **7**: 132–68.

Vincent, R. K. (1975) The potential role of thermal infrared multispectral scanners in geological remote sensing, *Proceeding of the IEEE* **63**: 137–47.

Vincent, R. K. (1980) The use of radar and Landsat data for mineral and petroleum exploration in the Los Andes Region, Venezuela, in *Radar Geology: An Assessment*. Jet Propulsion Laboratory: Pasadena, California, pp. 367–84.

Vincent, R. K. and Thomson, F. J. (1971) Discrimination of basic silicate rocks by recognition maps processed from aerial infrared data: *In Proceedings of the Seventh International Symposium on Remote Sensing of Environment*, Vol. 1. University of Michigan: Ann Arbor, Michigan, pp. 247–52.

Vincent, R. K., Thomson, F. and Watson, K. (1972) Recognition of exposed quartz sand and sandstone by two-channel infrared imagery, *Journal of Geophysical Research* **77**: 2473–77.

Von Bandt, H. F. (1962) *Aerogeology*. Gulf Publishing Company: Houston, Texas.

Waite, W. P. and MacDonald, H. C. (1970) Snowfield mapping with K-band radar, *Remote Sensing of Environment* **1**: 143–50.

Watson, K. (1975) Geologic applications of thermal infrared images, *Proceedings of the IEEE*, **63**: 128–37.

Welsted, J. (1979) Air-photo interpretation in coastal studies – examples from the Bay of Fundy, Canada, *Photogrammetria* **35**: 1–27.

Williams, Jr., R. S. *et al.* (1973) Iceland: preliminary results of geologic, hydrologic, oceanographic and agricultural studies with ERTS-1 imagery. In Anson, A. (ed.) *Symposium Proceedings of Management and Utilization of Remote Sensing Data*, American Society of Photogrammetry: Falls Church, Virginia, pp. 17–35.

Williams, Jr., R. S. (1976) Dynamic environmental phenomena in southwestern Iceland: In Williams, Jr., R. S. and Carter, W. D. (eds) *ERTS-1: A New Window on our Earth*. Geological Survey Professional Paper 929, US Government Printing Office: Washington, DC, pp. 109–12.

Williams, Jr., R. S. *et al.* (1983) Geological Applications: In Colwell, R. N. (ed) *Manual of Remote Sensing* Second Edition, Volume II, American Society of Photogrammetry: Falls Church, Virginia, pp. 1667–1953.

Woldai, T. (1979) Geomorphology and land use of the Jiang Plain and surroundings, Hubei Province, China, from Landsat imagery, *ITC Journal* **No. 4**, 519–33.

Wolfe, E. W. (1971) Thermal IR for geology, *Photogrammetric Engineering* **37**: 43–52.

Wood, C. R. (1972) Ground-water flow, *Photogrammetric Engineering* **38**: 347–52.

Zernitz, E. R. (1932) Drainage patterns and their significance, *Journal of Geology* **40**: 498–521.

Chapter 5   The biosphere: vegetation, crops and soils

The biologically inhabited part of the lithosphere, atmosphere and hydrosphere is generally known as the *biosphere*. Different forms of plant and animal life can be found living in this relatively shallow but densely populated zone of the interface between the atmosphere and lithosphere or hydrosphere. We have already discussed in Ch. 2, remote sensing applications to study the most complex life form on earth – human population. In this chapter, the focus is on plant life and its related phenomena which have a direct impact on the well-being of the human population. It is worthy to note that plants comprise the greatest bulk of the total world *biomass* (i.e. volume of living material) both above and below the lithosphere and in the hydrosphere.

Information requirements

In the study of vegetation, crops and soils, there is invariably the need to carry out surveys with a view to discover their spatial distribution, structure and type. This information is indispensible for the purpose of management in agriculture and forestry, for informed decision making in planning, for feasibility studies in land development projects and many engineering works. The spatial distribution emphasizes the locational variations in the characteristics of soils and plants. The pattern refers to the layout and arrangement of the soil units and plants in a particular location. The identification of the specific types of soils and plants permits individual homogeneous units to be defined and their extent delimited. All these information requirements can be effectively met with the use of conventional and modern remote sensing techniques coupled with minimal ground survey.

Applications using aerial photography

As in cases of geological, geomorphological and hydrological applications, the role of aerial photography remains dominant in providing information on the distribution, structure and type of vegetation, crops and soils. The importance of the third dimension affordable by the stereoscopic coverage, the distinctiveness of tonal and textural variations of the photographic images and the superior spatial resolution are among the factors which have contributed significantly to the success of aerial photography applied to the study of the biosphere. In vegetation studies, aerial photography is particularly valuable as it permits accurate metric data relating to heights and volumes to be obtained photogrammetrically. The employment of colour infrared film also aids greatly in extracting qualitative data of the vegetation.

Case Study 1:

Forest inventory with large-scale aerial photographs in temperate and tropical environments

In countries with an abundant forest cover, an inventory of the forest to determine the volume of the timber resources is necessary for economic and mangement purposes. The Forest Management Institute of the Canadian Forestry Service has developed techniques to collect tree data with large-scale aerial photographs. These generally involve two stages of work: (1) tree species recognition; (2) volumetric inventory. The method of identification of tree species was developed by Sayn-Wittgenstein (1961, 1978) who advocated the use of morphological characteristics of trees, such as crown shape, branching habit and foliage characteristics for identification clues. This approach is believed to be more objective than the use of indirect indicators such as topography, drainage, aspect and association, which are useful but not always reliable. The success of this approach is largely determined by the scale of photography because morphological characteristics of trees become progressively less distinct as the scale is decreased until they are inseparable from photographic tone, texture and shadow pattern. The most useful morphological characteristics from the vertical aerial photographic point of view are crown shapes, texture and branching habits. The stereomodel of the aerial photographs facilitates recognition of crown shapes. *Crown shapes* can be generally described as oval, dome, cylindrical, conical, rounded, widespreading and flat-topped (Fig. 5.1). One should note that although there is usually one typical crown shape for each species, there are, however, many deviations because crown shape varies greatly according to the influences of the environment and genetics. *Branching characteristics* of trees can be described according to the length and thickness of branches; variation in branch size, the branch direction, whether ascending, horizontal or drooping; the branch form, whether straight, crooked or twisted; the arrangement of branches and twigs; the density and coarseness of twigs and the colour of the bark. The appearance of tree crowns is also influenced by *foliage characteristics*, such as the tone and colour of trees in leaf. Trees with large and glossy leaves tend to appear in lighter tones and sometimes produce highlights as a result of specular reflection. Shadows visible inside the crown and shadows cast on the ground indicate crown characteristics (Fig. 5.2). The shadow density also provides clues to the degree of the crown and foliage: the darker the shadow the more compact.

With the use of very large scale aerial photographs (1:500) most tree species can be recognized by morphological characteristics alone. At scales of 1:2,500 and 1:3,000, small and medium branches are still

Fig. 5.1
Some common crown types
(*Source*: Sayn-Wittgenstein
1978)

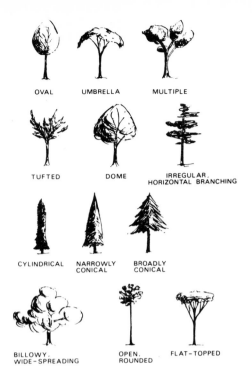

OVAL   UMBRELLA   MULTIPLE

TUFTED   DOME   IRREGULAR, HORIZONTAL BRANCHING

CYLINDRICAL   NARROWLY CONICAL   BROADLY CONICAL

BILLOWY, WIDE-SPREADING   OPEN, ROUNDED   FLAT-TOPPED

Fig. 5.2
Stereogram illustrating the
difference between balsam fir
(1) and white spruce (2). Note
the pointed top and more
symmetrical shape of balsam
and the more open tops and
coarse branching of spruce. (1a)
shows the 'soft' branches of
small, densely-crowned balsam
fir (*Source*: Sayn–Wittgenstein
1978)

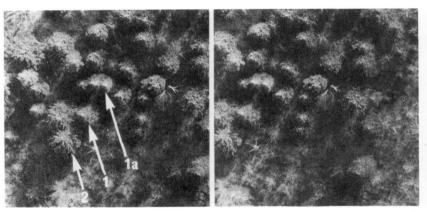

visible and individual crowns can be clearly distinguished. As the
scale decreases to 1:8,000, one cannot describe crown shape although
individual trees can still be separated. In smaller scale photography
such as 1:15,840, indirect evidences such as shadows have to be
employed to determine crown shapes. For scales even smaller than
this, individual trees growing in stands cannot be recognized. Tree
type identification depends on tone and texture but not crown shape.

Finally, attention should be drawn to the seasonal variations which
affect the appearance of trees on aerial photographs. An understanding of the phenological change in a forest is essential in determining the right season of photography for tree species identification.

On the whole, summer is a widely used season for photography in temperate latitudes because of the good weather and the lack of variation in appearance of a single species in a large area, unlike in spring and autumn. However, the colour differences of the leaves in different species as they mature are not as greatly emphasized as in spring photography. Based on the morphological approach, Sayn-Wittgenstein (1978) has been able to develop detailed photo-keys for approximately 40 tree species in the Canadian environment.

The next stage of a forest inventory is to collect tree data. Aldred and Lowe (1978) reported the application of a photographic measurement system to a test area of 3,000 km² just to the northwest of Manning, Alberta. The site is well drained and is covered by a wide variety of boreal forest types such as white spruce, aspen, balsam poplar, lodgepole pine, black spruce, balsam fir and tamarack. The aim of the inventory was to provide estimates of gross total volume, softwood sawlog volume (white spruce and lodgepole pine) and pulpwood volume (all species). Aerial photography was specially flown using reconnaissance cameras of 70 mm format (focal lengths 75 mm and 300 mm) from a height ranging from 200 m–1,000 m above the ground, thus giving nominal photographic scales of 1:4,000 and 1:1,000. A radar altimeter was used onboard the aircraft to measure the flying height. Fixed-area sample plots of 0.03 ha were established on the resultant stereo pairs of photographs (Fig. 5.3). The area under study had already been demarcated into four strata known as L (lumber), R (roundwood), H (high uncommercial) and U (low uncommercial) by the Alberta Forest Service. A stratified random sampling

Fig. 5.3
Large-scale 70 mm stereo-photographs (contact scale 1:840) showing the 0.03 ha photo-plot and the airborne data recorded on the edge of the photographs during the exposures. Note that the flying heights displayed were 257 m (left) and 261 m (right). The series of dots at the lower right corner of the left photograph is a binary code showing the amount and direction of tilt at the instant of exposure. The time of photography is also displayed (e.g. 09 h 39 m 42 s on the right photograph; *Source*: Aldred and Lowe 1978)

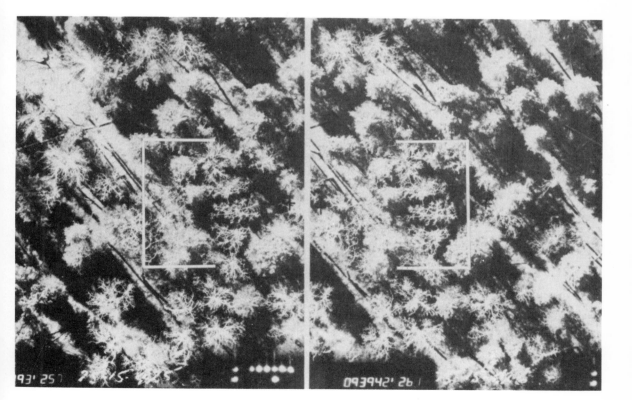

procedure was adopted to select sample plots in each of the four strata for photographic measurement. The study area was first divided into long, parallel strips spaced at 25 m apart (spacing dependent on plot size). A number of strips to be photographed was randomly selected. The length of the strip covering the four strata was divided into 20 m sections (plots). The required number of photographs was then selected at random. The sample size was estimated with the following equation:

$$n = \frac{(t.cv)^2}{e} \qquad [5.1]$$

where n is the number of photographs required to achieve a given accuracy e to a confidence level specified by t for a stratum with an estimated coefficient of variation cv. The optimum shape and size of the plot were found to be square-shaped of 300 m² in area or 17.32 m on the side. Between 25–50 trees per plot seemed best. These square sample plots were located in the centre between the principal points of the stereo pair of aerial photographs (Fig. 5.3).

The measurement system consisted of a mirror stereoscope and a parallax bar directly attached to a programmable desk calculator. The stereo pair of aerial photographs was examined and the photographic coordinates of the principal points, plot corners and parallax were all measured with the parallax bar under the mirror stereoscope. The measured values were digitized and stored in the calculator. In addition, data on focal length, flying height, tilt, tip and print enlargement factor were stored. Corresponding ground control points were also digitized. Computer programs were developed to carry out orientation, to calculate the exact plot area and to fit a plane or curved surface to the points previously digitized at the ground level. The heights of trees over any location could be measured with the parallax bar. The fitted ground surface provided a datum from which the height difference between the top and the base of the tree could be obtained. The perimeter of the tree crown could also be traced and digitized, thus determining the crown area (Fig. 5.4). By applying volume and diameter regression models developed for the tree species in the study area, the tree height and crown area data were input to estimate diameters at breast height (dbh) and volumes of trees. The basic model for tree volume estimation (V) is:

$$V = b_1 (H) + b_2 (H^2) + b_3 (e^{H/100} - 1) \qquad [5.2]$$

where b_1, b_2 and b_3 are regression coefficients empirically determined, H is photo-measured tree height and e is the exponential function. The basic model for estimating *dbh* (D) from photo-measured tree height (H) is:

$$D = b_0 + b_1 (H) + b_3 (SINH(H)) \qquad [5.3]$$

where *SINH* is a hyperbolic sine function and b_0 was set to 0 to force the regression through the origin.

The accuracy of the whole approach of plot volume estimation was determined by random and systematic errors in species identification, exclusion and inclusion of the number of trees within the photo-plot,

Fig. 5.4
A stand map showing the plot boundaries, tree locations and crown perimeters (*Source*: Aldred and Lowe 1978)

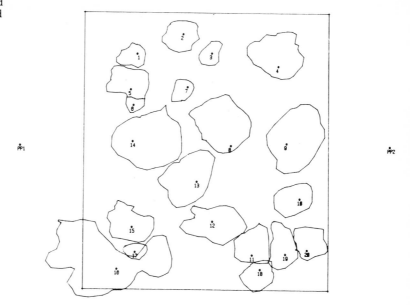

flying height, plot size determination, tree-height measurement and the regression equations employed. The combined effect was a standard error of ±9.5 per cent (Table 5.1). This met the required accuracy performance of the Alberta Forest Service. It is interesting to note in connection with the forestry inventory that individual tree species were correctly identified 95 per cent of the time from large-scale aerial photographs.

Will this approach of forest inventory be applicable to the tropical environment? This approach was tried in Surinam in South America, (Aldred 1976). It was anticipated that this would present difficulty because in the tropical forest several hundred species could occur within a few kilometres and thousands of species could occur in a large region, quite unlike the temperate zones where no more than twenty species had to be dealt with in one area. Other difficulties include high crown density, multiple crown layers, the presence of epiphytes and

Table 5.1
Combined effect of measurement errors in accuracy of plot volume estimates
(*Source*: Aldred and Lowe 1978)

Source	Standard error of plot-volume estimate (per cent)
Species identification	trace
Omissions and commissions	7.5
Radar altimeter	1.0
Plot-size determination	2.1
Tree-height measurement	2.0
Regression equation	5.0
Combined effect	9.5

lianas, and the poorly understood patterns of phenological changes (Fig. 5.5). As a result, photo-interpretation had to be carried out with vague or indefinite clues. In the test area of Surinam, which lies west of the Nassau Mountains and east of the Van Blommesteinmeer, the rolling and occasionally dissected terrain is covered by a tropical rainforest with at least 40 important species. The main tree canopy was about 25–30 m high on average. The aerial photography was acquired with an RMKA 15/23 survey camera and one or two simultaneously operating Vinten 492 reconnaissance cameras (70 mm format). Different film types were tried, which included black-and-white panchromatic, colour negative, colour infrared and colour positive films. A Canadian photo-interpretator who had no tropical experience was asked to identify the tree species of 58 trees. The result of the interpretation was compared with that of another interpreter who was familiar with the Surinam environment. It was found that there was about 75 per cent agreement between the two in their species identification. It was revealed that the description of the appearance of trees in the photo-key should be very clear and well illustrated to avoid misinterpretation. Good colour photographs at scales of 1:4,000 or larger were particularly useful in identifying several important species. Colour infrared film was less useful because of the large number of red and pink hues which could not be related to the natural colours observed on the ground. The dark shadow effect of colour infrared was another disadvantage. Black-and-white panchromatic photography was totally unsatisfactory, unlike in its application to temperate zones. True colour positive transparencies were superior to colour negatives because of their higher spatial resolution which permitted detailed observation of branching habits and foliage. Given good photo-keys and proper training of the photo-interpreter, a satisfactory recognition of tree species in the tropical rainforest should be possible. The usefulness of large-scale true colour aerial photographs for rainforest species identification was echoed by Myers and Benson (1981) in connection with their experiment in north Queensland, Australia. They reported that 50 per cent of the dominant and co-dominant species in the study area were correctly identified in the test stands at least once. Twenty-four species were identified with more than 75 per cent accuracy and 11 of these were correctly identified in every case examined by the five interpreters. They recommended that

Fig. 5.5
Stereotriplet showing the large number of tree species in the tropical environment (Surinam). (1) Djadidja – rounded crown in upper storey, fine branching; (2) Tamaren prokoni – flat, shallow crown, fine branches; (3) Donceder – dominant, rounded, dark crown, coarse texture; (4) Zwarte foengoe – large, spreading crown, coarse texture; (5) Rode foengoe – large, spreading crown, coarse texture; (6) Bolletri – single, solid crown, grainy texture; (7) Hoogland gronfolo – large crown, foliage in clumps, coarse texture with openings in crown (*Source*: Wittgenstein, Milde and Inglis 1978)

colour transparencies at 1:2,000 scale in 70 mm format should best be employed and that photographs should not be exposed at solar altitudes of less than 45°–50° in order to achieve the best results.

As for the applicability of the photographic measurement system to the tropical environment, Aldred (1976) evaluated the accuracy of tree-height measurement in the tropical high rain forest in Surinam, using large-scale aerial photographs (1:500). It was found that the accuracy of parallax bar readings near the tree base was largely determined by the combined effect of forest cover and topography. Where the terrain was flat and the visibility of the tree base was good, tree-heights could be measured with a precision of less than ±2 m at 95 per cent level of confidence. This kind of accuracy is comparable to that obtained under temperate forest conditions. Where openings in the forest occurred 20 m or so from the trees in closed forest canopies, the heights could be measured with a precision of about ±6 m. Under close forest canopy conditions with no reference to artificial openings or visible ground, the precision of height measurement fell to ±11 m. Crown area could be measured fairly consistently on the large-scale aerial photographs. Its measurement was affected by the tendency of the branches of trees in the tropical rain forest to intermingle or overlap and of some trees to divide into separate sections appearing as separate crowns in the general canopy. This tendency also adversely affected the tree count. Several film types were also tried and again true colour transparency 70 mm film was found to be superior. A flying height of 300 m–600 m above the ground (nominal scale 1:2,000 to 1:4,000) and photography under light-overcast conditions were recommended as ideal. The low-altitude photography increases the chance of obtaining coverage of tropical areas which are perennially cloud-covered.

One major limitation of the application of aerial photography to forest inventory or woodlands survey is that it cannot adequately take into account the well developed structure in terms of layering, i.e. the layers of canopy, understorey, shrub, herb and ground (Fig. 5.6). The aerial photographs focus attention to the uppermost layer – the tree canopy. As a result, the subordinate layers tend to be ignored (Shaw 1971). Fortunately, correlations exist between the canopy and the

Fig. 5.6
Diagrammatic representation of vegetation stratification (*Source*: Tivy 1982)

other layers of the forest. Although the degree of correlation is not as strong as one would expect, one can still make a limited degree of inference of the layers beneath.

Case Study 2:

Wetland vegetation mapping with high-altitude colour infrared

The use of aerial photography to map wetland vegetation is more or less a standard practice because of the difficulty of ground access to the wetland areas (Carter 1982). According to Anderson's land use and land cover classification scheme (1976), wetlands are defined as 'those areas where the water table is at, near, or above the land surface for a significant part of most years.' Any seasonal fluctuations in water level therefore will affect the extent of the wetland. In mapping the vegetation in the wetland, it is essential to establish first a wetland classification scheme. A typical example is the project undertaken by Carter, Malone and Burbank (1979) to map the wetland vegetation in western Tennessee, USA using high-altitude colour infrared photography. A new wetland classification was developed based primarily on vegetation and on frequency and duration of inundation. Special attention was given to the mapping ability and interpretability of the classes from the high-altitude colour infrared photography while at the same time the scheme should be useful for ground inventory or habitat evaluation. The resultant scheme (Table 5.2) showed 6 wetland classes and 12 subclasses which fitted well into the land use and land cover classification system of the United States Geological Survey (USGS). The colour infrared photography was taken with a Wild RC-10 150 mm focal length camera (format size 23 × 23 cm) at a nominal scale of 1:130,000 in February (high water, no leaves), May (early growing season) and November (low water, no leaves) of 1975 (Pl. 7, page 212).

The mapping scale, however, was 1:24,000 which is the scale of the USGS topographic map series. These geometrically accurate maps contain detailed information on topography, cultural features and drainage patterns which are needed to help mapping the wetland from the aerial photographs. In addition, the high altitude photography gave a mappable areal unit of 0.5 ha which was too small for some wetland classes and a larger mapping scale was desirable. The delineation of the wetland classes and subclasses was carried out manually by an experienced photo-interpreter who had participated in field checking. A variety of photographic mapping criteria was established for distinguishing among these wetland classes and subclasses, from which a photo-interpretation key based on the image characteristics of tone, texture and shape was developed (Table 5.3). With the aid of the Bausch and Lomb Zoom Transferscope, the interpreted data were transferred to stable base drafting film keyed to the appropriate 1:24,000 map. The final map product (Pl. 7) was extensively checked in the field in order to resolve ambiguities. It was found that the most variable boundaries in terms of yearly or seasonal fluctuation were those between upper and lower bottomland hardwood; between

Table 5.2
Comparison of Wetland Classes and Subclasses for the Tennessee Valley Region with USGS Level II categories (Anderson *et al.* 1976)
(*Source*: Carter, Malone and Burbank 1979)

| Tennessee Valley Region | | |
Wetland Classes	Wetland Subclasses	Level II Class
FW-1 Bottomland Hardwood	(FW-1a) Upper Bottomland Hardwood	Forested Wetland
	(FW-1b) Lower Bottomland Hardwood	
FW-2 Swamp	(FW-2a) Forested Swamp	
	(FW-2b) Shrub Swamp	
	(FW-2c) Dead, Woody Swamp	
M-1 Marsh	(M-1a) Wet Meadow	Non-forested Wetland
	(M-1b) Emergent Marsh	
	(M-1c) Seasonally Emergent Marsh	
M-2 Seasonally Dewatered Flats	(M-2a) Vegetated	
	(M-2b) Non-vegetated	
M-3 Agriculture Subject to Flooding		Agriculture
OW-1 Open Water	(OW-1a) Vegetated	Open Water
	(OW-1b) Non-vegetated	

bottomland hardwood and forested swamp; between seasonally emergent marsh and open water; and around agriculture subject to flooding. However, the technique developed was found useful to map wetland vegetation with an emphasis on vegetation physiognomy, physical site characteristics and hydrologic regime.

Case Study 3:

Vegetation damage assessment with normal colour and colour infrared aerial photography

Another approach in the use of aerial photography in vegetation study is to detect damage caused by diseases or pests to the vegetation. According to Murtha (1978), vegetation damage is 'any type and intensity of an effect, on one or more types of vegetation, produced by an external agent, that temporarily or permanently reduces the financial value, or impairs or removes the biological ability of growth and reproduction, or both.' Aerial photographs are used to detect vegetation damage, if any, and to find out the possible damage agents. This is usually done with reference to a host vegetation which is healthy. There are two major manifestations of damage: either the

Type	Tone	Texture	Shape
Upper Bottomland Hardwood FW-1a	Oct – bright Feb – lt. brown Nov.– dk. blue–lt. brown	Very rough	Variable – boundary indistinct between FW-1a and FW-1b
Lower Bottomland Hardwood FW-1b	Oct – bright red Feb – dk. brown–lt. grey Nov – lt. brown–lt. blue	Very rough	Variable – boundary indistinct between FW-1b and FW-2a
Forested Swamp FW-2a	Oct – blue-green with some red and yellow Feb – dk. brown Nov – dk. blue	Very rough	Elongated-variable from large to extra small
Wet Meadow M-1a	Oct – pink Feb – not visible Nov – lt. pink–lt. blue	Medium to fine textured	Small variable shape
Emergent Marsh M-1b	Oct – red, blue, green Feb – lt. pink-grey Nov – lt. pink-grey	Fine to medium textured	Variable – square to elongated.
Seasonally Emergent Marsh M-1c	Oct – pink Feb – not visible Nov – not visible	Fine to smooth texture	Variable – small to elongated
Seasonally Dewatered Flats Vegetated M-2a	Oct – reddish blue–blue Feb – not visible Nov – not visible	Fine to smooth texture	Variable – small to medium
Seasonally Dewatered Flats Non-vegetated M-2b	Oct – white–lt. blue Feb – not visible Nov – not visible	Smooth	Large elongated
Agriculture Subject to Flooding M-3	Oct – pink–red Feb – not visible Nov – pink–bright blue	Coarse to smooth	Variable
Open Water Vegetated OW-1a	Oct – pink–red Feb – not visible Nov – not visible	Very fine to smooth	Variable – small
Open Water Non-vegetated OW-1b	Oct – dark blue Feb – light blue Nov – light blue	Uniform smooth	Variable – elongated

Table 5.3
Reelfoot Lake Photo-interpretation Key
(*Source*: Carter, Malone and Burbank 1979)

tree has suffered from a change in morphology, or a change in physiology, or sometimes both. Morphological damage involves a change in the shape or outline of the vegetation. In the case of a tree, this means defoliation. Physiological damage involves a change in function as expressed by a deviation from a normal pattern, such as a decrease in photosynthates, deterioration of chloroplasts, interruption of translocates, including water, etc. The damages on the vegetation have an effect on spectral reflectance as the vegetation is recorded by aerial photography. A first visual symptom of physical damage is yellowing of the foliage. This means a shift in the peak of reflectance of a normal green leaf in the green region (0.5–0.6 μm) towards the red region (0.6–0.7 μm) (Fig. 5.7). On the other hand, morphological damage affects spectral reflectance only when new surfaces are exposed. Any morphological changes are best described on the basis of form, texture and boundary patterns. The changes in spectral reflectance caused by physiological damage, therefore, will affect normal colour or colour infrared film. On a normal colour film, the colour of the leaves of the damaged vegetation will change from green to yellow and finally to

Fig. 5.7
Generalized spectral reflectance patterns for (a) a normal green leaf; (b) leaf with incipient damage indicated extra-visually by a change in the level of near-infrared reflectance; (c) a yellowed leaf after a period of chronic damage and (d) a dead red-brown leaf. Arrows indicate the deviation from the normal reflectance pattern (*Source*: Murtha 1978)

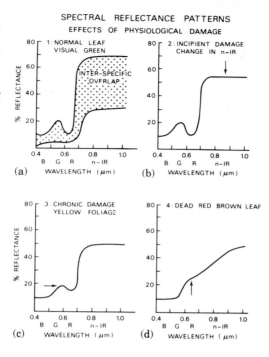

red. On a colour infrared film, the colour will change from magenta to darker magenta, to mauve and finally to yellow. Although one may detect the damaged vegetation it is extremely difficult to find out the causes of the damage from aerial photographs alone. Murtha (1978, 1982) suggested three approaches to the problem: the use of *a priori* knowledge, photo-keys, and enhancement techniques which should be combined together for the best results to identify the *damage syndrome*. The damage may be the result of an environmental stress, such as air pollution (sulphur dioxide) or a biotic stress, such as insects. They do exhibit differences in damage characteristics as revealed by normal colour and colour infrared photography at scales between 1:1,000 and 1:4,000. Table 5.4 summarizes these strain symptoms of trees caused by environmental and biotic stress agents. Finally, in the assessment of the impact of the damage, five approaches are generally used: (1) counting the individuals affected; (2) delineating the areal extent of the damage; (3) multiplying the affected area by ground surveys of crop production estimates to obtain a damage volume estimate; (4) stratifying the area into damage intensity levels; or (5) multiplying the area of damage intensity levels by predetermined volumes to get stratified loss volume.

Case Study 4:

Crop identification and discrimination

Aerial photography of varying scales in black-and-white panchromatic, colour and colour infrared films has been extensively employed

Table 5.4
Selected stress agents and key
for noticeable strain symptoms
(*Source*: Murtha 1982)

Stress	Key Strain Symptoms
Abiotic – Environmental	
Water deficit (drought)	Small foliage, wilting
Water excess (flood)	Topography-related, discoloured to dead foliage
Air pollution (e.g. SO_2)	Affects many species, interveinal necrosis and chlorosis, decrease from source
Wind (storm)	Broken stems
Fire (ground)	Burn scar, blackened ground
Acid rain	Top dieback, lower foliage appears healthy
Biotic – Insects	
Bark beetles (Douglas fir)	Red-brown and faded foliage
Defoliators (spruce budworm)	Bare branches, seasonal red-brown foliage
Terminal feeders (white pine weevil)	Dead leader
Sucking insects (woolly aphid)	Yellowed foliage, thin crown
Leaf miners (birch leaf miner)	Discoloured foliage
Biotic – Diseases	
Stem rusts (blister rust)	Dead top or leader
Root rots (Fomes, Poria)	Thin foliage, dead trees in pockets
Leaf wilts (Dutch elm)	Entire or part of crown with no dead or discoloured foliage

to identify farm crops in different climatic environments. As early as 1959, Goodman made use of sequential black-and-white panchromatic aerial photographs to study farm crops in northern Illinois, USA. She found that tone, texture and association were three major clues which helped in the identification of the farm crops. Fields of dense stands of alfalfa were registered black whereas fields of ripe wheat appeared nearly white. In order to standardize images, a densitometer was used to measure quantitatively the tone values of all objects, which were then adjusted relative to the tone of a reference object. On the other hand, the texture of the photographic image of individual crops varied with the external characteristics of the crops and the way in which they were planted, cultivated and harvested. Textural characteristics were also, to a certain extent, associated with variations of soil moisture. One could therefore describe textures of fields as 'lined', 'plaid-like', 'corduroy', 'striped', 'swath' or 'mottled'. The use of association of objects such as straw stack, hay stack, herds of cows, etc. has helped photo-interpreters to identify a field of oats, a field of hay and a rotation pasture respectively. The time of photography affects the accuracy of the identification. In the case of northern Illinois, photography taken during the second half of July was found to be the optimum for the purpose. More recently, Philipson and Liang (1975; 1982) developed an aerial photo-key for crop identification in the tropics in medium-scale photography (1:10,000–1:30,000). They specifically indicated that in view of the tropical environment only black-and-white panchromatic single-date photography was to be used to develop the key. The key placed emphasis on: (1) field characteristics, such as density, size, shape, assemblage and appearance; (2) management characteristics, such as relief, presence of sub-units, irrigation or drainage; (3) crop characteristics, such as intercropping, cropping pattern or density, tone, form and texture. These were auxiliary features for direct and indirect recognition. They were related to the actual planting and harvesting procedures. In this way, some major tropical crops, namely, sugar cane, lowland rice, maize, tobacco, pineapple, banana, rubber, coconut, coffee and cacao could be identified with the aid of the photo-key.

The identificiation of farm crops was also attempted using colour and colour infrared photography which was normally taken at a much higher altitude than the black-and-white panchromatic photography. The accuracy of the resultant crop identification could be improved with the use of multi-season photography. In an application to Switzerland, Steiner and Maurer (1969) found that in combining June and July true colour photographs (1:8,000 scale) the crop identification accuracy was 75 per cent as compared with 40 per cent (June) and 29 per cent (July) accuracy on single photography sets. With the addition of crop height as a discriminating variable, the crop identification accuracy was further raised to 88 per cent. Steiner (1970) also observed that by selecting the best three dates of photography, an overall identification accuracy of 90 per cent could be achieved. In this context, one should note that an increasing number of photographic variables is being employed for the crop identification purpose. A computer-assisted approach is more appropriate and will be discussed under the section dealing with applications using satellite data.

Case Study 5:

Detection of crop conditions using colour infrared photography

The photographic infrared can record crop conditions that are not visible to the human eyes. These conditions refer to the vigour of the crops. Variations in the infrared reflection from the crops are the result of chlorophyll breakdown in the mesophyll tissue before visible symptoms of stress or infection appear. The behaviour of infrared reflectance or absorptance is largely dependent on the optical characteristics of the air-water interface within the cells. It has been demonstrated that the density levels of colour infrared photographs can be related to the infection level of diseases of crops in fields (Jackson and Wallen 1975). On the whole, the higher the level of infection, the lower will be the infrared reflectance from the crop (De Carolis and Amodeo 1980). This line of application is best demonstrated by Henneberry *et al.* (1979) who attempted to determine the effectiveness of plant growth regulators on cotton cultivation in Arizona, USA using colour infrared photography. The photography was obtained by a KA-2 aerial camera (f = 305.5 mm and 23 × 23 cm format) with Kodak Aerochrome infrared film type 2443, through

Wratten 15 and CC40B filters. The photography was flown between 10 a.m. and 2 p.m. to obtain maximum sun-angle and light intensity at altitudes between 609.6 m and 3,048 m, giving photographic scales of 1:2,000 and 1:10,000 respectively. The growth regulators were chemicals applied to remove late-season fruiting forms of cotton and the associated larvae of the bollworm. The colour infrared photography was found to be most effective in detecting treated and untreated cotton fields (Pl. 8, cf (A), (B) and (C), page 212). The treated fields appeared much darker red than the untreated fields and became brownish-red as the treatment time got longer, thus indicating the loss in vigour of the crop. This application further indicated that the difference in the amount of nitrogen fertilizer each plot received could be detected. A plot receiving no nitrogen fertilizer appeared brown. A plot of cotton infested with root rot was also clearly distinguished (Pl. 8, Plot (D)). This permitted the extent of damage and the magnitude of the problem to be determined.

More down-to-earth applications of colour infrared photography for crop management were reported by Lamb (1982) and Baber, Jr. (1982). Both of them have made use of low-altitude colour infrared photography to monitor crop growth in the arid region of the USA using electrically driven, centre-pivoted irrigation systems and the application of fertilizers. The colour infrared photography proved to be an effective tool to detect pre-visual crop stress caused by the defects of the irrigation system. The nature and pattern of the colour variations of the crop (magenta colour) indicate the probable source of the stress.

Case Study 6:

Soil mapping

A soil consists of unconsolidated materials formed as a result of the decomposition, alteration and organization of the upper layers of the earth's crust under the influence of water, climate, plant and animal activities. Generally, one can distinguish three layers of soil: the topsoil (A Horizon), the subsoil (B Horizon) and the parent material (C Horizon). The topsoil is the most extensively weathered horizon and contains the most organic matter. In applying aerial photography to soil mapping, one may or may not be able to see the surface of the soil. It depends on whether there is any vegetation cover on the soil. Even if it is visible, the aerial photographs are not actually used to identify soil types but to locate changes in the land surface patterns that may relate to different soil properties (White 1977). In other words, much of the photographic interpretation is to make indirect inference of the soil properties. The photographic tone has been an important clue, but the geomorphological features as revealed by stereoscopic viewing are particularly useful in helping to delineate soil boundaries. There is a close relationship between the kinds of soil and the nature of the parent material, relief, climate, vegetation and age of the landform (Soil Conservation Service 1966).

On the whole, the photographic tone of a soil is largely determined by its water content and organic matter content. A coarse–textured soil, such as a sandy soil, is not able to retain as much water as a fine-textured soil, such as a clay soil. Hence, a sandy soil reflects more radiant energy than a clay soil. Similarly, soils with high organic matter give low reflectance (Girard 1980). However, surface roughness can also play a part in affecting the reflectance. In general, the reflectance is lower for a soil with a rough surface (i.e. more diffused reflection). Evans, Head and Dirkzwager (1976) cautioned the use of photographic tone to identify soil properties. In their study of the Fens of East Anglia, UK, they found that tonal patterns were good indicators of different soils, but the correlations of photo-tone with organic matter content, moisture content, calcium carbonate content and siltiness of the soils were poor especially over large areas. At two other sites of study in Bedfordshire, UK, they also found that the photographic tones were related to the lithological pattern and soil surface characteristics. Their study suggested that in some areas such as the Fens where soils merged because of their depositional history, the photographic tones reflected differences in the soil profile at depth whereas in other areas the photographic tones reflected only the properties of the plough horizon or at the surface. In the former case, soils could be mapped as soil series, associations or complexes. In the latter case, only the phase differences of soils could be mapped. In many cases, photographic tones were found to have little meaning and could be resulted from such transient features as surface waterlogging, stubble remaining on the surface after ploughing or different dates of cultivation. Although the predictability of soil types based on photographic tones is poor, field experience and landform analysis can adequately supplement the use of aerial photographs in soil mapping. Apart from the use of black-and-white panchromatic film which is standard for soil mapping, one can make use of normal colour and colour infrared films. More soil differences can be revealed by subtle changes in hue of the normal colour photography (Anson 1970; Kuhl 1970). The colour infrared displays a greater contrast between bare soil (blue) and vegetation (red). A higher degree of wetness is reflected by a darker blue colour.

An example of soil mapping from aerial photographic interpretation based on landform and tone is given below. The area is in the Atlantic Coastal Plain, Charleston County, South Carolina, USA (Fig. 5.8a). The interpretation was carried out by an experienced soil scientist. He identified seven types of soils from the photograph: (1) Rutlege loamy fine sand; (2) Seabrook loamy fine sand; (3) Kiawah loamy fine sand; (4) Stono fine sandy loam; (5) Tidal marsh, high; (6) Tidal marsh; (7) Eustis fine sand (Fig. 5.8b). Field checking of the map revealed an accuracy of about 95 per cent. The study area revealed a ridge-slough landform characterized by parallel lines of beach dune sands modified by wind and water with time and shoreline recession. These ridge areas which supported only sparse vegetation and had less organic matter in the topsoil exhibited white or chalky tones. Low areas having the most organic matter showed up dark, whereas areas intermediate in elevation had an intermediate amount of organic matter and showed up as greyish-white. In this way, the Eustis soil series (chalky-white toned ridges), the Seabrook and Kiawah series (intermediate areas) and the Rutlege series (low areas) were delineated. The Seabrook and Kiawah series were separated by the clue that the former had more white than grey-white tone and the latter had more grey-white tone than white (Soil Conservation Service 1966).

(a)

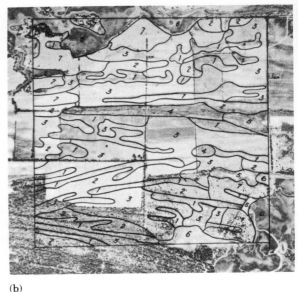

(b)

Applications using thermal infrared imagery

Thermal infrared sensors in the 'windows' at 3–5 μm or 8–14 μm at aircraft altitude are specifically applied to measure the surface temperature of the vegetation, crops and soils either in the form of high-resolution imagery or in digital format, amenable to computer processing. It has already been demonstrated in Ch. 1 that the magnitude of the signal recorded by the sensor is a function of the emitted thermal radiation from a surface, which is proportional to $\varepsilon\sigma T^4$ where ε is the emissivity, σ the Stefan-Boltzmann constant, and T the absolute temperature. It may be assumed that the emissivity of all natural surfaces is very close to one (actually 0.9–0.95). Thus, the signal is solely a function of surface temperature. From a knowledge of surface temperature, inference can be made of the soil moisture. A knowledge of these two environmental factors of temperature and moisture will permit vegetation or crop vigour to be determined. This explains why thermal infrared sensors are particularly useful in crop yield prediction, as demonstrated by the following case studies.

Case Study 1:

Soil moisture determination

Soil moisture information is important to the agriculturalist. A deficit in moisture may lead to the wilting of plants, and timely remedial actions through irrigation can save the crops. However, remote sensing methods of monitoring soil moisture are still in the experimental stage (Hardy 1980). The thermal infrared linescan sensor

(IRLS) provides the greatest potential for promising development in this direction. The sensor (preferably recording signals in the 8–14 μm waveband from the aircraft) directly measures the surface temperature which is a function of the energy balance between evaporation and transpiration on the one hand, and the emitted radiation on the other. If most of the incoming energy is used to supply latent heat to support the evaporative processes, both surface temperature and emissivity will be low. Thus, a distinction can be made between evaporating and non-evaporating surfaces. Examples of evaporating surfaces are water, plant cover or moist soil, while non-evaporating surfaces are dry soil or dry impervious surfaces. The moisture content detected in this way refers only to a very thin layer which has little or no relation to that of the deeper zone of the soil.

For the estimation of subsurface soil moisture, one has to employ the technique of thermal inertia mapping already discussed in Ch. 4. A soil consists of varying portions of water and air in the voids according to the moisture content. Air which has a very low thermal conductivity and specific heat is an effective thermal insulator. Therefore, a soil with a high proportion of air will exhibit large day and night contrasts of surface temperature, while a soil containing a low proportion of air will show relatively small temperature variations. This thermal property is expressed by the thermal inertia, P, which is defined as:

$$P = \sqrt{k\rho c} \qquad [5.4]$$

where k is the thermal conductivity, c is the specific heat, and ρ is density (see Ch. 4). By acquiring thermal infrared data of an area in the same 24 hours at approximately the times of minimum and maximum temperatures, i.e. pre-dawn and early afternoon, the thermal inertia of the area can be found and employed as an indicator of subsurface moisture content (Fig. 4.18). It should be noted, however, that thermal inertia is not a unique indicator because different combinations of k, c and ρ can produce identical thermal inertia values.

Case Study 2:

Crop temperature and crop yield prediction

The thermal infrared sensor can be used to measure the temperature of a crop canopy from which the water tension within the plant itself can be inferred. Two approaches are possible. The first measures the difference between the crop maximum temperature (1–1.5 hours after local solar noon) and minimum temperature (pre-dawn), which is then normalized by ambient air temperatures acquired regionally within a few km of the field site. These values are added together to produce a value known as the 'Stress Degree Day' (SDD) (Idso *et al.* 1977a). The second approach is to measure the difference between crop canopy temperatures and ambient air temperatures at the time of maximum solar heating. The daily sum of this temperature difference is then a value for the 'Stress Degree Day' (SDD). Millard *et al.* (1978) experimented with these two methods in the study of a wheat canopy in Arizona, USA. Both aircraft and ground measurements of crop canopy temperatures were made. Plant water tensions were also measured. For future comparison between the two sets of temperature data, a plot of field was selected for use to adjust the aircraft-measured temperatures with the ground-measured temperatures. The temperature of a tank of water (1.3 m diameter) was also used as a standard to calibrate the thermal infrared scanner. The pre-dawn and afternoon scanner data were converted into digital form and were then registered with reference to control points (low-emitting aluminum spray paint) randomly placed in the area before the thermal infrared imagery was flown. With the use of a computer and a proper software package, the difference in pre-dawn and afternoon temperatures could be displayed as an image where subtle variations in the diurnal range of temperatures were easily detectable (Pl. 9, page 212). It was clearly seen that well-irrigated fields of wheat and alfalfa showed very small variations of diurnal temperatures; the bare soil showed a great variation; and the severely water-stressed wheat showed diurnal temperature variations of about 10°C greater than the well-irrigated wheat. The SDD was computed by the following equation:

$$SDD = \left[\frac{(\text{p.m.}-\text{a.m.}) \text{ crop temperature}}{(\text{p.m.}-\text{a.m.}) \text{ air temperature}} \times 18 \right] - 18 \qquad [5.5]$$

The correlation between the aircraft-acquired and ground-acquired canopy temperatures was very high (0.99). The second approach which involved measuring the difference between canopy and air temperatures at the time of maximum solar heating only was found to be a simpler but an equally effective measure of SDD. The results of the research confirmed that canopy temperatures acquired about an hour and a half past solar noon were well correlated with pre-dawn plant water tension, a parameter directly related to plant growth and development. The standard deviation between airborne and ground-acquired canopy temperatures was 2 °C or less.

The possibility to accurately measure crop canopy temperatures opens up ways to predict crop yields. Idso *et al.* (1977a) developed a formula which linked the difference between leaf temperature and air temperature $(T_L - T_A)$ and crop yield together. This temperature difference is the SDD. Crop yield (Y) was hypothesized to be linearly related to the total SDDs accumulated over some critical period:

$$Y = \alpha - \beta \left(\sum_{i=h}^{e} SDD_i \right) \qquad [5.6]$$

where SDD_i is the mid-afternoon (about 2 p.m.) value of $(T_L - T_A)$ on day i, h is the time of appearance of heads and awns (i.e. the beginning of above-ground growth), e is the day of head growth (i.e. the end of growth), α is the intercept and β is the gradient of the slope of the line. To determine the period $h \rightarrow e$, Idso *et al.* (1977b) also advocated the use of albedo measurement of the reflected solar radiation from the wheat canopy. They discovered that the albedo of plots declined as the season progressed. As the crop grows, leaves proliferate and progressively shut out more and more of the high-albedo soil from view. As the crop matures, the plant heads grow and proliferate, thus leading to an increase in albedo again. This heading time for wheat was determined to

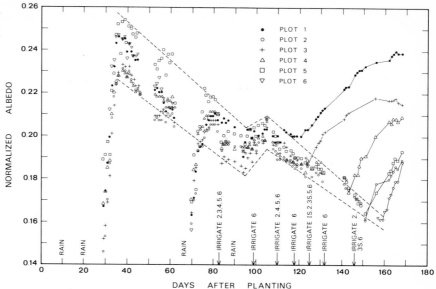

Fig. 5.9
Seasonal progressions of normalized clear-day albedo values for the north halves of the six plots of wheat, with notations on times of rainfall and irrigation (*Source*: Idso *et al.* 1977)

be at about day 100 (Fig. 5.9). However, this period of beginning and end of growth is subject to climatic variations as well. In a recent study, Idso *et al.* (1979) refined the model by including 'Growing Degree Day' (GDD) as an extra parameter, which is defined by the equation:

$$GDD = \frac{T_{max} + T_{min}}{2} - T_b \qquad [5.7]$$

where T_{max} and T_{min} are the daily maximum and minimum air temperatures respectively, and T_b is a base temperature below which physiological activity is assumed to be inhibited. However, the time prior to the appearance of heads and awns, i.e. the pre-heading period, determined the *potential* grain yield according to the climatic conditions, and similarly the period $h \rightarrow e$ was also affected by the climatic conditions prevailing, which determined the *actual* yield. They found that the only other climatic parameter of importance was light. It was proposed that the potential yield of a grain crop was proportional to the summation of daylight minutes over the beginning and end of the heading period during which daily average air temperature was greater than the base temperature, T_b. By incorporating this assumption, the crop yield prediction model (Fig. 5.9) could accommodate more varied environmental conditions and give more realistic estimates.

Applications using radar imagery

The use of imaging radar, an active microwave remote sensing system, to study vegetation, crops and soils is somewhat limited because much of these can be done better with the conventional black-and-white aerial photography. However, in tropical and equatorial regions which tend to be cloud-covered most of the time, imaging radar is the only sensor which has the capability of penetrating thick layers of clouds to obtain timely land cover information. The following case studies will typify the approaches normally taken in this field of application.

Case Study 1:

Vegetation mapping in Nigeria

According to Morain (1976), radar images can reveal three main kinds of vegetational information: (1) geographic pattern; (2) gross structure and physiognomy; (3) type identification. A good knowledge of geography, biology and ecology is essential to improve the level of reference of the interpreters. The image tone and texture are two major parameters employed to identify and delineate vegetation types. They are affected by surface roughness and dielectric properties of the terrain features as well as the incidence angle, polarization and frequency of the radar system. Morain (1976) has presented a matrix key which incorporates texture, tone and polarization very well for use in identifying vegetation types at Horsefly Mountain, Oregon, USA using K-band imagery (Fig. 5.10). It is interesting to note that tone is a function of the wavelength or frequency of the system being used and the viewing angle (Fig. 5.11). Leaf size, leaf area, leaf/twig ratio,

Fig. 5.10
Matrix key for identifying natural vegetation using two-polarization K-band imagery. HH tone and texture form the 'X-axis'; HV tone and texture form the 'Y-axis'. The intersections of these attributes define the location of vegetation types within the matrix (*Source:* Morain 1976)

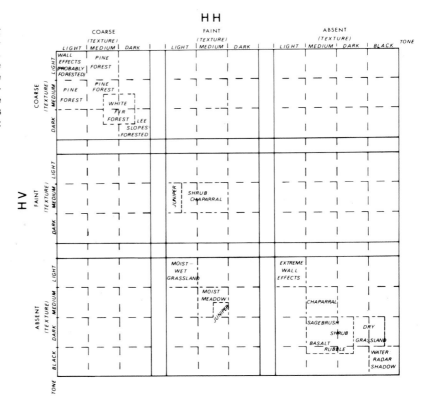

percentage of ground cover and moisture condition can affect the radar image tones. Texture is affected by height variability of vegetation. Therefore, the forest exhibits more spatial variation than the grass or shrub types of vegetation. Hence, forests display coarser textures than grass or shrubs. On the other hand, system resolution as determined by the system frequency will also affect texture: the finer the resolution, the more prominent the texture. Polarization affects tone and texture of vegetation. The horizontally transmitted and horizontally received (HH) polarization is dominated by reflectances arising from *surface* scattering whereas the horizontally transmitted and vertically received (HV) polarization contains most of the information on *volume* scattering. In other words, signals have penetrated to some depth within the image scene. A bright return on HV imagery indicates the heterogeneous nature of the imaged material. For homogeneous materials there should be little difference between HH and HV reflectances (Fig. 5.12).

A large-scale project of radar mapping of vegetation in Nigeria was successfully undertaken by Hunting Technical Services Ltd. in Britain (Parry and Trevett 1979). The choice of radar was necessitated by the perennial cloud-cover condition particularly over the wet tropical forest region of South Nigeria. The Nigerian government attempted to obtain aerial photographic coverage of the country, which was not completed even after three years of operation because of the cloud cover. The project aimed at producing vegetation maps at 1:250,000 scale for the whole country, which would form the basis for a reliable assessment of forest cover and timber resources of Nigeria. The Motorola real-aperture radar system with a horizontal plane polarization was employed in the SLAR acquisition of data. Rectified SLAR images at 1:250,000 scale were used for interpretation. By consulting all the available ancillary information, a team of experienced interpreters tried to relate the various radar signatures to different vegetation types. Tone and texture were the basic criteria employed in the interpretation. It was found that the arid, Sahel zone of North Nigeria proved to be more difficult to interpret because the sparse vegetation permitted most of the radar return to come from the soil surface. With large areas of mixed farmland, several dissimilar areas gave the same signal response. However, as soon as the first rains came, the changed soil moisture content had altered the signal response completely.

Fig. 5.11
SLAR signature differences in a uniform vegetation unit on level terrain (*Source*: Parry and Trevett 1979)

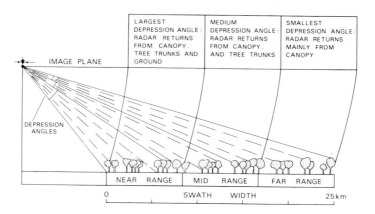

(a)

(b)

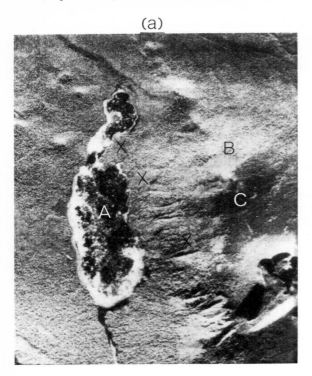

Fig. 5.12
K-band HH polarization (a) and HV polarization (b) images of Horsefly Mountain, Oregon, USA. On the HH image, area A is a swampy meadow in which the lighter tones are wet spots and the darker tones are dry: area B is a pine forest dominated by *Pinus ponderosa*, exhibiting a medium grey tone but relatively coarse texture; and area C is a fir forest dominated by *Pseudotsuga menziesii* distinguished by both its dark tone and more or less prominent texture. There are chaparral communities (marked by X) detectable by their finer texture in comparison with the surrounding pine forest. On the HV image, the swampy meadows all appear dark like a lake and the forest shows brighter tone, indicating a significant contribution of volume scattering from the forest (*Courtesy*: S. A. Morain; *Source*: Morain 1976)

Another difficulty was the development of a suitable vegetation classification system. Finally, a classification of vegetation units based on physiognomy was adopted. This classified vegetation in three tiers: (1) main formations such as grassland, woodland, forest and farmland, (2) subformations such as mature forest, immature forest, riparian forest or swamp forest; (3) species for the subformations. Field checks formed an essential part of this project which took up 35 man months of work spread over 18 months. The initial interpretation classification was then replaced by that determined in the field. In all, 69 map sheets to cover the whole of Nigeria showing land use and vegetation in colour were produced in 18 months. The cost was estimated to be about US $3.00 per km². Some preliminary experiments in digital analysis of SLAR data were also attempted. It was found that tone in SLAR imagery was an unreliable feature to use alone for classification of land use whereas texture was proved to be more useful. However, the influence of relief could have adversely affected the characteristic textures of vegetations. This case study is presented here with the intention to show the limitations of SLAR which can at best be a reconnaissance vegetation survey tool.

Case Study 2:

Crop identification in Indiana

The use of imaging radar data for crop identification has attracted a lot of attention in recent years because of the prospect of monitoring crop growth from space at any time. Ulaby *et al.* (1980) reported an experiment with the synthetic aperture radar in L-band (1.6 GHz) to identify crops in Huntington County, Indiana, USA. The test site was covered by two adjacent passes flown at 2,170 m from Mean Sea Level with a swath width of 4,500 m each. L-band imagery with HH and HV polarizations was obtained. Depression angles ranged from 31° in the near range to 15° in the far range. The radar return values were also digitized as grey levels (128 steps) for classification purposes. The crop types and the number of fields imaged are show in Table 5.5. Fallowed fields and small grains were combined with pasture for classification purposes because of the similar radar returns for all these three categories. The corn and soyabeans were mature and ready for harvest.

Because of the tonal variation across the image from the near range to the far range (Fig. 5.11), the data had to be normalized. This involved breaking down the image into five strips parallel to the flight line, and from each strip all corn and soyabean data were averaged separately and plotted as a function of range. It was found that the curves were quite similar in shape for both passes, for HH and HV polarizations and for both corn and soyabeans, although corn consistently gave a higher return than soyabeans. A range correction curve was later calculated by dividing each test site into 250 parallel strips. All soyabean pixels (which were the most extensive) within each strip were averaged and plotted and a curve was fitted to their distribution with range (Fig. 5.13). This curve was then used to normalize the pixels which were aggregated into fields. Means and standard deviations were calculated for each field. These provided the parameters required for computer classification using a linear discriminant analysis. For all of the analyses, 50 per cent of the samples were randomly selected for training and the remaining samples were used for testing. The classification analysis was performed using each of the two discriminating variables, like polarization (HH) and cross-polarization (HV) returns, singly and in combination. The results revealed that HH yielded an overall classification of 65 per cent, but with the additional HV component, this was raised to 71 per cent.

The accuracy of crop discrimination with radar can be further improved with multidate data. A follow-up study by Shanmugan *et al.* (1983) using L-(1.6 GHz) and C-(4.75 GHz) band data obtained in late

Table 5.5
Number of fields per category for Pass 1 and Pass 2 (*Source*: Ulaby *et al.* 1980)

Pass 1*			Pass 2**		
Crop Type		*Number of Fields*	*Crop Type*		*Number of Fields*
Continuous Cover:			Continuous Cover:		
Fallow	6		Fallow	0	
Grains	5	21	Grains	6	23
Pasture	10		Pasture	17	
Woods		10	Woods		16
Corn		40	Corn		35
Soyabeans		42	Soyabeans		42

* 40°33.3′N to 40°58.4′N, 85°26.4′W to 85°29.5′W.
** 40°41′N to 40°58.4′N, 85°29.8′W to 85°33′W.

Fig. 5.13
Mean radar image grey tone
level for soyabeans as a function
of range. The test site was
divided into 250 strips parallel
to the flight line (*Source*: Ulaby
et al. 1980)

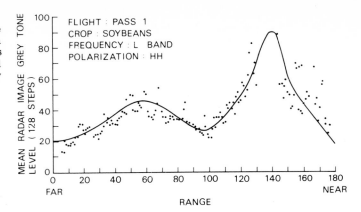

July and early August 1978 from aircraft over a test site in western
Kansas, USA (crop types: corn, pasture or bare soil) indicated a
discrimination accuracy in the range of 97–100 per cent, using a
maximum likelihood classifier. There was no need to use
multi-frequency and/or multipolarization data at all because the
multidate data alone were more informative than the spectral and
polarization data. However, soil moisture tended to adversely affect
the overall accuracy of classification by over 20 per cent with the like-
polarized systems. In such a case, the cross-polarized systems should
be used.

There are other factors that need to be considered, such as the
choice of a proper frequency. It has been suggested that frequencies
above 8 GHz (wavelengths shorter than 3.75 cm) were best for crop
discrimination (Ulaby *et al.* 1980). Row direction of the crops in the
fields is another factor because this tends to affect radar return more
significantly at longer wavelengths. The HH polarization has been
found to be more sensitive to row direction than HV.

Applications using multispectral remote sensing techniques

Multispectral sensing makes use of different spectral bands of the
electromagnetic spectrum to differentiate between objects or to determine the physical condition of objects based on the amount of reflectance in each band. Computer processing is normally required because
of the large number of multispectral data involved. A typical early
example was the recognition of tree species in a forest area in Michigan, USA using data collected in six spectral regions between
0.4–1.0 μm (Rohde and Olson, Jr. 1972). With the aid of a computer
it was found that approximately 85 per cent of all trees within the
forest area could be correctly identified. Discrimination among conifers
was not as successful as for broadleaved species, but spruce could
consistently be separated from pine. Another early application was to
detect diseased citrus trees in Florida, USA (Edwards, Schehl and
DuCharme 1975). The spectral bands used were 0.53–0.58, 0.72–0.76,
0.77–0.81, 0.82–0.88, 1.05–1.1 and 1.2–1.3 μm. Reflectance intensity

from sensors recorded on magnetic tape was employed in a computer for analysis using the Laboratory for Applications of Remote Sensing (LARS) software developed at Purdue University. The overall accuracy of the method for determining healthy and diseased trees was found to be 89 per cent using data collected at 0.82–0.88 μm from aircraft at an altitude of 460 m.

More recently, applications of multispectral remote sensing have focused on the estimation of the amount and distribution of vegetation. A knowledge of the vegetation amount permits vegetation yield to be estimated. When applied to crop growth, crop yield can be predicted (Curran 1980). The amount of green vegetation can be measured by the green Leaf Area Index (LAI), green biomass or per cent cover. The LAI is the area of green leaves per unit area of ground. The biomass is the mass of the green vegetation per unit area, usually measured by harvesting and weighing vegetation. Per cent cover of green vegetation can be measured by photographing upwards into the vegetation canopy from the forest floor. The percentage cover was calculated with a dot grid.

Case Study 1:

Estimation of green leaf area index (LAI)

The estimation is based on reflectance from a vegetation canopy. The intensity of the reflectance is dependent on the wavelength used and the three components of the vegetation canopy, viz., *leaves*, *substrate* and *shadow*. Leaves reflect weakly in the blue and red wavelengths but strongly in the near infrared wavelengths (Fig. 5.14). The reflec-

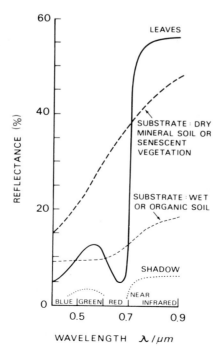

Fig. 5.14
Reflectance properties of three components of a vegetation canopy, leaves, shadow and either a light-toned mineral substrate or a dark-toned organic subtrate (*Source*: Curran 1983)

tance of vegetation substrates (or soil background) varies according to the nature of the substrate, i.e. senescent vegetation; light-toned mineral soil; or dark-toned organic soil (Curran 1983). The shadow component in a vegetation-canopy is very dark in the visible wavelengths with radiation strongly absorbed by leaves, but appears fairly dark in the near infrared wavelengths with radiation slightly absorbed by leaves. The relative area of these three spectrally dissimilar canopy components determines the reflectance of the total canopy. It was found that green LAI is *negatively* related to red reflectance but positively related to near-infrared reflectance (Fig. 5.15). A ratio of red to near infrared reflectance therefore expresses the increasing difference between red and near-infrared reflectance with increasing green LAI. A simple form of this ratio is just IR/R, where IR is the near-infrared and R is the red reflectance (Tucker 1979). Another popular form of the ratio is (IR–R)/(IR+R) which is also known as the *vegetation index*. The relation between the vegetation index and the green LAI is generally curvilinear, and is dependent on the species and conditions of measurement (Fig. 5.16). The curve reaches an asymptote when the substrate is covered by several layers of leaves. It is noteworthy that the detection of reflectance change depends on the reflectance contrast between green leaves and substrate (i.e. the soil background). Reflectance in near-infrared wavelengths is more sensitive to changes in the green vegetation on dark-toned substrates than on light-toned ones, whereas reflectance in red wavelengths is more sensitive to changes in the green vegetation on light-toned substrates than on dark-toned ones (Fig. 5.15). Other environmental factors that can affect reflectance of vegetation canopy include sun and sensor angles. The angle of the sun controls the area and the darkness of the shadow. The look-angle of the sensor determines the amount of substrate (soil) visible to the sensor. As the look-angle moves from the vertical, less soil and more vegetation will be visible.

Fig. 5.15
Relation between Leaf Area Index (LAI) and red and near-infrared reflectance for a vegetation canopy growing on a light-toned and a dark-toned substrate, derived from modelled data (*Source*: Curran, 1983)

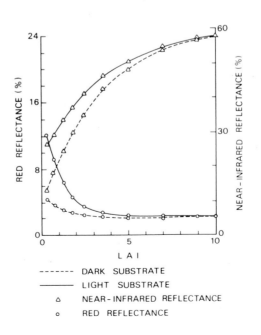

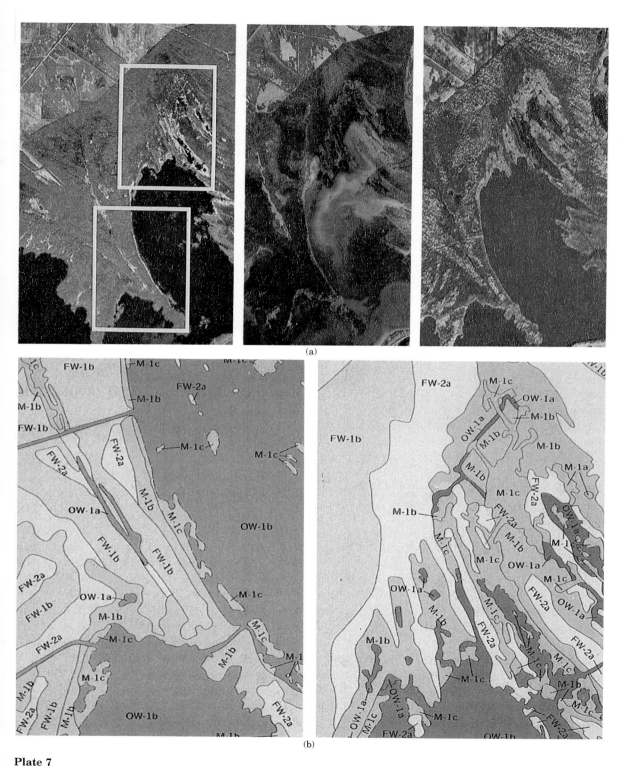

Plate 7
Seasonal high-altitude colour infrared photographs of Reelfoot Lake, Tennessee. (a): Left: October 1974 growing season, the two mapped areas are framed; Middle: February 1975: high water with no leaves; Right: November 1975: low water with no leaves. These were used to produce wetland maps. (b) all marsh (M), open water (OW) and forested wetland (FW) classes (*Source*: Carter *et al*. 1979)

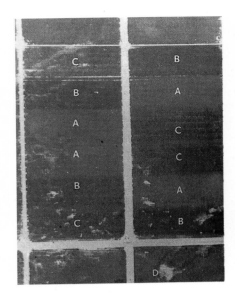

Plate 8
Colour infrared photograph of cotton
fields: (A) – untreated with plant growth
regulators; (B) and (C) – treated 13 and 26
days respectively; before the photograph
was taken; (D) – plants infested with cotton root rot (*Courtesy*: W. G. Hart; *Source*:
Henneberry *et al.* 1979)

Plate 9
Pseudo-coloured imagery of the difference between afternoon and morning surface temperature measurements, 1 April 1976, Phoenix.

	°C		°C
White	41–43	Dk Green	26–28
Red	38–40	Lt Blue	23–25
Orange	35–37	Dk Blue	20–22
Yellow	32–34	Violet	17–19
Lt Green	29–31	Black	14–16

(from Millard *et al* 1978)

Plate 10
(a) Black and white thermal infrared image (8–14 µm) of Farmington, New Mexico, USA acquired on 18 January 1978 at 11.30 p.m. Note that warm objects appear dark while cold objects appear bright after a digital reversal process. (b) The same scene colour enhanced to emphasize contrast. Note that ten colours were used and each colour represented 0.5 °C (*Courtesy:* T. K. Budge)

(a)

(b)

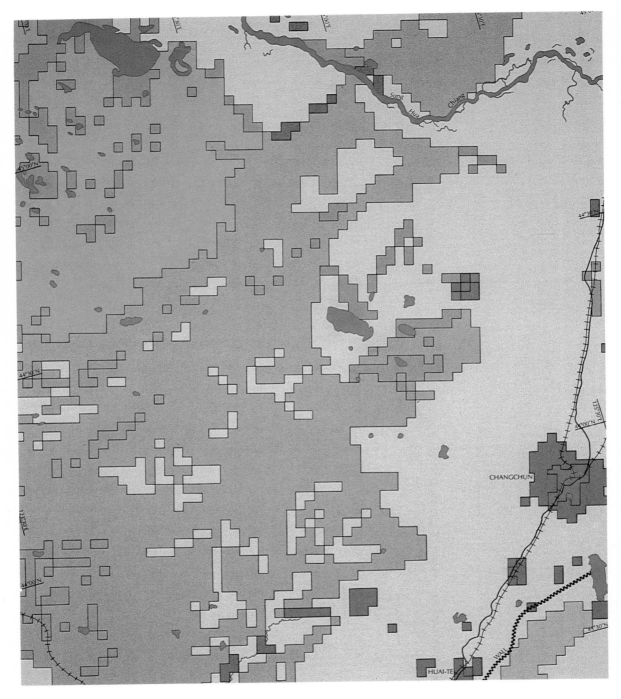

Plate 11
A sample of the land use map of northeast China 1973 interpreted from Landsat-1 data. Yellow = extensive field crop-land; dark green = intensive market garden; light brown = new agricultural land; dark brown = steppe, scrubland and grassland; light green = mixed deciduous and evergreen forest; red = urban or built-up land (*Source*: Welch, Pannell and Lo 1975)

Fig. 5.16
The relation between the Vegetation Index and Leaf Area Index (LAI) for different vegetation canopies: (a) asymptote not reached; (b) asymptote reached at a low LAI; (c) asymptote reached at a high LAI (*Source*: Curran 1983)

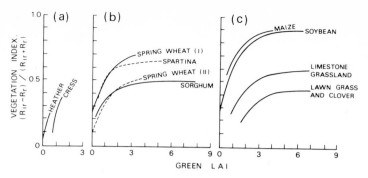

To estimate green LAI from remotely sensed multispectral reflectance, two stages are involved: (1) the determination of a calibration curve; (2) the use of multispectral reflectance data collected from aircraft or spacecraft as input to this calibration curve (Curran 1983). This is best demonstrated by the work of Curran (1981, 1983) in his study of Snelsmore Common, a heathland in Berkshire, UK where four vegetation types were studied: young *Calluna* (heather), mature *Calluna, Pteridium* (bracken) and *Pteridium-Calluna* (bracken-heather association). Ground data were first collected at each random point selected in the field. Radiometric red and near-infrared reflectance measurements or photographic red and near-infrared reflectance measurements or both, were taken, and vegetation samples were collected at each sample point. Because the near-infrared reflectance data were found to be affected markedly by sun angle, they were corrected to a constant solar angle. These corrected data were then used to calculate a vegetation index. In order to remove the effect of the substrate (soil) from the canopy reflectance, the following form of vegetation index known as *Perpendicular Vegetation Index* (PVI) developed by Richardson and Wiegland (1977) was used:

$$PVI = \sqrt{[(R_{s,r} - R_{v,r})^2 + (R_{s,ir} - R_{v,ir})^2]} \qquad [5.8]$$

where $R_{s,r}$ and $R_{s,ir}$ are substrate reflectances in red and near-infrared wavelengths respectively, and $R_{v,r}$ and $R_{v,ir}$ are vegetation reflectances in red and near-infrared wavelengths respectively. The computed PVI was regressed against the green LAI of the four vegetation types (Fig. 5.17). These provided calibration curves for the different vegetation types.

The second stage involved the use of multispectral aerial photography taken at about 11 a.m. local time on 20 dates, from June 1980 to August 1981, from a light aircraft with a 35 mm camera, a near-infrared film and filters combined to obtain red and near-infrared photography at 1:5,500 scale. At each random sample point the vegetation was harvested and the green LAI determined. For each flight the aerial photographic red and near-infrared reflectance data were transformed to PVI values which were then used to estimate the LAI of each of these areas with the aid of the calibration curves established previously (Fig. 5.17). The accuracy of the estimation is shown in Table 5.6. An estimate of green LAI could be made with a low error of the estimate to an overall accuracy of 42 per cent; at a medium error of the estimate it was raised to 74 per cent; and at a high error of the

Fig. 5.17
The linear relation between the Perpendicular Vegetation Index (PVI) and the green Leaf Area Index (green LAI) for a *Pteridium-Calluna* association. The solid line is described by the equation $x = 0.24 + 0.22y$. The standard error (SE) of estimate is 0.66; $n = 60$; $r = 0.83$; significant level, 1 per cent (*Source*: Curran 1983)

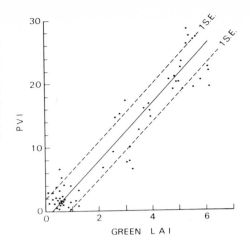

Table 5.6
Relation between the error of a green LAI estimate and the accuracy of that estimate (For example the green LAI of *Pteridium* can be estimated with an error of ±1.0, 73 times out of every 100.)
(*Source*: Curran 1983)

Vegetation Association	Error of Green LAI Estimate	Accuracy (%) of Green LAI Estimate at given Error Level
Young *Calluna*	±0.05 (low)	34
	±0.1 (medium)	62
	±0.2 (high)	92
Mature *Calluna*	±0.05 (low)	41
	±0.1 (medium)	73
	±0.2 (high)	100
Pteridium	±0.5 (low)	39
	±1.0 (medium)	73
	±2.0 (high)	100
Pteridium-Calluna	±0.5 (low)	53
	±1.0 (medium)	89
	±2.0 (high)	98

estimate, 98 per cent. The accuracy was found to vary with season. It was highest in winter and spring (from December to early May) with an average of 84 per cent; lowest in summer and autumn (from late May to November) with an average of 52 per cent only.

Such an approach can also be applied to estimate green LAI of crops such as spring wheat (Ahlrichs *et al.* 1979).

Case Study 2:

Estimation of grass canopy biomass

Biomass of short grass prairie vegetation is traditionally measured by hand clipping plots of known area and weighing the vegetation removed to determine the wet or dry biomass per unit area. This destructive method is time consuming and inefficient to use. A remote sensing method has been developed to accurately estimate grass canopy biomass (Pearson, Tucker and Miller 1976). A calibration curve has to be established first. In order to speed up the ground truth collection of biomass data in the field, a homemade, hand-held biomass meter was constructed which was capable of measuring the canopy reflectance at 0.68 μm (red) and 0.80 μm (near-infrared). The ratio of the red and near-infrared reflectances was then related to the dry biomass data collected at 25 sample plots of blue grama grass (*Bouteloua gracilis*), each plot being 1/4 m² in area. A linear relationship was established between the two (Fig. 5.18), showing a correlation of 0.98 for all 25 plots measured. This was found adequate for biomass estimation of shortgrass to midgrass grass canopies. Subsequently, one flight line of 12-band multispectral imagery over the Pawnee National Grassland which is located about 56 km northeast of Fort Collins, Colorado, USA was analysed with the aim of mapping the spatial distribution of biomass directly using the knowledge and results obtained in the basic field spectrometry experiments discussed above. At the time of the overflight, a total of 87 biomass measurements of 1 m² areas were made for control and 35 of them occurred in the flight strip analysed. The results indicated that with the use of a simple ratio of the two spectral bands most closely representing the red and near-infrared bands used in the hand-held biomass meter, which happened to be 0.66–0.72 μm (Channel 8) and 0.80–1.00 μm (Channel 10) respectively, a correlation coefficient (R) of 0.79 was obtained for 26 ground-truth biomass samples. A dry biomass equation in the following form was obtained:

$$\text{Dry biomass} = -327.7 + 429.2 \text{ (ch 10/ch 8)} \qquad [5.9]$$

This result compared favourably with that obtained by multiple regression of all scanner channels (R = 0.79) and that obtained by spectrum matching using maximum likelihood computations and training sets of known biomass (R = 0.98). The biomass for each ground cell imaged could then be obtained in this way. The total biomass of grass for the entire flightline was estimated to be 1.696

Fig. 5.18
Dry standing crop biomass for 25 sample plots of blue grama grass (*Bouteloua gracilis*), each plot being ¼ m² in area, as a linear function of the ratio of the solar induced radiance of the plots at 0.80 μm referenced to that at 0.68 μm (*Source:* Pearson *et al.* 1976)

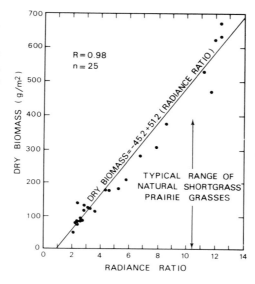

$\times$ 10^6 kg, thus giving an average biomass of 1,812 kg/ha. The spatial distribution of biomass in four classes (50, 100, 250 and 400 g/m^2) could also be displayed on the image.

Applications using satellite data

The availability of satellite data, notably Landsat MSS data, has made applications which take advantage of wide areal coverage, computer data processing capabilities and sequential coverage capabilities of these data possible. A specific example each taken from the fields of vegetation, crops and soils will serve to illustrate these approaches well.

Case Study 1:

Coniferous tree species indentification and mapping using Landsat data

The Landsat MSS data which are available in four spectral bands can be visually interpreted with the aid of a colour additive viewer to manually delineate forest from non-forest land, or to separate between water, conifers, muskegs and hardwood stands, as has been done in the boreal forest region of Alberta, Canada (Kirby 1974). However, it would be difficult to visually differentiate tree species from the small-scale coarse-resolution colour composite. Walsh (1980) demonstrated an analytical approach to identify and map coniferous tree species in Crater Lake National Park, Oregon, USA using Landsat digital MSS data and a special computer image analysis system known as Inter-active Digital Image Manipulation System (IDIMS).

His approach requires the collection of ground truth data, the development of training statistics, an understanding of the environmental factors that affect spectral reflectance variation of surface-cover types and the selection of an appropriate classification algorithm for use by the computer. The study area which has an area of 405.6 km^2 is located in southwestern Oregon between the latitudes of 43°04'N and 42°46'N and between the longitudes of 122°15'56"W and 121°-59'50"W. It has 12 surface-cover types including seven classes of coniferous tree species. The study area is quite diverse in topography and has numerous volcanic landforms. The collection of ground truth data has to be executed with care because the quality of these data will determine the success of the subsequent analysis. The location of ground truth sample sites was aided by consulting colour infrared aerial photography, topographic maps and a slope map of the Park so as to permit sampling of at least the major slope aspects, slope angles and the main forest-cover types. The size of the sampling unit which consisted of about 25–30 Landsat pixels (or approximately 13.6 ha) was large enough to be field checked. At each sample site, the following data were observed: tree species or brush type; percentage of tree species in mix; distribution of tree species in mix; tree height – single or multiple layered canopy; crown size; crown density; diameter of the trees; breast height; slope angle; slope aspect; tree species branching characteristics; sample site understory; time of year of sampling; vegetation vigour; moisture characteristics of the site; and sample site location on Landsat and aerial photographic data. The large clusters of pixels are desirable for the supervised classification technique employed. In addition to the Landsat digital data, a digital terrain tape of the area was obtained from digitized topographic maps to provide information on slope angle and slope aspect. The 10 September 1974 Landsat CCT was the optimal data set to be used because of the phenological conditions of the Park vegetation and the lower sun angle. The IDIMS was used to delimit training areas in 32 × 32 pixel blocks called Intensive Study Areas (ISA). The ISA (training areas) were placed in such a way that they tapped as much spectral variation of surface-cover types within the park as possible. The spectral variation was caused by the different vegetation communities and environmental parameters such as slope angle, slope aspect, crown size and crown density which tended to influence the reflectance values of vegetation types within the park. Altogether 25 ISAs were designated and with the aid of the ground-truth information the following data were collected: photo-interpretation of 1:122,000, 1:30,500 and 1:7,600 colour infrared aerial photography; photo-interpretation of terrestrial photography; photo interpretation of a Landsat colour composite displayed on the IDIMS CRT terminal. The individual ISAs were later enlarged ten times for examination to discard unwanted areas. Training statistics were then obtained from all the delimited areas within the ISA. These were input to the IDIMS which generated naturally occurring clusters of pixels with the function called ISOCLS. This simply made use of distance measurements and statistical parameters to help distinguish between classes. The mean values and standard deviations of these naturally occurring clusters were computed using a function called STDMAX. All pixels were assigned to the clusters they were nearest and the process continued until the standard deviation of each cluster was less than a threshold value (2.20 in this case study). A maximum likelihood classifier (MLC) was then used in the 'CLASFY' function to classify the entire study area (512 × 512 pixels) with reference to the training statistics. Altogether 59 clusters or classes were generated. These were combined to 12 classes only since the major concern was coniferous tree species separation (Table 5.7). The accuracy of the classification was checked by randomly selecting a sample of pixel clusters (5 × 5 pixels) within each class, which were then checked against ground truth data. The result indicated an average accuracy of 88.8 per cent.

The application described above is typical of a supervised approach to vegetation or land-cover mapping with the use of digital image data, although the type of computer system employed may have differed. The speed of computer mapping for Landsat data can be illustrated by another example of application carried out in the Philippines using the General Electric Image-100 system (Lachowski *et al.* 1979). A nationwide forest land-cover assessment project resulted in the production of 30 Landsat vegetation maps for the whole country at 1:500,000 scale in six months, with five major forest-associated land-cover types and an overall accuracy that varied between 85 and 95 per cent. In addition, the land areas occupied by each forest-related land-cover type were also generated. It has to be stressed again that

Table 5.7
Major surface cover types of Crater Lake National Park mapped by Landsat (*Source*: Walsh 1980)

Surface cover type

1. Ponderosa pine, large, densely stocked and/or Shasta red fir, large, moderately stocked.

2. Ponderosa pine, large, poorly stocked and/or Shasta red fir, large, poorly stocked.

3. Ponderosa pine (65%) and White fir (35%).

4. White fir (75%) and Shasta red fir and/or Ponderosa pine.

5. Mountain hemlock (80%) and Shasta red fir (20%).

6. Shasta red fir (80%) and Mountain hemlock (20%).

7. Water.

8. Pumice fields, rock outcrops, and bare soil.

9. Lodgepole pine (80%), densely stocked with scattered amounts of related pines, firs, and hemlock.

10. Lodgepole pine (80%), poorly stocked with scattered amount of related pines, firs, and hemlock.

11. Shadow and boundary pixels of land-water interface; scattered deciduous and brush associated with steep moist areas.

12. Scattered tree and light brush or pumice underlying canopy openings.

invariably, much field work had to be done to obtain ground truth information.

Some workers do not agree that the supervised approach is the best for image classification. Strahler (1981) demonstrated the use of an unsupervised approach to inventory natural vegetation and rangeland using tone, texture and terrain data derived from Landsat digital imagery and collateral data. He also attempted to do timber volume inventory with the Landsat data, having collected ground truth information on tree height, tree stocking and regional type. The importance of terrain in affecting the reflectance of the tree species was also observed. Townshend and Justice (1980) have experimented with the unsupervised classification of MSS Landsat data for mapping spatially complex vegetation in Southern Italy and achieved a disappointing 60 per cent overall accuracy. They concluded that the unsupervised approach was not superior to the supervised one as suggested by other workers in areas of complex terrain. It is clear therefore that many methodological issues await to be settled.

Case Study 2:

Identification, area estimation and yield prediction of agricultural crops by computer classification of Landsat MSS data

The Landsat data also open up opportunities to forecasting and estimating crop production. One approach is to identify agricultural crop types and estimate the area covered by the crops. Another appproach is to estimate the vegetation amount using a Vegetation Index involving red and near-infrared ratios.

(a) Identification and area estimation

For the first approach, Bauer *et al.* (1979) presented a case study of a three-county area in northern Illinois, USA where approximately 81 per cent of the total land area is cropland. About 71 per cent of the cropland was planted to corn and soyabeans. Other land uses included: urban (5.3 per cent), pasture (5.1 per cent), forests and wood lots (3.4 per cent) and others (4.4 per cent). The Landsat MSS data acquired on 9 August, 1972 were analysed with the LARSY Version 3 multi-spectral data analysis system developed by the Laboratory for Applications of Remote Sensing, Purdue University, West Lafayette, Indiana, USA. The procedures involved were similar to those described above, viz., (1) defining a group of spectral classes for training; (2) specifying these to a statistical algorithm for the computation of a set of defined parameters (statistics) for each class; (3) applying the calculated statistics to 'train' a pattern recognition algorithm; (4) classifying, using a maximum likelihood classifier, into one of the training classes; and (5) displaying the classification results in map or tabular forms. The analysis only attempted to identify three classes of crops: 'corn', 'soyabeans' and 'other'. Five hundred fields were visited on the ground or examined from large-scale aerial photography to record the crop type or land use. They were used to train the classifier. Only 'pure' pixels (i.e. only one crop or cover type) around the centre of a field were used to train the classifier.

The result of analysis indicated an overall accuracy of 82.8 per cent which was compatible to the accuracy achievable with the use of large-scale aircraft MSS data collected at 3,000 m altitude. The results were considered significant for Landsat data, considering the limited number of spectral bands, limited dynamic range and coarse spatial resolution. Most significantly, they covered an area of 3,000 km², unlike the limited 20 km² coverage of the aircraft data. Crop areas were also estimated by counting the number of pixels in each class in the computer. Comparison with the US Department of Agriculture (USDA) county area estimates indicated a close correspondence between the two (a root mean square difference of 9.3). This, however, could be improved by applying the error matrix based on the test data collected in the field. From this one could determine the proportions of corn classified as corn and non-corn and the proportions of soya-beans classified as soyabeans and non-soyabeans. Thus, one could obtain more accurate crop area estimate by using the following equation:

$$A = CP^{-1} \qquad [5.10]$$

where C is the classification vector with n crops or classes; P^{-1} is the inverse of the $n \times n$ classification performance matrix and A is a $1 \times n$ vector of the crop areas. This helped to raise the root mean square difference between the USDA and Landsat estimates to 0.6.

One point which was experimented on was the extendability of training statistics from one county to the other two counties. Equally good performance was revealed. This, however, was dependent on many environmental factors such as atmospheric conditions and the similarity of the agricultural practice. It was also important to note the need to have more training data in dealing with areas having greater variations in crops and soils.

The research also experimented with three methods of improving further the computer classification: (1) use of prior probability information; (2) use of spatial feature in classification; (3) use of multitemporal data. Prior probability information on the occurrence of individual crop types could be obtained from an earlier survey or other secondary sources to establish 'weights' for each crop type. This helped to increase the overall classification accuracy by 2.3 per cent at the expense of decreased accuracy for soyabeans. The use of spatial dimension of the Landsat data implied the use of the whole field rather than an individual data point for classification. This made use of the fact that points in close proximity to one another were likely to be members of the same class. An improvement of 5 per cent in the performance was achieved. Finally, the use of multidate data was believed to have great potential for improving the recognition of crop species because of different stages of growth of the crops. Analysis was carried out with three sets of Landsat data acquired on 9 August, 14 September and 2 October 1972 which were registered to achieve geometric coincidence of the picture elements of the three dates. The results seemed to suggest that the combined use of spectral and temporal features could maintain an overall accuracy between 71 and 74 per cent. On the other hand, the correct choice of an optimum scene alone for classification could give rise to an excellent accuracy of 76 per cent. For this study area, the optimum scene was obtained on 9 August 1972 well before crop maturation, thus enhancing crop spectral separability. Therefore, multitemporal data did not necessarily improve the overall results of the classification. It was also noted in this study that the Landsat MSS bands were not optimum for crop identification. Kumar and Silva (1977) suggested that one channel from the visible, near-, middle- and thermal-infrared could give the highest separability of agricultural crops.

(b) Band ratioing

The second approach to crop yield prediction is to make use of the near-infrared/red ratio method to estimate crop biomass (Tucker 1979). Rouse *et al.* (1974) made use of Landsat MSS data to develop vegetation indices to monitor rangelands and wheat crops. The simple ratio of MSS7/MSS5 which best approximated near-infrared/red ratio could be used as a measurement of the relative amount of green, but a better index to use was the *Vegetation Index* (VI) computed as:

$$VI = \frac{MSS7 - MSS5}{MSS7 + MSS5} \qquad [5.11]$$

To avoid working with negative ratio values and the possibility that the variances of the ratio would be proportional to the mean values, the constant of 0.5 was added and a square root transformation was applied to give a *Transformed Vegetation Index* (TVI):

$$TVI = \sqrt{VI + 0.5} \qquad [5.12]$$

It was found that there was a close relationship between TVI and the green biomass so that one could follow crop development as ground cover, biomass and leaf area indices increased. Tucker and Miller (1977) proposed the use of MSS6/MSS5 rather than MSS7/MSS5 for rangeland biomass estimation because of the low biomass situations based on the spectral contrasts of soil compared to green vegetation. All these indices can be used to forecast agricultural crop yields and to monitor crop vigour.

(c) Satellite-aided crop forecasting

The prediction of crop yield on a global basis has become a very important area of application of Landsat data in recent years. Since 1974 the Large Area Crop Inventory Experiment (LACIE) developed methodology for area, yield and production estimation of wheat with the cooperation of three agencies in the USA, – the National Aeronautics and Space Administration (NASA), the US Department of Agriculture and the National Oceanic and Atmospheric Administration (NOAA). The performance goals were: (1) to achieve an accuracy for estimates at harvest to be within ±10 per cent of true country production 90 per cent of the time; (2) to reduce Landsat data to production area information within 14 days after acquisition in an operational environment; (3) to base on objective and repeatable procedures (Erb 1980). In the early phases, the experiment concentrated primarily on a 'yard-stick' wheat-growing region of the USA – the nine-state wheat region in the Great Plains. As the experiment progressed, two additional major producing regions (Canada and the USSR) were included. Later on, exploratory studies were also made in five other major producing regions India, Peoples' Republic of China, Australia, Argentina and Brazil. The experiment extended over three overlapping global crop seasons. Technically, the approach was to estimate production of wheat on a regional basis where production is the product of area and yield. Area was obtained by computer classification of Landsat MSS data and yield estimates were obtained from statistical regression models relating wheat yield to local meteorological conditions, notably precipitation and temperature. Only sample unit sites selected by a stratified random sampling method were analysed in this way in each region. Each sample unit was a 5 × 6 nautical mile segment which was analysed for wheat percentage. Weather data were acquired from the World Meteorological Organization's (WMO) Global Telecommunications System, the US Air Force Enviromental Technical Applications Center and the NOAA Environmental Satellite Service. The wheat yield models were statistical regression models based upon recorded historical wheat yields and weather. These regression models forecast wheat yield for fairly broad geographic regions (called yield strata) using calendar monthly values of average temperature and cumulative precipitation over the yield strata. Thus, monthly updated yield estimates during the growing season were obtained. The wheat production estimate for a region required a process of upward aggregation of the segment level wheat percentages to the yield strata regions where the aggregate area estimates and yield model estimates were multiplied to provide estimates of production. Estimates of production for larger regions were the sum of the appropriate strata level production estimates.

The accuracy of the LACIE met the performance goals set. As an example, the 1977 LACIE wheat production final estimate for the USSR was 91.4 million metric tons which compared favourably with the official USSR figure of 92 million metric tons, a difference within 1 per cent of error. The LACIE technology is continuously being modified and improved. It is generally agreed that the yield models developed perform adequately if no significant changes in trend occur and if the average weather conditions for a region are not drastically different from the historical data used in their development. The major contribution of LACIE is its integrative approach of extracting information from Landsat MSS data in a timely and objective manner.

A more recent development of satellite-aided crop forecasting is the Inventory Technology Development Project in the joint program for Agriculture and Resources Inventory Surveys through Aerospace Remote Sensing (AgRISTARS) which aims at the development of automated information extraction techniques for non-US crop analysis of multitemporal Landsat data without using ground-based observations (Erickson *et al.* 1982). Area estimate technology was experimented for spring small grains, summer crops, corn and soyabeans. A high degree of automation was stressed. Initially experiments were carried out over the USAs Northern Great Plains (North Dakota, South Dakota, Montana and Minnesota) and Saskatchewan, Canada for spring small grains; and the USAs Central Corn Belt states of Iowa, Illinois, Indiana and Missouri for summer crops, corn and soyabean experiments. Preliminary results for a more fully objective automated approach were promising. The availability of the high-resolution Thematic Mapper data from Landsat 4 should further improve the accuracy of the technology developed.

Case Study 3:

Delineation of soil unit boundaries with Landsat MSS data

The Landsat MSS data have been found to be particularly suited for the delineation of soil boundaries caused by climatic and vegetative differences which otherwise are not observable on conventional aerial photographs with restricted areal coverage (Westin and Frazee 1976). Soil boundaries can also be revealed by differences in parent materials and relief, which are all detectable visually from Landsat images (band 5 and/or band 7). The use of the computer to discriminate soil types by digital analysis of Landsat data has also been attempted (Podwysocki, Gunther and Blodget 1977). A typical approach is to use image enhancement and classification techniques to produce a spectral map of cover types which are then correlated with soil characteristics. Obviously, such an approach is more successful in the arid or semi-arid environment where the vegetation cover is sparse.

A good case study to illustrate this approach is the work reported by Imhoff *et al.* (1982) who made use of the digital processing system of the Office for Remote Sensing of Earth Resources (ORSER) at the

Pennsylvania State University. The study area was arid and semi-arid rangeland near the town of Green River in each central Utah, USA. The soils in the area were mostly Aridisols, Entisols and Inceptisols. The natural vegetation consisted of desert shrubs and bunch grasses. The Landsat data for the study area were first geometrically corrected for earth rotational skew, orientation to true north and overlapping pixels. Two Landsat image products were generated: (1) a spectral thematic map where both supervised and unsupervised procedures were used to obtain in the classification categories and (2) an enhanced image created by a contrast stretch and an equal area density slicing technique. The Landsat images were also registered cartographically to a soil map drawn on the US Geological Survey topographical quadrangle. This required the use of control points selected on the map and the images. It was hoped to overlay soil boundary and other ancillary information such as transportation networks, a map projection graticule or survey grids onto the Landsat image as well. In other words, all those map data had to be digitized first and special computer programs had to be written to match the digitized ancillary data with the Landsat digital data together. The resultant image could be viewed in a colour CRT display. By using the method of spectral signature classification, this application succeeded in defining ten spectral categories based on a variety of soil surface characteristics such as terrain roughness, per cent of area of rock outcrop, parent material type, presence of rock fragments, and vegetation type and canopy densities. An unsupervised approach was recommended as being more suitable for soil mapping in view of its fairly objective processing method applicable over a large area (unlike the supervised approach which was also more expensive). The image enhancement procedure also resulted in an image product with 15 spectral classes on the basis of variation in grey level over all four spectral bands. The enhanced imagery was found to be particularly helpful in locating soil unit boundaries in the field. There was a good match between the spectral pattern and the mapped soil boundary lines. The spectral patterns in the enhanced image appeared to be more similar to the field data than those produced simply by spectral signature classification.

Conclusions

The application of remote sensing to the study of the biosphere has particular economic significance in view of our increasing concern about population growth and resource. The great variety of approaches and the high degree of sophistication in the methodology developed are astonishing. Many of the techniques discussed in this chapter overlap those examined in connection with the lithosphere. To recapitulate, one sees again the dominant role played by aerial photography, especially colour infrared, in vegetation and crop inventory. However, there are signs of growing importance of non-photographic remote sensors, notably radar, in an environment where the conventional aerial camera cannot operate efficiently, such as in the perennially cloud-covered tropical or equatorial regions of the world where great resource potentials await to be developed. Information on soil moisture and temperature which are important to crop growth and yield is collected by these non-photographic remote sensing systems as well. The role of multispectral remote sensing in monitoring vegetation growth and crop yield is also growing in importance. One sees the development of band ratioing technique being applied to the vegetation field with the intention to develop an effective vegetation index. The availability of satellite data, especially Landsat MSS data, has made the development of a global crop yield forecasting system (LACIE) possible, which integrates resources satellite data with environmental satellite data. All these only seem to be possible with the support of computers and sophisticated image analysis techniques. It appears that with the more advanced satellite remote sensing system, such as the Thematic Mapper of Landsat-4 and the high resolution visible imaging sensors of the SPOT satellite to be launched by the French, an even higher degree of accuracy and reliability in extracting resources information from the biosphere can be achieved.

References

Ahlrichs, J. S., Bauer, M. E., Hixson, M. M., Daughtry, C. S. T. and Crecelius, D. W. (1979) Relation of crop canopy variables to the multispectral reflectance of small grains, *Proceedings of the Internatioal Symposium on Remote Sensing for Observation and Inventory of Earth Resources and the Endangered Environment,* Vol. 1: International Archives of Photogrammetry Vol. XXII-7, pp. 629–47.

Aldred, A. H. (1976) *Measurement of Tropical Trees in Large-Scale Aerial Photographs,* Forest Management Institute, Ottawa, Ontario, Canada.

Aldred, A. H. and Lowe, J. J. (1978) *Application of Large-Scale Photos to a Forest Inventory in Alberta.* Forest Management Institute: Ottawa, Ontario, Canada.

Anderson, J. R., Hardy, E. E., Roach, J. T. and Witmer, R. E. (1976) *A Land Use and Land Cover Classification System for Use with Remote Sensor Data.* Geological Survey Professional Paper 964, US Government Printing Office: Washington, DC.

Anson, A. (1970) Color aerial photos in the reconnaissance of soils and rocks, *Photogrammetric Engineering* **36**: 343–54.

Baber, Jr., J. J. (1982) Detecting crop conditions with low-altitude aerial photography: In C. J. Johannsen and J. L. Sanders (eds) *Remote Sensing for Resource Management.* Soil Conservation Society of America: Ankeny, Iowa, pp. 407–12.

Bauer, M. E., Cipra, J. E., Anuta, P. E. and Etheridge, J. B. (1979) Identification and area estimation of agricultural crops by computer classification of LANDSAT MSS data, *Remote Sensing of Environment* **8**: 77–92.

Carter, V. (1982) Applications of remote sensing to wetlands: In Johannsen, C. J. and Sanders, J. L. (eds) *Remote Sensing for Resource Management.* Soil Conservation Society of America: Ankeny, Iowa, pp. 284–300.

Carter, V., Malone, D. L. and Burbank, J. H. (1979) Wetland classification and mapping in Western Tennessee, *Photogrammetric Engineering and Remote Sensing* **45**: 273–84.

Curran, P. (1980) Multispectral remote sensing of vegetation amount, *Progress in Physical Geography* **4**: 315–41.

Curran, P. J. (1981) The estimation of the surface moisture of a vegetated soil using aerial infrared photography, *International Journal of Remote Sensing* **2**: 369–78.

Curran, P. J. (1983) Multispectral remote sensing for the estimation of green leaf area index, *Philosophical Transactions of Royal Society of London, Series A* **309**: 257–70.

De Carolis, C. and Amodeo, P. (1980) Basic problems in the reflectance and emittance properties of vegetation: In Fraysse, G. (ed) *Remote Sensing Application in Agriculture and Hydrology.* A. A. Balkema: Rotterdam, pp. 69–79.

Edwards, G. J., Schehl, T. and DuCharme, E. P. (1975) Multispectral sensing of citrus young tree decline, *Photogrammetric Engineering and Remote Sensing* **41**: 653–7.

Erb, R. B. (1980) The Large Area Crop Inventory Experiment (LACIE) methodology for area, yield and production estimation: results and perspectives: In Fraysse, G. (ed) *Remote Sensing Application in Agriculture and Hydrology.* A. A. Balkema: Rotterdam, pp. 285–97.

Erickson, J., Dragg, J., Bizzell, R. and Trichel, M. (1982) Research advances in satellite-aided crop forecasting, Paper presented at the *International Society for Photogrammetry and Remote Sensing*, Commission VII International Symposium: In Toulouse, France, 13–17 September 1982.

Evans, R., Head, J. and Dirkzwager, M. (1976) Air photo-tones and soil properties: implications for interpreting satellite imagery, *Remote Sensing of Environment* **4**: 265–80.

Girard, C. M. (1980) Application of photo-interpretation technique to the classification of agricultural soils, choice of sensor, use of the results: In Fraysse, G. (ed) *Remote Sensing Application in Agriculture and Hydrology.* A. A. Balkema: Rotterdam, pp. 37–51.

Goodman, M. S. (1959) A technique for the identification of farm crops on aerial photographs, *Photogrammetric Engineering* **25**: 131–7.

Hardy, J. R. (1980) Survey of methods for the determination of soil moisture content by remote sensing methods: In Fraysse, G. (ed) *Remote Sensing Application in Agriculture and Hydrology.* A. A. Balkema: Rotterdam, pp. 233–47.

Henneberry, T. J., Hart, W. G., Bariola, L. A., Kittock, D. L., Arle, H. F., Davis, M. R. and Ingle, S. J. (1979) Parameters of cotton cultivation from infrared aerial photography, *Photogrammetric Engineering and Remote Sensing* **45**: 1129–33.

Idso, S. B., Hatfield, J. L., Jackson, R. D. and Reginato, R. J. (1979) Grain yield prediction: extending the Stress-Degree-Day approach to accommodate climatic variability, *Remote Sensing of Environment* **8**: 267–72.

Idso, S. B., Jackson, R. D. and Reginato, R. J. (1977a) Remote sensing of crop yields, *Science* **196**: 19–25.

Idso, S. B., Reginato, R. J. and Jackson, R. D. (1977b) Albedo measurement for remote sensing of crop yields, *Nature* **266**: 625–28.

Imhoff, M. L., Petersen, G. W., Sykes, S. G. and Irons, J. R. (1982) Digital overlay of cartographic information on Landsat MSS data for soil surveys, *Photogrammetric Engineering and Remote Sensing* **48**: 1337–42.

Jackson, H. R. and Wallen, V. R. (1975) Microdensitometer measurements of sequential aerial photographs of field beans infected with bacterial blight, *Phitopathology* **65**: 961–68.

Kirby, C. L. (1974) Forest and land inventory using ERTS imagery and aerial photography in the boreal forest region of Alberta, Canada: In Freden, S. C., Mercanti, E. P. and Becker, M. A. (eds) *Third Earth Resources Technology Satellite-1 Symposium, Vol. I: Technical Presentations: Section A.* National Aeronautics and Space Administration: Washington, DC, pp. 127–43.

Kuhl, A. D. (1970) Color and IR photos for soils, *Photogrammetric Engineering* **36**: 475–82.

Kumar, R. and Silva, L. F. (1977) Separability of agricultural cover types by remote sensing in the visible and infrared wavelength regions, *IEEE Transactions on Geoscience Electronics* **15**: 42–9.

Lachowski, H. M., Dietrich, D. L., Umali, R., Aquino, E. and Basa, V. (1979) Landsat assisted forest land-cover assessment of the Philippine Islands, *Photogrammetric Engineering and Remote Sensing* **45**: 1387–91.

Lamb, F. G. (1982) Agricultural uses of low-altitude aerial photography: In C. J. Johannsen and J. L. Sanders (eds) *Remote Sensing for Resource Management.* Soil Conservation Society of America: Ankeny, Iowa, pp. 402–6.

Millard, J. P., Jackson, R. D., Goettelman, R. C., Reginato, R. J. and Idso, S. B. (1978) Crop water-stress assessment using an airborne thermal scanner, *Photogrammetric Engineering and Remote Sensing* **44**: 77–85.

Morain, S. A. (1976) Use of radar for vegetation analysis: In Lewis, A. J. (ed) *Geoscience Applications of Imaging Radar Systems, Remote Sensing of the Electromagnetic Spectrum* **3**: 61–78.

Murtha, P. A. (1978) Remote sensing and vegetation damage: a theory for detection and assessment, *Photogrammetric Engineering and Remote Sensing* **44**: 1147–58.

Murtha, P. A. (1982) Detection and analysis of vegetation stress: In Johannsen, C. J. and Sanders, J. L. (eds) *Remote Sensing for Resource Management.* Soil Conservation Society of America: Ankeny, Iowa, pp. 141–58.

Myers, B. J. and Benson, M. L. (1981) Rain forest species on large-scale color photos, *Photogrammetric Engineering and Remote Sensing* **47**: 505–13.

Parry, D. E. and Trevett, J. W. (1979) Mapping Nigeria's vegetation from radar, *Geographical Journal* **145**: 265–81.

Pearson, R. L., Tucker, C. J. and Miller, L. D. (1976) Spectral mapping of short-grass prairie biomass, *Photogrammetric Engineering and Remote Sensing* **42**: 317–23.

Philipson, W. R. and Liang, T. (1975) Airphoto analysis of the tropics: crop identification, *Proceedings of the Tenth International Symposium on Remote Sensing of Environment.* Environmental Research Institute of Michigan: Ann Arbor, Michigan, pp. 1079–92.

Philipson, W. R. and Liang, T. (1982) An airphoto key for major tropical crops, *Photogrammetric Engineering and Remote Sensing* **48**: 223–33.

Podwysocki, M. H., Gunther, F. J. and Blodget, H. W. (1977) *Discrimination of rock and soil types by digital analysis of Landsat data*, Report X-923-77-17. NASA/Goddard Space Flight Center: Greenbelt, Maryland.

Richardson, A. J. and Wiegand, C. L. (1977) Distinguishing vegetation from soil background information, *Photogrammetric Engineering and Remote Sensing* **43**: 1541–52.

Rohde, W. G. and Olson, Jr., C. E. (1972) Multispectral sensing of forest tree species, *Photogrammetric Engineering* **38**: 1209–15.

Rouse, Jr., J. W., Haas, R. H., Schell, J. A. and Deering, D. W. (1974) Monitoring vegetation systems in the Great Plains with ERTS: In Freden, S. C., Mercanti, E. P. and Becker, M. A. (eds) *Third Earth Resources Technology Satellite-1 Symposium, Vol. I: Technical Presentations Section A.* National Aeronautics and Space Administration: Washington, DC, pp. 309–17.

Sayn-Wittgenstein, L. (1961) Identification of tree species by crown characteristics at temperate area, *Photogrammetric Engineering* **27**: 792–809.

Sayn-Wittgenstein, L. (1978) *Recognition of Tree Species on Aerial Photographs.* Canada Forestry Service: Department of the Environment.

Shanmugan, K. S., Ulaby, F. T., Narayanan, V. and Dobson, C. (1983) Identification of corn fields using multidate radar data, *Remote Sensing of Environment* **13**: 251–64.

Shaw, M. W. (1971) The use of aerial photography in the survey of woodlands: In Goodier, R. (ed.) *The Application of Aerial Photography to the Work of the Nature Conservancy.* The Nature Conservancy: Edinburgh.

Soil Conservation Service (1966) *Aerial-Photo Interpretation in Classifying and Mapping Soils*, United States Department of Agriculture, Superintendent of Documents, US Government Printing Office, Washington, DC.

Steiner, D. (1970) Time dimension for crop survey from space, *Photogrammetric Engineering* **36**: 187–94.

Steiner, D. and Maurer, H. (1969) The use of stereo height as a discriminating variable for crop classification on aerial photographs, *Photogrammetria* **24**: 223–41.

Strahler, A. H. (1981) Stratification of natural vegetation for forest and rangeland inventory using Landsat digital imagery and collateral data, *International Journal of Remote Sensing* **2**: 15–41.

Tivy, J. (1982) *Biogeography*, Longman, London and New York, p. 171.

Townshend, J. R. G. and Justice, C. O. (1980) Unsupervised classification of MSS Landsat data for mapping spatially complex vegetation, *International Journal of Remote Sensing* **1**: 105–20.

Tucker, C. J. (1979) Red and photographic infrared linear combinations for monitoring vegetation, *Remote Sensing of Environment* **8**: 127–50.

Tucker, C. J. and Miller, L. D. (1977) Contribution of soil spectra to grass canopy spectral reflectance, *Photogrammetric Engineering and Remote Sensing* **43**: 721–6.

Ulaby, F. T., Batlivala, P. P. and Bare, J. E. (1980) Crop identification with L-band radar, *Photogrammetric Engineering and Remote Sensing* **46**: 101–5.

Walsh, S. J. (1980) Coniferous tree species mapping using LANDSAT data, *Remote Sensing of Environment* **9**: 11–26.

Westin, F. C. and Frazee, C. J. (1976) Landsat data, its use in a soil survey program, *Soil Science Society of America Journal* **40**: 81–9.

White, L. P. (1977) *Aerial Photography and Remote Sensing for Soil Survey.* Clarendon Press: Oxford.

Chapter 6 Land use and land cover mapping

The problems of land use and land cover mapping

The mapping of land use and land cover is closely related to the study of vegetation, crops and soils of the biosphere examined in the preceding chapter. Since land use and land cover data are most essential to planners who have to make decisions concerning land resource management, they are strongly economic in nature. These data are usually presented in map form accompanied by area statistics for each category of land use and land cover. The use of remotely sensed imagery is particularly appropriate for the production of these maps. This explains the great interest focused on this field.

Land is a raw material of a site, which is defined in terms of a number of natural characteristics, namely, climate, geology, soil, topography, hydrology and biology (Aldrich 1981), *Land use* is man's activities on and in relation to the land, which are usually not directly visible from the imagery. Land use has been studied from many diverse viewpoints so that no one single definition is really appropriate in all different contexts (Campbell 1983). It is possible, for example, to look at land use from the land capability point of view by evaluating the land in relation to the various natural characteristics mentioned above. From this, we move to the *land cover* which describes the 'vegetational and artificial constructions covering the land surface' (Burley 1961). These are all directly visible from the remotely sensed imagery. Three general classes of of data are included in land cover: (1) physical structures built by human beings; (2) biotic phenomena such as natural vegetation, agricultural crops and animal life; (3) any types of development. Thus, based on the observation of land cover as a proxy, one hopes to infer human activities and land use. However, there are human activities that may not be directly related to the type of land cover, such as recreational activities (Anderson *et al.* 1976). Other problems include multiple use which may occur simultaneously or alternately; vertical arrangement of uses; and the minimum area size of mapping. Therefore, land use and land cover mapping require some arbitrary decisions to be made and the resultant maps inevitably contain some degree of generalized information according to their scales and the purposes of applications.

Classification schemes

One crucial factor in determining the success of land use and land cover mapping lies in the choice of an appropriate classification scheme designed for an intended purpose. A good classification scheme should be easy to use with no ambiguity in defining each land use and land cover category. It must also be able to generate the degree of details required. In other words, the level of accuracy of the resultant maps is closely related to the classification scheme which should take into account the final map scale. This in turn is related to the method of data acquisition. The use of ground survey methods allows the adoption of a smaller land parcel as the mapping unit and the production of more detailed maps on a large scale. This approach is normally required in a built-up area or an intensively developed area where human activities are extremely complex. Nowadays, most land use and land cover surveys make use of aircraft or satellite imagery, and ground surveys are restricted to the field checking of interpreted results only. The scale and spatial resolution of the aerospace imagery together determine the amount of detail that can be interpreted with certainty, thus determining the minimum size of the land parcel as a mapping unit, which may vary from one category to the other. On the whole, Clawson (1966) recommended that a single or pure line concept should be used in a single classification scheme and an inductive approach be adopted. This implies the use of the smallest recognizable parcel of land for interpretation in as much detail as possible so that the uses can be grouped into the categories most appropriate to the needs for the researchers and planners. The design of the land use and land cover scheme can take either a functional approach, a morphological approach or both. The functional approach is activity-oriented, as evidenced by the use of such terms as agricultural, grazing, forestry, urban activities, etc. The morphological approach emphasizes land cover with the use of such terms as arable land, grassland, woodland, moorland, built-up area, etc. as exemplified by Stamp's scheme for the First Land Utilization Survey of Britain in 1930 (Stamp 1960; Campbell 1983). Anderson (1971) suggested that the activity-oriented or functional approach would be more appropriate as a general-purpose classification scheme for use with aerospace imagery. At this point, it is instructive to examine the USGS (United States Geological Survey) land and land cover classification system for use with remote sensor data (Table 6.1) devised by Anderson and his colleagues (1976), which is the most commonly used scheme in the US. The scheme has been designed for use with high-altitude or orbital remote sensor data and was intended to meet the following criteria: (1) a minimum level of interpretation accuracy of at least 85 per cent; (2) equal accuracy for different categories; (3) repeatable results; (4) applicability over extensive areas; (5) categorization permitting land cover to be used as surrogate for activity; (6) possibility for use with remote sensor data acquired at different times; (7) integration with ground surveyed data or large-scale remote sensor data possible through the use of sub-categories; (8) aggregation of categories possible; (9) comparison with future data possible; (10) multiple uses of land recognizable. The scheme emphasized a rather coarse level of mapping – Levels I and II which were appropriate for maps at the scales of 1:250,000 and 1:100,000 used in the USGS

Table 6.1
Land use and land cover classification system for use with
remote sensor data
(*Source*: Anderson *et al.* 1976)

Level I	Level II
1 Urban or Built-up Land	11 Residential
	12 Commercial and Services
	13 Industrial
	14 Transportation, Communications, and Utilities
	15 Industrial and Commercial Complexes
	16 Mixed Urban or Built-up Land
	17 Other Urban or Built-up Land
2 Agricultural Land	21 Cropland and Pasture
	22 Orchards, Groves, Vineyards, Nurseries, and Ornamental Horticultural Areas
	23 Confined Feeding Operations
	24 Other Agricultural Land
3 Rangeland	31 Herbaceous Rangeland
	32 Shrub and Brush Rangeland
	33 Mixed Rangeland
4 Forest land	41 Deciduous Forest Land
	42 Evergreen Forest Land
	43 Mixed Forest Land
5 Water	51 Streams and Canals
	52 Lakes
	53 Reservoirs
	54 Bays and Estuaries
6 Wetland	61 Forested Wetland
	62 Nonforested Wetland
7 Barren Land	71 Dry Salt Flats
	72 Beaches
	73 Sandy Areas other than Beaches
	74 Bare Exposed Rock
	75 Strip Mines, Quarries and Gravel Pits
	76 Transitional Areas
	77 Mixed Barren Land
8 Tundra	81 Shrub and Brush Tundra
	82 Herbaceous Tundra
	83 Bare Ground Tundra
	84 Wet Tundra
	85 Mixed Tundra
9 Perennial Snow or Ice	91 Perennial Snowfields
	92 Glaciers

programme. The scheme was also aimed to provide compatibility with
other classification schemes being used by other government agencies
involved in land use inventorying and mapping. The whole system is
'resource-oriented' rather than 'people oriented'. Land cover is used as
the principal surrogate for interpreting activity. The minimum
mapping unit adopted by USGS for this scheme is 4 ha for urban or
built-up uses, water areas, transitional areas in an urban situation,
confined feeding operations, certain other types of agricultural land,
and strip mines, quarries and gravel pits. All other categories are
delineated with a minimum unit of 16 ha (Place 1977). The success of
the USGS land use and land cover classification system is proved by
the fact that it has been used not only within the US but also in other
countries of the developed and developing world with slight modifications (Allan 1980).

Applications using aerial photography

The use of aerial photography for land use and land cover mapping follows more-or-less standard procedures as explained in Campbell (1983) and Baker *et al.* (1979). First of all, the aerial photographs have to be assembled together in the form of a mosaic to define the area of interest. One can carry out a 'trial' interpretation of three to five frames of photographs from widely spaced parts of the study area. This is then followed by a search for collateral materials such as maps, documents and literature relating to the study area. It is advisable to visit the study area and carry out a ground check of the 'trial' interpretation. A classification scheme is then designed with the purpose of a specific application in mind. The minimum area to be interpreted should also be defined. It should be the smallest area that one can interpret with certainty. The manual interpretation of the land use and land cover categories according to the classification scheme is performed. This requires the application of the following image characteristics: size, shape, shadow, texture, tone, pattern, site and association. Texture, tone and pattern are particularly useful to land cover interpretation whereas association helps to infer human activities (land use). The land use and land cover categories can be delineated directly on the photographic prints or better on a plastic overlay. In case of doubts, the interpretation should also be checked in the field. The interpreted details are then transferred to a base map with the aid of such instruments as reflecting projectors, sketch masters or zoom transferscopes. The accuracy of the resultant map is determined by field checking a sample of land parcels allocated by a sound statistical sampling design. The area of each category of land use and land cover is measured by using a dot grid, a planimeter or a digitizer. Finally, a report has to be written to provide the regional setting, to explain the method of data acquisition, to assess the accuracy of the maps and to summarize the land use patterns of the study area. In the following case studies, the emphasis has been placed on how this standard approach has been modified to tackle the problems in practice.

Case Study 1:

Rural land use mapping in a hilly country

A team of Swiss specialists were sent to Nepal to study the development of the Jiri area and to recommend directions for its future development. The team was faced with the problem of inavailability of land use maps without which they could not correctly evaluate agricultural production and land capability (Schmid 1971). In order to produce the needed land use maps of the area (700 km²) within a period of four months; existing aerial photography flown for the forest survey of Nepal and handheld vertical photography together with topographic maps at 1:50,000 scale and at 1:25,000 scale (original survey) had to be used. The aerial photography was taken with a Zeiss camera (f = 305.42 mm) with a 23 × 23 cm format from an altitude of 8,200 m, thus giving an average scale of 1:200,000. The hand-held photography was taken with a Hasselblad camera ($f = 38.62$ mm) in a 6×6 cm format, enlarged to 13×13 cm from altitudes of 4,600 m and 5,200 m, thus giving scales of 1:34,000 and 1:45,000 respectively. These photographs were in no way standard or of excellent quality. The study area is predominantly an agricultural area specializing in the cultivation of food grains, mainly paddy, wheat, barley and millet, supplemented by potatoes. In view of the hilly nature of the area, all these are grown on terraced fields along slopes. Above the field agriculture, areas of forest and shrub occur. They are used as grazing grounds for buffalo, goats and cattle. Dense forests are only found in the higher remote parts of the mountains. Much of the original forest land had been damaged as a result of agricultural activities and the traditional practice of cutting tree branches for use as firewood. In the land use mapping, a detailed 13-category land use scheme was adopted with an emphasis on the land cover (Table 6.2). The aerial photographs were interpreted with a stereoscope and the land use categories delineated on the photographs. Continuous field checking was made with the aid of binoculars from a point overlooking the valleys and hillsides. It was found that mistakes tended to occur in the areas of deep valleys where dark shadows obscured features. However, most of the categories could be correctly identified on the aerial photographs. The results of the survey were transferred to the 1:25,000 scale topographical map. Since no optical instrument was available, the transference of details had to be done slowly by eyesight with reference to map details. The areas of the various land use categories were measured on the original land use map with the aid of a 1×1 mm grid. The counting was done independently by two people, each counting twice. The averaged results were grouped according to altitudinal zones and to village *panchayats* (a kind of municipal community with 37,410 inhabitants). The altitudinal zones used were the average upper limits of rice cultivation (1,800 m), agriculturally used land (2,600 m) and forests (3,800 m). In order to measure terraced field areas accurately from the map, a correction factor had to be computed for each land use category according to

Table 6.2
Land use classification scheme for air photograph interpretation of East Nepal
(*Source:* Schmid 1971)

Category	Description
1	Settlements and corresponding non-agricultural territories
2	Irrigable crop land, annually cultivated
3	Not properly irrigated or unirrigated crop land, annually cultivated
4	Temporarily used crop land
5	Abandoned crop land
6	Horticulture and experimental fields
7	Orchards
8	Uncleared and uncleaned pastures
9	Cleared and cleaned pastures
10	Attended meadows, improved
11	Bushes, less than 3 metres
12	Dense tall forests of broadleaf and coniferous trees
13	Waste land

the slope angle. On the basis of the land use statistics, agricultural production was estimated with a knowledge of the crop rotation cycles in categories 2 and 3, referred to in Table 6.2. The land use map also permitted the forest resources in Nepal to be assessed. This case study illustrates the use of aerial photography for land use and land cover mapping in a less than ideal situation in a developing country.

Case Study 2:

Urban land use mapping in an old industrial city

Due to the complexity of human activities and the high density of buildings, mapping the land use of an urban area is always difficult, both from the ground or from aerial photography. The photography required should normally be of large scale, say not smaller than 1:50,000. The success of the photo-interpretation depends on a good correspondence between function (human activities) and form (morphology) of the built-up area (Collins and El-Beik 1971). One major problem is the prevalence of multistorey building which are characterized by multiple use, especially in the city centre or the Central Business District (CBD). These difficulties necessitate a much greater amount of field work than that required in the rural environment (Pollé 1974). Mapping land use of the city of Leeds, UK from 1:10,000 scale 'split-vertical' aerial photographs illustrates the problems very well (Collins and El-Beik 1971). Three aerial photographs which covered an extensive area with a dominant land use: commercial, industrial or residential were selected and a key of the urban land use was compiled by carrying out a field survey of the land use covered by these three photographs (Table 6.3). Clearly, the key emphasized

Table 6.3
Key to urban land use in the city of Leeds
(*Source*: Collins and El-Beik 1971)

Key to urban land use				
1 Commercial	C		*5 Transportation*	T
(a) offices	C_1		(a) railways	T_1
(b) shops	C_2		(b) roadways	T_2
			(c) waterways	T_3
2 Industrial	I		*6 Public buildings*	B
(a) offices	I_1		(a) educational	B_1
(b) factories	I_2		(b) hospitals	B_2
(c) petrol storage tanks	I_3		(c) churches	B_3
(d) gas storage tanks	I_4			
(e) power stations	I_5			
3 Residential	R		*7 Open improved land*	P
(a) detached houses	R_1		(a) parks	P_1
(b) semi-detached houses	R_2		(b) cemeteries	P_2
(c) terraced houses	R_3		(c) sports	P_3
(d) back to back houses	R_{3b}		(d) parking places	P_4
(e) single storey houses	R_4		(vehicles)	
(f) block of flats	R_5		(e) allotments	P_5
4 Bodies of water	W		*8 Open unimproved land*	V
(a) streams and rivers	W_1			
(b) lakes	W_2			
(c) reservoirs and water works	W_3			

Fig. 6.1
The split-vertical aerial photo-
graph of the commercial area of
Leeds acquired on 30 July 1963
(*Courtesy*: W. G. Collins; *Source*:
Collins and El-Beik 1971)

land cover and building types. The minimum unit of interpretation for
the built-up area was a building unit, the function of which had to
be inferred. The photo-interpreter then examined the three photographs carefully under the stereoscope in order to become familiar
with the photo images of the various items in the key (Figs 6.1, 6.2).
When all these items could be consistently and accurately identified,
the other photographs covering the city were interpreted. The land use
categories were delineated on a transparent overlay for each photograph. Since the photographic scale roughly matched the Ordinance
Survey map scale of 1:10,560, details could be transferred directly
from the overlays to the map. In addition, an attempt was made to
interpret the industrial activity in the city in more detail. This necessitated the compilation of an industrial air photo key by using a field
survey based on the types of premises visible on the aerial photographs (Table 6.4). The photo-interpreted urban land use map was
checked against one compiled from field work. It was found that an
accuracy of 88.5 per cent was achieved for the overall map with eight
major categories of land use. The error was mainly caused by the

Fig. 6.2
Interpreted land use map of the commercial area of Leeds (see also Fig. 6.1) (*Source*: Collins and El-Beik 1971)

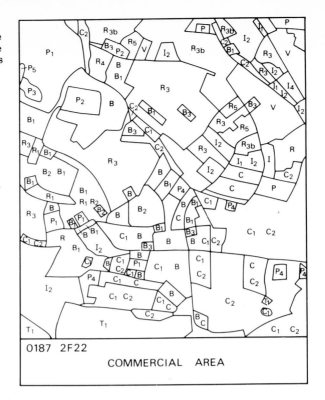

0187 2F22

COMMERCIAL AREA

Table 6.4
Key to industrial land use of city of Leeds (*Source*: Collins and El-Beik 1971)

Symbol	Type of Industry
I	General industrial area
I.1	Industrial offices
I.2	Factories
I.21	Extraction industry factories
I.22	Processing industry factories
I.23	Fabrication industry factories
I.23a	Heavy fabrication industry factories
I.23b	Light fabrication industry factories
I.3	Petrol storage tanks
I.4	Gas storage tanks
I.5	Power stations
I.6	Construction industry
I.7	Warehouses
I.8	Sewage works
I.9	Water works

mismatch between form and function. The industrial land use was particularly difficult to interpret correctly because of the limited relationship between form and function of many industrial buildings. As for the industrial land use types, the extraction industry could be interpreted with certainty whereas processing and fabrication (manufacturing) industries exhibited low accuracies in the interpretation. It was concluded that for detailed urban land use studies, especially those concerned with industrial activity, the optimum photo scale was 1:2,500.

The approach described above has mostly been applied to urban land use mapping in developing countries where aerial photography is the more easily available form of remote sensing data. However, the cities in the Third World are characterized by a high degree of multiple use and complexity so that the correlation between form and function is even weaker. Gautam (1976) presented an example of using 1:25,000 scale aerial photography to map the land use of Bikaner City in India, which illustrated very well the complexity of the land use classification scheme designed. Unfortunately the accuracy of the resultant map was not evaluated.

Case Study 3:

Land use change detection

Land use change detection using aerial photography has been applied to both rural and urban areas and found to be an efficient method of discovering the development trends of a region. Change detection involves the use of sequential aerial photography over a specific region from which the land use map for each date was mapped and compared. A land use change map between two time periods is usually produced (Campbell 1983). A good example is Avery's (1965) use of US Department of Agriculture aerial photography at 1:20,000 scale in 1944 and 1960 to evaluate land use changes in Clarke County, Georgia. A six-category land use scheme was designed for use in the interpretation. The six categories were: cultivated land, pine forest, hardwood forest, urban land, idle land and water. A land use map for each period was obtained and the area for each category was determined with the aid of a dot grid (16 dots per in^2), each dot representing about 4 acres at the nominal 1:20,000 scale. Two practical problems were noteworthy. The first one was to ensure that the two maps corresponded exactly in scale and in areal extent. The second problem was to ensure the compatibility of the land use interpretation. Ideally, it required the same person to interpret and prepare the two maps. The apportionment method to allocate area to each category of land use was the best method to use since the total area of the county under study was known (Paine 1981). This simply means that:

$$\text{Type area} = \frac{\text{No. of dots in type}}{\text{Total dots counted}} \times \frac{\text{Total land area of}}{\text{the study area}}$$

Avery's study discovered a shift of the agricultural pattern of Clarke County, Georgia from a heavy emphasis on cotton to poultry production, livestock and farm woodlot management while cropland was reverted back to forest land as a result of the influx of manufacturing industry to provide employment. It is also possible to carry out land use change detection in a more simplified way: just to match two land use maps compiled for the two different dates over a light table and to only delineate any changes. In this way, a land use change map can be obtained, from which the change area can be measured. This has been done by the present author for a more up-to-date study of land use change in Clarke County, Georgia between 1970 and 1983, using aerial photographs at the nominal scale of 1:12,000. The land use scheme adopted was the Level II USGS system shown in Table 6.1. An example of the resultant change map for a part of the study region is shown in Fig. 6.3. An increase in the 'urban or built-up land' category, particularly in residential, commercial and services, as well as a decrease in the 'agricultural land' and 'forest land' categories were noted.

A modern approach to land use change detection in a more complex and rapidly changing environment involves the use of the computer. Adeniyi (1980) presented a good example of how change detection using sequential aerial photography was carried out in Lagos, Nigeria. A major problem was the great difference in scale of the aerial photography available: 1:40,000 for the year 1962 and 1:20,000 for the year 1974. The study area encompassed urban built-up areas, urban vacant land and non-urban land (Fig. 6.4). A land use classification scheme of nine major land use and land cover categories with 45 sub-categories was devised. The nine major categories were residential, commercial, industrial, institutional, transportational and utilities, recreational and open spaces, vacant land, non-urban land, and water. A minimum mapping unit of one ha (100 × 100 m) was used as the basis of interpretation and for subsequent storage of land use data by the computer. On each stereomodel, the built-up areas were identified and delineated first, followed by vacant and non-urban land uses. The interpretation was carried out under a mirror stereoscope with 3X

Fig. 6.3
Land use change map of southwestern Athens, Georgia, USA, 1970–1983.

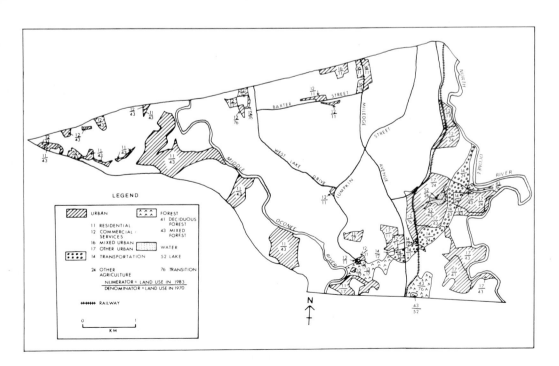

Fig. 6.4

Stereotriplet of part of central mainland Lagos, 1974, illustrating the residential types with the following class codes: 10 = large plot, 1–2 storeyed buildings with vegetated open space; 11 = medium plot, mostly two-storeyed buildings without vegetated open space; 12 = medium plot, mixed, 1–2 storeyed buildings with small individual open space; 13 = single storeyed row houses with moderate common open space; 14 = mixed traditional and modern 1–3 storeyed buildings; 16 = traditional single-storeyed rooming buildings; 17 = apartment buildings (four storeys and above). Also indicated are undeveloped (dry) vacant land (72) and undeveloped (wet) forested vacant land (74) (*Courtesy*: P. O. Adeniyi; *Source*: Adeniyi 1980)

magnification onto an acetate overlay. The interpretation was limited to the central portion of each photograph only, to avoid excessive relief displacements. The interpreted land use classifications were field checked or field completed in cases of any doubt. These were subsequently transferred to a base map of 1:20,000 scale with the aid of the Bausch and Lomb Zoom Transfer Scope. The land use data on each map was later manually encoded on the one ha cell size. A clear and stable Mylar sheet with 100×100 m grids was placed on top of the land use maps. The dominant land use falling within the grid would be coded with the appropriate land use category code. The encoded data were carefully checked before they were key-punched onto a disk and then transferred onto a computer tape for processing by the computer. Special computer programs were written to: (1) compare each major land use category for the two time periods on a cell by cell basis; (2) provide information about the locations, types and amounts of changes; (3) produce land use and change maps with the drum plotter (Fig. 6.5a, b, c). The computer approach permitted a high degree of flexibility and the generation of a large volume of land use data. A data bank was created, in which the land use data could be easily updated and integrated with other types of data, such as population data, for urban research and planning. The land use change revealed the rapid increase of the residential land use and a strong lateral

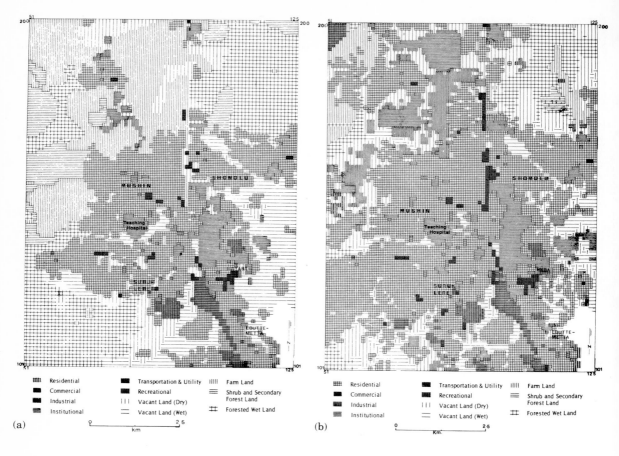

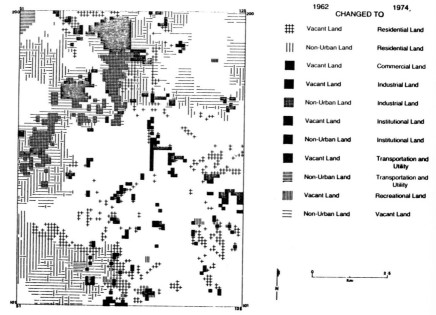

Fig. 6.5
A sample of the computer generated land-use map of part of the Lagos mainland in Nigeria for (a) 1962 (b) 1974, and (c) land use change, 1962–1974 (*Source*: Adeniyi 1980)

expansion of the urban area of Lagos. This computer approach is certainly more desirable for land use change detection. With the recent advances in computer technology, the laborious encoding and digitizing of the land use data for computer input can be done much more quickly and easily.

Applications using thermal infrared imagery

There is a paucity of applications of thermal infrared imagery to land use and land cover mapping. This is perhaps due to the fact that the spatial resolution of the thermal line scan system is rather coarse and that its emphasis has been to record thermal emission from the land cover. The tonal variations of the thermal infrared imagery are closely related to the emissivities of the materials that make up the land cover. It would not be possible to classify subtle land uses based on emissivities alone. An evaluation of the applicability of thermal infrared scanning imagery for land use classification using the USGS classification system has been carried out by Brown and Holz (1976). The scanner was an AN/AAS-18 linescan infrared detecting set in the wavelength range from 8–14 μm. It was a photo conductive quantum detector utilizing mercury-doped germanium (GeHg) as a detector crystal material. The imagery was obtained in the early evening hours of 11 November 1974 from an altitude of 754 m above the water surface of a reservoir in Oak Creek Lake study area, Texas. This is an artificial water inpoundment in a tributary of the Texas Colorado River. Since the scale in thermal infrared image remains constant along the flight direction and decreases outwards from the nadir line in the scan direction (see Ch. 1), only the scale of the imagery parallel to the flight direction was determined with reference to the 1:24,000 topographic map of the study area, which was found to be 1:25,840. The 1:500,000 aeronautical charts and 1:24,000 USGS topographic maps were also consulted in the delineation of land use categories on the thermal infrared imagery. The results suggested the following (Fig. 6.6):

1. *Residential* use along the lakeside could be identified by the dark structures, indicating that they were cool. These were vacation homes and coolness suggested that they were unoccupied.

2. *Commercial services* were found to occur as a complex or cluster which included a motel, restaurant, grocery store and gasoline station.

3. *Transportation, communications and utilities* were detectable. A railroad could be differentiated from the road because of the lower emissivity of the metal of the railroad compared to that of the macadam highway which was lighter in tone. A thermal power plant was revealed by the discharge of heated effluents into the reservoir.

4. *Cropland* was revealed by contour ploughing which resulted in a difference in temperature caused by changes in moisture because of

Fig. 6.6
Thermal infrared images
(8–14μm) of the Oak Creek
Lake, West Texas, USA (*Source*:
Brown and Holz 1976)

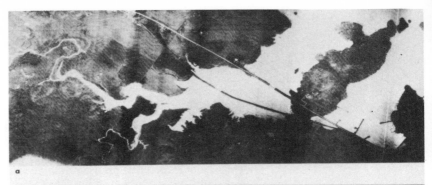

ploughing. Orderly rectangles could also be seen, thus strongly
suggesting that they were croplands rather than pasture lands.

5. *Grass rangeland* exhibited a rougher texture in the image because
of the complexity of rangeland flora where the grass was short with
occasional patches of bunch grass and larger dry-climate shrubs.

6. *Savanna rangeland* was much darker in tone than the grass
rangeland in the thermal infrared imagery because of the thicker
savanna cover which provided shade to retard ground absorption of
solar energy in the daytime and hence the reradiated longer-wave
energy emitted from the earth surface was weaker.

7. *Deciduous forest* appeared as a fairly light-toned patch of extremely
granular texture as a result of the solar heating of the trees during
the daytime and their different heights and shapes.

8. *Streams and waterways* could be easily detected because of the
great contrast between the water and the land. Even fairly narrow
linear water bodies (less than 1.5 m) could be recognized.

9. *Reservoirs* here implied a multipurpose utilization with a strong
emphasis on recreation.

10. *Other water bodies* exhibited light-toned returns on the imagery.
They could be circular, triangular and linear in shape. Some of them
were found to be earth-dammed stock tanks, or water storage basins
for range cattle.

This study concluded that the interpretation of thermal infrared imagery required an understanding of the intensity of infrared energy emitted by an object according to its absolute temperature, thermal capacity, texture, background and emissivity. It provided a wide variety of information unavailable from the photographic system alone. The USGS land use and land cover classification system was found to be sufficiently inclusive to serve as an adequate framework for land-use surveys, utilizing thermal infrared scanner imagery, although some slight modifications in the redefinition of certain land use categories (such as rangeland, savanna and streams) were required to meet the needs of the advancing technology of remote sensing. Clearly, the extraction of land use information from thermal infrared imagery requires more inference or deduction than in the case of aerial photography.

A more common application of thermal infrared imagery is to map or detect heat loss from a specific type of land use, such as residential, commercial or industrial. This makes use of the fact that any surface, such as a roof, radiates energy the intensity of which is dependent on the temperature of the surface, the wavelength of the radiation and the physical properties of the material. In particular, the physical properties of the material determine the efficiency of the surface in emitting energy, i.e. its emissivity. A typical heat loss detection survey normally acquires night-time thermal infrared, daytime thermal infrared and daytime photographic imagery from a low altitude (between 300–500 m) which ensures an acceptable ground resolution. The night-time imagery should be collected four or more hours after sunset to minimize the effects of solar heating, ideally with snow and ice-free roofs, air temperatures below 0° C and light wind conditions. These environmental factors can affect the interpretation of heat loss from the roofs of buildings. The night-time thermal imagery was examined to detect excessive bright tone surfaces. The daytime thermal imagery permits differentiation between clear or ice-covered snow while the higher resolution daytime photograph facilitates identification of snow or ice patches, roof vents, chimneys and other roof structures. They provide supplementary information to the night-time thermal imagery to resolve cases of doubts in the interpretation (Brown 1979). Plate 10 (page 212), illustrates an example of the thermal infrared scanner imagery of Farmington, New Mexico, USA on 18 January 1978. The flight took place between 10.30 p.m. and 12.00 p.m. from an altitude of 365.8 m (1,200 ft) using a Daedalus D5-1260 multispectral scanner to record emitted thermal radiation in the 8–14 μm range. On a thermal image warm objects normally appear bright while cold objects appear dark. In Pl. 10a this has been digitally reversed to permit easier detection of temperature changes on roof tops so that warm spots became dark and the cold spots registered bright. This is revealed by the temperature scale below the picture which shows an 0.5° C step from −5° C to −9.5° C. The colour rendition of the black-and-white thermal image (Pl. 10b) permits easier visual discrimination of rooftop temperature changes (Budge and Inglis 1978). However, the detection of heat loss from buildings is complicated by many environmental factors such as sky temperature, wind speed, cloud cover, and roof conditions. The emissivities of the roofs tend to be confined to a very narrow range so that the observed variation in rooftop emissivities has only a minimal effect on the temperatures depicted in the thermal image (Artis and Carnahan 1982). Ground surveys and interviewing of the residents are normally required if accurate mapping of heat losses is to be achieved.

Applications using radar imagery

The use of radar imagery to map land and land cover has attracted a great deal of attention in recent years because the radar imagery is an all-weather system which complements the aerial photography well. Also, the radar image appears visually similar to the aerial photograph and familiar image characteristics such as tone, texture, pattern, shape and association can be applied to the interpretation of the radar imagery. In addition, the areal coverage of the radar imagery obtained by Side-Looking Airborne Radar (SLAR) is much wider than that of aerial photography, thus facilitating the detection of the overall land use pattern. The following case studies illustrate the typical microscopic and macroscopic approaches of land use mapping using SLAR imagery.

Case Study 1:

Small-scale rural land use mapping

Henderson (1975; 1979) attempted to evaluate the usefulness of SLAR for general land use mapping at a small scale (1:250,000 or smaller). His approach was to delineate land use or landscape regions – a process called regionalization normally used by geographers. To establish the regions, physical and cultural features of the environment had to be interpreted from the radar imagery. By using the Westinghouse AN/APQ-97 real-aperture K-band SLAR imagery at an approximate scale of 1:180,000 over the West and Midwest of the US (from Minnesota to North Utah), he was able to detect five major components of a land use region (Henderson 1975): (1) surface configuration; (2) natural vegetation; (3) field patterns and size; (4) settlement pattern; (5) transportation and communications network. Differences in each of these characteristics were noted on the radar image. A land use region should be one within which the above mentioned physical and cultural characteristics are similar. In other words, the combination and intermixing of the relevant environmental characteristics had to appear homogeneous and distinct if compared to the adjacent areas. The radar imagery was in the form of a continuous film strip which facilitated the description of the areas and detection of changes. Each of the five characteristics mentioned above was delineated on a transparent overlay and then composited again to determine the boundaries of the level of the land use regions. This qualitative approach permitted 14 land use regions to be delineated:

I. Commercial cropland and livestock
II. Commercial cropland with pasture and coppice woodlands
III. Subhumid cropland with pastures

IV. Subhumid pasture with irrigated and dryland farming
V. Semi-arid limited grazing and pasture–dissected hill
VI. Forest, meadow and limited grazing–low mountains
VII. Semi-arid pasture and grazing–tablelands
VIII. Semi-arid limited grazing–dissected low and high hills
IX. Limited grazing with sparse scrub–high hills
X. Mostly ungrazed semi-arid foothills with mining
XI. Semi-arid tablelands with limited grazing and river valley cropland
XII. Mining and mostly ungrazed scrub–semi-arid tablelands and low mountains
XIII. Scrub, scrub forest with limited grazing–low mountains and high hills
XIV. Mountain valley hay and pasture

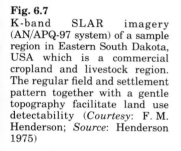

Fig. 6.7
K-band SLAR imagery (AN/APQ-97 system) of a sample region in Eastern South Dakota, USA which is a commercial cropland and livestock region. The regular field and settlement pattern together with a gentle topography facilitate land use detectability (*Courtesy*: F. M. Henderson; *Source*: Henderson 1975)

In delineating these regions, collateral materials have to be used. Both the presence and absence of the features were noted. Two examples are shown in Figs 6.7 and 6.8. The results of the delineation were compared to the traditional land use map by J. R. Anderson (1970) published in the *National Atlas* and the land-resource region map by M. E. Austin (1965). It was found that eight of the ten land use divisions (80 per cent) compiled by Anderson and seven of the eleven divisions (74 per cent) compiled by Austin were similar to the land use divisions or combinations of divisions created from radar. It was further noted that more detailed regions were created using radar imagery than using the traditional approach. It should be possible, therefore, to revise existing land use maps with radar imagery.

In a follow-up application of radar imagery for small-scale land use

Fig. 6.8
K-band SLAR image of Western South Dakota, USA – a sample region of semi-arid pasture and grazing with plateau-like topography containing a few gentle hills and some steep relief. Vegetation cover is scarce and is predominantly brush and grasses. This presents a contrast to Fig. 6.7. (*Courtesy*: F. M. Henderson; *Source*: Henderson 1975)

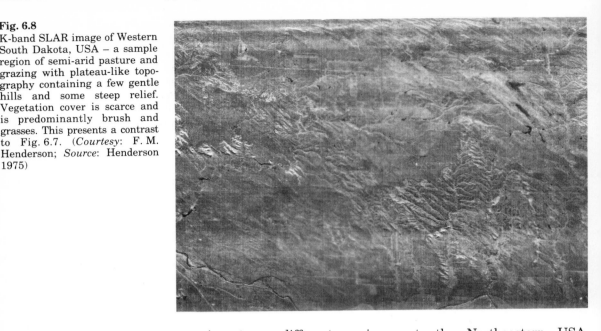

mapping to a different environment, the Northeastern USA, Henderson (1979) encountered less favourable results than those for the Midwest and West, USA. The approach was not to delineate land use regions but to delineate Level I and Level II land-use categories of the USGS system (Table 6.1). The radar imagery used was a Westinghouse K-band imagery flow over portions of Pennsylvania, New York and New England in July 1966 at an approximate scale of 1:225,000. A transparent grid of one-kilometre cells overlay was placed on top of the image, and the land use of each cell was interpreted with the aid of a Bausch and Lomb 240Z stereoscope in monomode. Based on this interpretation the land use regions for the study area were delineated. It was found that because of the longer history of development of the Northeastern USA, the land use units visible were smaller and the land use pattern more complex. The spatial resolution of the radar imagery (about 15–16 m) became a limiting factor, making the interpretation difficult. The fields were generally small and appeared in different shapes and sizes interspersed with forest vegetation. This contrasted with the much larger fields of the Midwest and West. Natural forest stands were extensive in areas, which gave rise to strong returns that tended to obscure features on the radar image. The delimitation of the extent of built-up areas was more tenuous due to the interspersion of wooded areas, idle land or open spaces and scattered development of new residential areas under a forest canopy. Similarly, the visibility of transportation and communication elements was affected by the considerable topographic relief and natural forest stands. Although Level I categories of land use could be delineated, more detailed analysis at Level II was difficult. When the resultant land use regions were compared with Anderson's 1970 and Austin's 1966 maps, the agreement was poor. It was concluded that *environmental modulation* reduced the level of land use detail visible in the Northeast compared to that of the Midwest and West, USA as a result of the much more fragmented and

Fig. 6.9
K-band imagery of Connecticut River Valley, USA – a sample region of woodland with forest land. Note the rugged topography, the complexity of the land use pattern and the lack of precise borders (*Courtesy*: F. M. Henderson; *Source*: Henderson 1979)

complex landscape elements of the Northeast (Fig. 6.9). For the radar imagery, forest vegetation was an important environmental factor contributing to the variation in the amount of land use information. In the Northwest, forest stands were found virtually everywhere, thus concealing drainage, relief, transportation and settlement features. In such a situation, larger-scale radar imagery is required.

Case Study 2:

Large-scale urban land use mapping

The application of radar imagery to urban land use mapping is possible with the improved spatial resolution (10×10 m) and the availability of multiple wavelength and multiple polarization systems. A major problem of the application to the urban environment is the regular, grid-like layout of many Western cities which gives rise to lines of strong returns caused by multiple reflections of the radar beam in the four cardinal directions, North, South, East and West (called cardinal effect). The basic identification of the built-up area is facilitated by this. Applications using small-scale real aperture K-band SLAR imagery at 1:250,000 scale indicated the possibility of identifying gross patterns of industrial, residential and open space uses in the city, but their boundaries could not be mapped in great detail (Lewis 1968; Moore 1969). For the latter purpose, a larger-scale radar imagery has to be used to improve the accuracy of the interpretation of urban land use. Bryan (1975) observed the importance of the manual approach in this application because the image characteristics of texture, tone, shape, pattern, association, etc. were valuable clues to interpretation. He conducted an experiment with multi-channel synthetic aperture radar imagery to see how well inexperienced human interpreters (685 undergraduate students) could do to interpret urban land use. The images were obtained over Melbourne, Florida, USA using a synthetic aperture radar system of operating at two wavelengths: X-band (3 cm, 9.3 GHz) and L-band (23 cm, 1.165 GHz) with cross- and like-polarizations for each band (Fig. 6.10). This resulted in four images (X-HH, X-HV, L-HH and L-HV) with a spatial resolution of 10 × 10 m at a scale of 1:50,000. Each interpreter had access to only one of these images. The USGS land use classification system was adopted. The results revealed that several types of urban land use could be identified at the 5 per cent level of confidence by the inexperienced student interpreters. These included residential and transportation uses. On the whole, the L-band images were found to be more difficult to interpret. By combining the results of all four images, most of the urban land use categories could be identified correctly at the 1 per cent level of confidence. Major sources of error in identification were the heavily wooded areas, such as residential areas, cemeteries, parks and recreational grounds, which tended to give identical returns regardless of the underlying land use because the radar returns were all from the tree tops. Also, these trees cast shadows which obscured features on ground. In radar imagery, shadows are no signal areas. To compensate for this loss of information, multiple looks of the same scene at different angles should be carried out. As the radar is also sensitive to surface roughness, the availability of multiple wavelength data allows the discrimination of variations in the surface roughness within the urban area. Thus, the X-band image exhibited more sensitivity to smaller surface roughness variations than the L-band as revealed by the detection of airport runway grass and the golf course in the X-band (Fig. 6.10). The conclusion was that, given good quality synthetic aperture radar image, detailed land use mapping of the urban area was possible using a manual approach (Fig. 6.11). The accuracy of urban land use mapping using radar imagery is still inferior to that achievable with aerial photography, because of the lack of correspondence between the form (as detected by the remote sensor) and the function (the human activities implied in the classification scheme) − a point mentioned before when we examined applications of aerial photography to land use mapping. The advantage of the radar sensors seems to lie in giving *additional* information to that of the aerial photography by virtue of its exploitation of the microwave portions of the electromagnetic spectrum. In a recent paper, Bryan (1983) presented several new approaches to the use of radar imagery for land use classification of urban areas, having taken into account of the importance of the depression angle of the sensor's radiation and the orientation of the imaged feature. As a result, the variation in tone was found to be less important than texture, context and shape in manual interpretation. A new approach was to use a 'squint' mode of acquiring the synthetic aperture radar data as a means to eliminate the strong cardinal effect caused by the orientation of streets in the urban area. This involved positioning the antenna to an angle off broadside of the aircraft. It is also possible to process the synthetic aperture radar data by a squint processing technique. The data were found to effectively reduce the

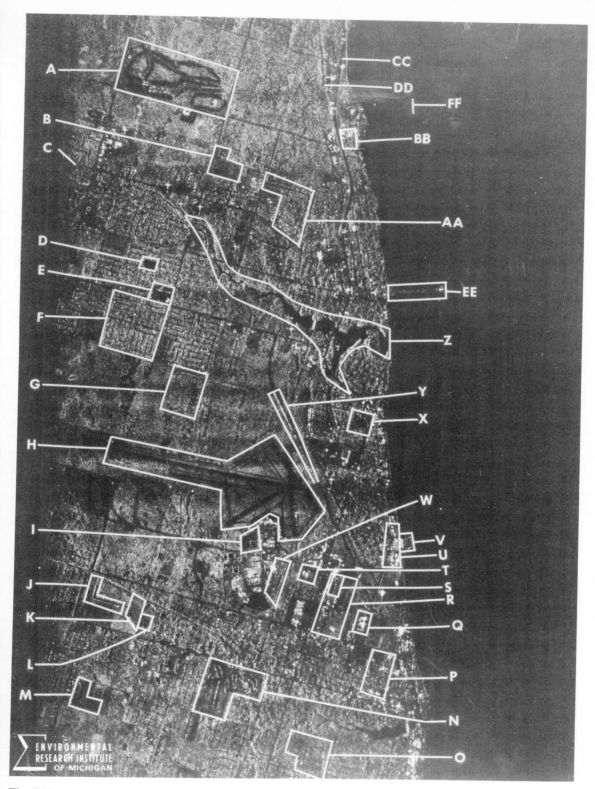

Fig. 6.10
X-band SLAR image (HH polarization) of Melbourne, Florida, USA. Interpretations of
the features marked can be found in Fig. 6.11 (*Courtesy*: M. L. Bryan; *Source*: Bryan
1975).

Fig. 6.11
Land use interpretation of
Florida, USA study area based
on SLAR imagery (see Fig. 6.10)
(*Source*: Bryan 1975)

high returns from streets and buildings, thus giving more textural and
tonal variations in the image (Fig. 6.12). Another new approach was the
acquisition of height information as a measure of surface roughness
of the urban area using synthetic aperture radar. Certainly, building
height information permits a better understanding of the morphology
of the urban environment as the author has demonstrated in his study
of the building heights of the city of Glasgow using aerial photography
(Lo 1970; 1971). Bryan (1983) drew attention to the need to understand the relationship between the traditional urban land use classification schemes and the type of ground truth required to properly
interpret the radar, i.e. an understanding of the morphology of the city
in relation to its function. All this of course is grounded on a good
understanding of the radar operation.

Applications using satellite imagery

The availability of satellite image data in different forms has attracted
a flood of applications in the land use and land cover mapping field.
The advantages of the satellite data are numerous. For the purpose
of land use mapping, the wide and repetitive coverage afforded by the
satellite platforms are specially important with regard to the cost-
effectiveness of collecting and the ease of updating the land use data.
Initially, the applications tended to concentrate on a manual approach
of extracting the land use data and producing maps. With the more general availability of software packages for the mainframe computer and

Fig. 6.12
L-band (HH) synthetic apera-
ture radar mosaics of Los
Angeles, California, USA
showing variations in an urban
scene due to 'squint' processing
at three different doppler frequences within the radar beam:
(a) 3.2° forward of the normal;
(b) normal processing at zero
doppler; (c) 3.2 aft of the
normal. Note the changes in the
patterns of high returns which
are the result of street and
building orientation relative to
radar look-direction (*Courtesy*:
M. L. Bryan; *Source*: Bryan
1983)

LOS ANGELES, CALIFORNIA

(a)

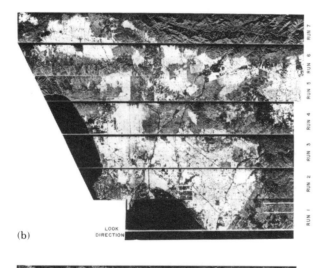

(b)

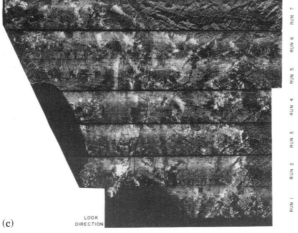

(c)

specialized computerbased image analysis systems, digital analysis of the satellite data, particularly those from Landsat, has become standard in recent years, although the role played by the manual approach continues to be important in developing countries. One also sees in recent years the availability of satellite radar data (Seasat) and thermal infrared data (HCMM, DMSP) which complement the multispectral data of Landsat. The procedures of manual and digital analyses of satellite data for land use mapping have been well established. The following case studies attempt to examine the problems and solutions when these procedures are applied under different environmental conditions.

Manual approach of analysis

This involved visual examining of the satellite image data which have been preprocessed to remove geometric distortions and radiometric errors. The original data are usually enlarged optically to permit the maximum detection of variations in tone, texture, pattern, shapes and other related image characteristics. Where multispectral data are available, as in the case of Landsat, false colour composites are normally employed for visual interpretation, although individual spectral band images of the scene will also be examined. Where a colour additive viewer is available, much more manipulation of the image data is possible through different combinations of spectral bands and colour filters. Land use interpretation accuracy here also depends on the land use classification scheme which has to be designed with regard to the cultural character of individual environments. The USGS scheme designed by Anderson *et al.* (1976) has provided a valuable guide for the design. The concept of different levels of land use or land cover details interpretable according to the different scales of the imagery, and hence, the spatial resolution quality is normally adopted.

Case Study 1:

Land use mapping using Landsat data

An example is the work of Welch, Pannell and Lo (1975) who made use of Landsat 1 MSS data acquired in 1972 and 1973 to produce a colour land-use map of Northeast China at a scale of 1:600,000 (Pl. 11, page 212). A false colour composite of each scene of the study area at a scale of 1:500,000 was projected onto a viewing screen from 70 mm band 5 (0.6–0.7 μm) images (nominal scale 1:3,369,000) using an I^{2}S colour additive viewer. A transparent overlay of 5 × 5 mm grids (equivalent to a ground cell of 2.5 × 2.5 km) was placed on top of the colour composite image for cell by cell interpretation of the land use or land cover. An eight-category land use and land cover scheme was employed as a result of the restrictions by the low spatial resolution of the Landsat data (IFOV = 80 m). These categories were:

1. Extensive field cropland
2. Intensive market garden

3. New agricultural land
4. Steppe, scrubland and grassland
5. Mixed deciduous and evergreen forest
6. Urban or built-up land
7. Bog and marshland
8. River, stream, lake and playa

These were equivalent to Level I of the USGS scheme. Apart from the macroscopic land use map, the urban land use within the cities of Shenyang and Anshan was also interpreted, using a much finer grid (2.5 × 2.5 mm or 1.25 × 1.25 km on ground). Because of the difficulty in obtaining ground truth data in China, old aerial photographs, maps and reports relating to the study area had to be consulted to improve the level of reference of the interpreter. Much greater difficulty was encountered in preparing the city maps at the scale of 1:100,000 because only tonal variation provided the major clue for the delineation of urban land use. It was only possible to delineate six categories of land use: (1) densely built-up city core area; (2) housing estate, administrative and military use; (3) industrial use; (4) park; (5) airfield; (6) intensive market garden. From these maps, one could obtain some insights into the urban structure of these cities. A methodology was developed that permitted mapping of areas where accurate up-to-date special data are unavailable or difficult to obtain.

Another example along this line of approach is the mapping of land use in the USSR using Landsat MSS imagery and collateral materials (Snyder 1981). Black-and-white positive prints of Landsat images at a scale of 1:500,000 were used. Most of these were band 5 prints but band 7 prints were also employed in resolving some ambiguities. The band 7 (0.8–1.1μm) prints being in the infrared spectrum helped to determine the extent of wetlands and water bodies. False colour composite prints were only used for verification within the detailed study area.

The *landscape* approach was adopted. In order to devise a 'landscape scheme', a study area of 41 scenes located in European USSR giving a combined surface area of about 1.4 million km² was selected. The band 5 Landsat prints were visually examined for landscape elements, namely, wetlands, woodland, cropland, grassland, built-up areas and linear features such as highways, railroads and waterways. The location and extent of these elements, and the degree and pattern of their intermixture provided the basis for classifying landscapes. From this study area, a landscape classification scheme consisting of 30 categories and subcategories of land use and land cover associations was devised (Table 6.5). The landscape approach delineated regions of uniform landscape elements. It was found that much of the USSR was dominated by a single element, such as woodland or cropland, i.e. in the uniform landscape category. The variations in tones of the *woodland* reflected differences in densities, cutting patterns and degrees of reforestation, with dense woodlands being the darkest areas in band 5 prints. By comparing images acquired during summer with those acquired during winter or early spring, one could easily differentiate deciduous woodlands from evergreen stands. *Grassland* gave an uneven appearance characterized by light-to-medium-grey tones. *Cropland* appeared dark-, light- or medium-toned, according to the crop type and field condition. Freshly ploughed fields appeared dark

Table 6.5
Classification scheme for landscapes of the USSR
(*Source*: Snyder 1981)

Uniform Landscapes

1. Water bodies

2. Barren land

3. Wetland
   A. Open
   B. Wooded

4. Woodland (excluding wooded wetland)
   A. Predominantly deciduous
   B. Mixed deciduous and evergreen
   C. Predominantly evergreen

5. Cut-over land

6. Grassland (including brushland)

7. Urban fringe and disturbed land

8. Urban and built-up land

9. Cropland

   A. More than 95 per cent of land surface cultivated
      (1) Average field size less than 150 ha
      (2) Average field size greater than 150 ha
      (3) Kuban type (irrespective of field size)
      (4) Irrigated (irrespective of field size)
   B. 85 per cent to 95 per cent of land surface cultivated
      (1) Average field size less than 150 ha
      (2) Average field size greater than 150 ha
   C. 65 per cent to 85 per cent of land surface cultivated
      (1) Average field size less than 150 ha
      (2) Average field size greater than 150 ha

Mixed Landscapes

10. Grassland/cropland
    A. Average size of cultivated fields less than 150 ha
    B. Average size of cultivated fields greater than 150 ha
11. Grassland/irrigated cropland
12. Grassland/cropland/woodland
    A. Average size of cultivated fields less than 150 ha
    B. Average size of cultivated fields greater than 150 ha
13. Grassland/woodland
    A. Woodland predominantly deciduous
    B. Woodland mixed or predominantly evergreen
14. Woodland/wetland/water
    A. Woodland predominantly deciduous
    B. Woodland mixed or predominantly evergreen
15. Woodland/cut-over land
    A. Woodland predominantly deciduous
    B. Woodland mixed or predominantly evergreen

while dry and bare earth was almost white. The intermixing of cropland and grassland was a characteristic of the land use pattern here, hence making it possible to differentiate the A, B and C types of croplands according to the extent of cultivation of the land surface (Table 6.5). The field size also showed variations in each type of cropland. *Urban areas* could be easily delineated because they were all sharply separated from the surrounding open space, but individual land uses within the urban area were usually impossible to visually identify. However, large parks and open areas, major streets and rail lines inside the urban area could be easily distinguished. The differentiation between older cities from the new cities was possible based on a difference in urban form, the former being more compact and irregular than the latter which exhibited distinct areas for residential and industrial uses. When no single landscape dominated, the landscape was delineated as a mixed one. The common types of mixed landscapes were: grassland and cropland; grassland and woodland; and grassland, cropland and woodland. In this way, a land use map at 1:1,200,000 could be obtained, and the same classification scheme could be adapted to map at scales as small as 1:2,500,000 without loss of detail.

The use of Landsat imagery for land use mapping has been particularly popular in the developing countries for rapid extraction of the needed data or to update old ones. Usually, multistage approach was adopted. This means the intercorrelation of all available data, ground truth data, aerial photography and satellite data. Landsat imagery provides an overall picture of the country which forms the basis for more detailed data collection using a combination of field work and aerial photography. Geiser *et al.* (1982) reported such a multistage land use mapping and change monitoring programme in Sri Lanka where the Swiss experts assisted to produce a new land use and land cover map at a scale of 1:100,000. Digitally enhanced colour composites enlarged to 1:100,000 from the best available MSS data were manually interpreted and corrected with aerial photography and field work. It was possible to interpret up to Level 3 of a land use classification with which the more important crop types, such as paddy, coconut, rubber, tea, were mapped. A similar application of Landsat for practical land use mapping with the manual approach was reported in Thailand (Omakupt 1978). North Thailand was mapped at a scale of 1:500,000 by interpreting 13 frames of black-and-white prints of Landsat–2 imagery of bands 5 and 7 at 1:500,000 and 1:250,000 scales. Diazochrome transparencies at 1:100,000 scale were also used. Planimetric details from the 1:250,000 scale topographic maps provided control for the land use mapping. The land use classification scheme consisted of five major categories and seven sub-categories as shown below:

    I. Urban land
   II. Agricultural land
      1. Horticultural land
      2. Perennial crop land
      3. Field crop land
      4. Paddy land
      5. Pasture and rangeland
  III. Forest land
      1. Dense forest
      2. Cut forest
  IV. Water body
   V. Miscellaneous land

This manual method was found to be the most cost-effective way of mapping land use. Research has confirmed that Landsat MSS data could produce generalized land use maps with percentages of accuracy ranging from 80 to 90.

Case Study 2:

Land use mapping using Skylab data

Skylab was a manned space station which orbited the earth from an altitude of 435 km at an orbital inclination of 50°. It carried an Earth Resources Experiment Package (EREP) which included three major imaging systems: the Multispectral Photographic Camera S-190A, the S-190B Earth Terrain Camera (ETC) and the Multispectral Scanner S-192 (NASA 1974). The photographic data acquired by S-190A and S-190B are of special interest because they are photographic data of high spatial resolution. The S-190B was a high resolution camera with a focal length of 460 mm and a film size of 125 mm, giving a nominal image scale of 1:950,000. Four types of films were used: Kodak 3443 colour IR, 3414 High Resolution Black-and-White, 3400 Panchromatic and SO-242 colour, giving an expected ground resolution (with object contrast 2:1 and exposure 1/500 second) of 38 m, 10 m, 24 m and 20 m respectively. An evaluation by Welch (1974) confirmed the high resolution of SO-242 colour to be 25 m. This type of colour photograph from S-190B was employed by Lins (1976) to map the land use of the city of Fairfax, Virginia, USA in greater detail. As indicated in the preceding case study, Landsat data's poor spatial resolution has limited their performance in urban land use mapping. The SO-242 colour photograph from S-190B was enlarged to a scale of 1:24,000 and an interpretation of the land use at Level III details based on the USGS scheme (Anderson *et al.* 1976) was carried out (Fig. 6.13). A

Fig. 6.13
Level III land use map of the city of Fairfax, Virginia, USA generated from a Skylab 3, S-190B colour photograph (*Source*: Lins 1976)

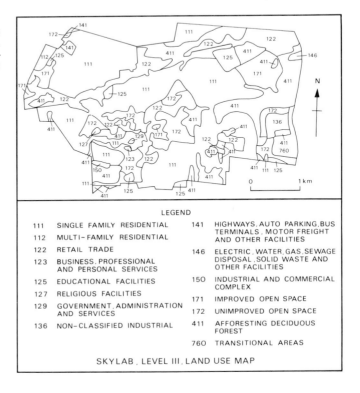

LEGEND

111 SINGLE FAMILY RESIDENTIAL	141 HIGHWAYS, AUTO PARKING, BUS TERMINALS, MOTOR FREIGHT AND OTHER FACILITIES
112 MULTI-FAMILY RESIDENTIAL	
122 RETAIL TRADE	146 ELECTRIC, WATER, GAS, SEWAGE DISPOSAL, SOLID WASTE AND OTHER FACILITIES
123 BUSINESS, PROFESSIONAL AND PERSONAL SERVICES	
125 EDUCATIONAL FACILITIES	150 INDUSTRIAL AND COMMERCIAL COMPLEX
127 RELIGIOUS FACILITIES	171 IMPROVED OPEN SPACE
129 GOVERNMENT, ADMINISTRATION AND SERVICES	172 UNIMPROVED OPEN SPACE
136 NON-CLASSIFIED INDUSTRIAL	411 AFFORESTING DECIDUOUS FOREST
	760 TRANSITIONAL AREAS

SKYLAB, LEVEL III, LAND USE MAP

similar interpretation was executed with high-altitude U2 photography of the same area obtained by a Wild RC-10 camera approximately two months later than that of the Skylab photography (Fig. 6.14). This land use map was field checked and corrected so that it was considered error-free for comparison with the land use map interpreted from Skylab S-190B photography. The two maps were compared first on a point by point basis with the sample points selected by a systematic aligned sampling strategy (Fig. 6.28; Berry and Baker 1968). Then, the area measurements of land use types generated by the two maps were compared. For the first comparison it was found that an accuracy of 83 per cent was achieved. The major problem here seemed to be the confusion between the tree-covered single-family residential areas and the forest land. This explained the great discrepancy in measured areas between these two uses. However, areas of commercial and services land use were readily detectable by S-190B colour photography because of the distinctive spectral and spatial characteristics of shopping areas. Less industrial area was interpreted because of the difficulty to discriminate it from the commercial and services category. On the whole, one could conclude that S-190B photography exhibited a great improvement over Landsat MSS data in the amount of land use details extractable. Welch (1982) observed that much higher spatial resolution was required for detailed studies of urban land use in Europe or Asia than were necessary in the USA or Canada because of the much higher building densities of the former. These higher building densities tended to reduce contrasts between urban land use or cover classes

Fig. 6.14
Level III land use map of the city of Fairfax, Virginia, USA generated from a RC-10 colour infrared photograph (*Source*: Lins 1976)

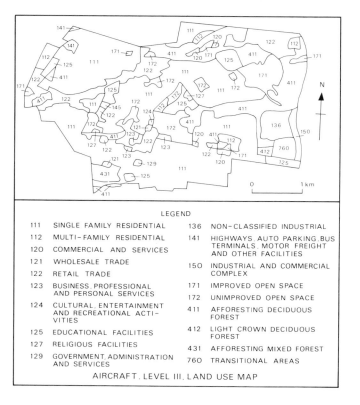

LEGEND

111	SINGLE FAMILY RESIDENTIAL	136	NON-CLASSIFIED INDUSTRIAL
112	MULTI-FAMILY RESIDENTIAL	141	HIGHWAYS, AUTO PARKING, BUS TERMINALS, MOTOR FREIGHT AND OTHER FACILITIES
120	COMMERCIAL AND SERVICES		
121	WHOLESALE TRADE	150	INDUSTRIAL AND COMMERCIAL COMPLEX
122	RETAIL TRADE		
123	BUSINESS, PROFESSIONAL AND PERSONAL SERVICES	171	IMPROVED OPEN SPACE
		172	UNIMPROVED OPEN SPACE
124	CULTURAL, ENTERTAINMENT AND RECREATIONAL ACTIVITIES	411	AFFORESTING DECIDUOUS FOREST
125	EDUCATIONAL FACILITIES	412	LIGHT CROWN DECIDUOUS FOREST
127	RELIGIOUS FACILITIES	431	AFFORESTING MIXED FOREST
129	GOVERNMENT, ADMINISTRATION AND SERVICES	760	TRANSITIONAL AREAS

AIRCRAFT, LEVEL III, LAND USE MAP

particularly in the developing countries of Asia. Therefore, spatial resolution requirements for urban studies would vary with region. For China and many other developing countries, an IFOV of 5 m or better (or about 12 m in ground resolution) was probably required for urban land use mapping. No civilian satellite imaging system available has yet met this requirement.

Case Study 3:

Land use mapping using Seasat SAR data

The use of satellite radar data, such as those from the now defunct Seasat, is aimed to provide *additional* or *complementary* information to that acquired by more conventional means. The spatial resolution of the satellite radar remote sensor is not high even with the synthetic aperture radar system. But in many circumstances radar is the only sensor that can yield the badly needed data. As an active system, the radar is weather independent and is more-or-less time-independent. Seasat, which was launched on 28 June 1978 into a nearly circular, 800 km orbit with an inclination angle of 108°, carried among other instruments a Synthetic-Aperture Radar (SAR) operating at the L-band (1.275 GHz, 23.5 cm) and HH polarization (Fu and Holt 1982). The SAR has a ground resolution cell of 25 × 25 m, and the imagery has a swath width of 100 km. The antenna was positioned on the spacecraft to image to one side only but, as a result of the orbital path of Seasat, each area could be viewed twice, with 'north passing' and 'south passing' imagery. The Seasat SAR operated for about 100 days until 10 October 1978 when a massive short circuit in the satellite electrical system ended the mission. Although Seasat was intended for monitoring the world's ocean surface, the SAR system also imaged land surfaces. An evaluation of the usefulness of the SAR data for land use mapping was conducted by Deane (1980) over the East Anglia area of the UK (Fig. 6.15). The choice of a flat area was to minimize the influence of relief on radar signature which was believed to be the important clue to land use identification. The field sizes in this area were large and a variety of land use types existed. The SAR images were blown up to 1:50,000 scale bromide prints from duplicate negatives of the optically-processed 1:700,000 scale imagery. A manual interpretation of these prints was carried out. On the basis of tone and texture, six units were identified (Table 6.6). They were correlated to the land use information on the 1:25,000 Ordnance Survey maps. It was seen that Unit 1 was generally identified as woodland, but the varying signatures of Units 2, 3 and 4 for the arable land needed to be checked in the field. In general, roads were not clearly defined on the imagery except where they were associated with an embankment, a line of trees or a boundary between two different radar signatures. Airfields could be detected by their uniform dark grey tone, and the Roman road was also identifiable probably because of its straightness. Field checks were carried out in two ground areas approximately 10 months after the acquisition of the imagery. The time discrepancy made checking difficult and questions had to be asked of the local farmers concerning the land use at the time of the imagery (10 August

Fig. 6.15
Seasat-A SAR image of the East Anglia study area in the UK (enlarged to 1:50,000 scale). The area mapped in Fig. 6.16 is framed (*Courtesy*: G. C. Deane)

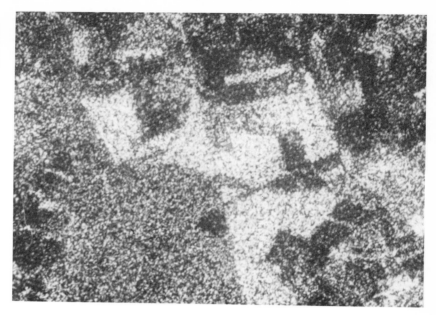

1978). Aerial photography and a land use map sheet from the Second Land Utilization Survey of Britain were also consulted. From these, the preliminary interpretation was corrected and finalized as shown in Table 6.7 and Fig. 6.16. The following observations were made:

Table 6.6
Seasat-A SAR: preliminary interpretation units in northern East Anglia UK (*Source*: Deane 1980)

1. Woodland in closed canopy was readily identified by its very bright returns but the distinction between coniferous and deciduous woodlands was impossible. Where there were scattered trees as in parkland and heathland, the signature was less bright.

Unit Number	Grey tone	Texture	Boundaries	Preliminary interpretation (*supported by information shown on OS maps*)
1	Very light	Coarse, strong	Distinct, straight	Woodland or hedges and trees at field boundaries
2	Dark	Coarse, very faint	Distinct, often coincide with field boundaries	Arable
3	Medium/dark	Coarse, faint	Clear, often at field boundaries	Arable
4	Medium	Coarse, moderately strong	Clear, often at field boundaries	Arable
5	Very light	None	Distinct (sometimes just bright spots)	Ploughed fields, harvested fields, buildings, barns, grain silos, towers, quarries and dense woodland
6	Light	Coarse, moderately strong	Distinct	Parkland/heathland

Fig. 6.16
Land use of the East Anglia study area in the UK, covering part of OS 1:25,000 scale sheet TL98, as interpreted from Seasat-A SAR imagery (see Fig. 6.15 and Table 6.7 for meanings of the code) (*Source*: Deane 1980)

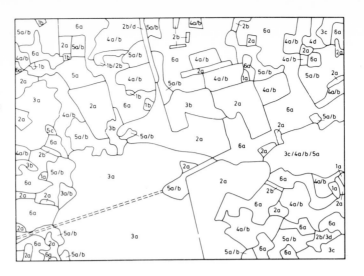

Table 6.7
Final classification of Seasat-A SAR signatures in northern East Anglia, UK (*Source*: Deane, 1980)

Description		Map Symbol
Very bright	recently harvested fields	1a
	large buildings (barns, aircraft hangars, etc.)	1b
	towns	1c
	dense clumps of mature trees	1d
	electricity pylons	1e
Bright	mature woodlands (coniferous and deciduous)	2a
	lines of trees	2b
	towns	2c
	embankments	2d
Very speckled	heathland, parkland	3a
	young woodland (without closed canopy)	3b
	pasture/meadow	3c
	villages	3d
Moderately speckled	root crops (beet, potato)	4a
	vegetables	4b
	orchards	4c
	meadow	4d
	agitated water surfaces	4e
	villages	4f
Faintly speckled	uneven grain crops	5a
	lower density vegetable and root crops	5b
	agitated water surfaces	5c
Dark, no speckle	unharvested mature grain crops	6a
	still water surfaces	6b
	airfield runways	6c

2. Arable land could be identified in broad categories only. Recently harvested fields normally gave very bright signatures also. The difference in tone was accounted for by the difference in crop types: grain crops, such as barley, wheat and rye, gave consistently dark signatures whereas vegetable and root crops produced a much more spec-

kled image. This depended on the degree of roughness of the fields imaged: an increased degree of roughness giving more speckled signature. The roughness of a field of crop could change with growth stages and environmental conditions, such as wind, rain, etc. at the time of imaging.

3. Villages could not be mapped accurately because they did not give strong radar responses or characteristic radar signatures. They were confused with the adjacent open woodland units. Major urban areas, however, could be delineated.

4. Roads and railways were identifiable only by association with other features and not by their own radar response.

From these observations, the conclusion was that with suitable field checking and supporting information from maps and aerial photographs, some land use categories could be mapped with reasonable reliability. The availability of a 'second look' image of the same area might aid to improve the accuracy of the land use map.

Despite the limitations of the Seasat SAR imagery for general land use mapping, its use for urban studies was also investigated by Henderson (1982). For this purpose, digitally processed imagery which is of superior quality to the optically processed imagery had to be used. The study area was the Denver, Colorado area in USA, an arid environment. A General Electric Image 100 interactive processing system was used to contrast stretch and enlarge the original image data to three scales: small (1:500,000), medium (1:131,000) and large (1:41,000). Black-and-white images at all these scales were visually interpreted. An automated machine/visual interpretation was also carried out for the image at the largest scale. The evaluation indicated that to obtain useful imagery at small and medium scales for interpretation, an average algorithm had to be used to reduce noise inherent in the data tapes; but for the large scale imagery, the raw data products were very satisfactory. It was found that Level I land cover classes of the USGS classification system could be delimited from the 1:500,000 images for the urban areas and the extent of the urban built-up area could be accurately defined, resulting in a very coarse land use map. With the use of the medium scale (1:131,000) images, more precise delimitation of the urban infrastructure was possible at Level II detail. Notably, newly developed residential areas and commercial-services/industrial core could be delineated. Open space was detectable but its use at times could only be inferred from its size, shape and location in relation to the urban area. Transportation network could not be identified. When the large scale (1:41,000) images were interpreted, more precise urban growth patterns could be measured. The growth on the urban fringe area was distinctly visible because of the sharp contrast between the recent residential development and the surrounding open space. Some small, isolated, low-density housing developments could also be detected. Open space could be identified but classification as to use remained difficult. More of the transportation elements could be seen at this scale. As a whole, an interpretation accuracy of 83.6 per cent was achieved. On the other hand, the automated machine/visual interpretation which involved density level slicing in discriminating land cover classes performed poorly, giving an overall accuracy of only 77.5 per cent. The conclusion was that digitally processed SAR imagery could be useful to urban studies at medium-to-large scale enlargements with a manual approach.

Digital approach of analysis

The digital approach of analysis involves the use of the computer and satellite data in digital form. The availability of the Landsat MSS data in Computer Compatible Tape (CCT) format encourages a lot of research in this direction. For land use and land cover mapping, there is a need to continuously update the data base to meet rapid changes in the environment caused by urbanization and industrialization. Simultaneous visual interpretation of numerous multispectral images becomes impractical especially over large areas. Only with the aid of computer techniques can we expect to carry out land use mapping with improved accuracy, that can be consistently repeated at regular time intervals. With the digital approach, the information content of the Landsat data can be fully utilized.

The digital approach is perhaps more appropriately used for land cover rather than land use mapping because it basically involves manipulation of the spectral data in number form to recognize or delineate patterns which have internal homogeneity. The procedures which originate from the method of 'pattern recognition' are in fact quite standardized (Swain 1978). First, the raw input data have to be refined before being employed for pattern recognition. There are digital preprocessing techniques to carry out: (1) *geometric adjustments* which aim to remove distortions caused by tilts of the spacecraft and the earth's rotation at the time of imaging, resulting in an image of desired scale or map projection; (2) *radiometric adjustments* which aim to change the grey level associated with each pixel of the original image according to a pre-specified scheme so that better contrast or feature definition can be achieved. The second stage is to subject the refined data for direct information extraction by the computer. This is the classification of the image for which a number of procedures are possible. The spectral data for each band of the image create a measurement space within which each pixel is located. A classification method has to be found to partition the measurement space approximately into *decision regions*, each corresponding to a specific land cover class (Fig. 6.17). There are many classification methods, the more commonly used ones being maximum likelihood, parallelepiped and minimum distance classification. It is possible to associate the pixel groups in advance to specific land cover classes before clasification. This requires field work or consultation of large scale aerial photography or reports to obtain the identities of these pixel groups. These data so collected are 'training' data. This is called the *supervised approach in classification*. The *unsupervised* approach does not make use of *a priori* information of the pixel groups at all and the pixels are allowed to be grouped naturally according to their spectral reflectance values. Such a procedure of grouping is called *cluster analysis* (Sheffield 1977). The cluster analysis therefore identifies spectrally distinct classes in the image data. These are then related to specific land cover types by doing field checks or consulting air photographs, maps and

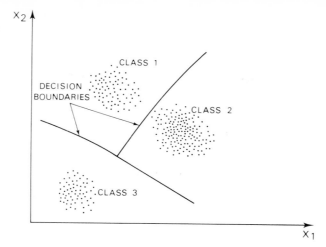

Fig. 6.17
Decision regions and boundaries
(*Source*: Swain 1978)

documents. The distinction between the supervised and unsupervised approaches tends to be blurred in recent years because a *hybrid* approach which makes use of an unsupervised classification to identify distinct spectral groups with their associated land cover classes and then a supervised classification by extrapolating over the whole area using the training statistics of these classes. The major problem of digital analysis lies in the choice of the correct method of classification for the problem in hand. However, the accuracy of the classification depends on the nature of the data: the pixel size in relation to the degree of homogeneity of the land cover type. The following case studies attempt to describe the digital approach in mapping the land use and/or land cover of an area and detecting any changes; they also emphasize the effort made to improve the accuracy of these maps.

Case Study 1:

Computer-assisted land use and land cover mapping using Landsat data

The author's experience in using the Office for Remote Sensing of Earth Resources (ORSER) pattern recognition package of the Pennsylvania State University in batch mode to map the land use and land cover of Hong Kong with Landsat digital data is given below to illustrate the procedures involved (Lo 1981). A full description of the ORSER system can be found in a paper by McMurtry *et al.* (1974). The first step involves examining the Landsat MSS images of the study area separately for each band in analogue form and to relate them with features identifiable on large-scale aerial photographs and maps. The study area is delimited on band 5 of the image with reference to the scanline number and element number at the four corners. The Landsat CCT digital data acquired from EROS Data Center in Sioux Falls, USA are radiometrically corrected but not corrected for variations in mirror-velocity, earth rotation or mapping projection (Taranik 1978). The delimited study area is extracted from the original data tape and stored in a disc file in a format which will facilitate further processing using a program called SUBSET (Fig. 6.18). To check that the study area is properly delimited, a program called NMAP is run to show the overall pattern of the data at the desired number of bands. This is equivalent to a brightness map. The Landsat data can be further geometrically corrected and scaled by running a program called SUBGM, which permits the matching of the Landsat data to a map of desired scale. This is followed by another run of NMAP which provides the basis for the selection of training data for a supervised classification. For Hong Kong, a land use classification scheme is determined with reference to existing land use maps and aerial photographs. For mapping at the scale of 1:100,000, eight land use and/or land cover classes are thought to be appropriate: (1) residential; (2) residential-commercial mixed; (3) reclaimed land; (4) cropland; (5) woodland; (6) grass and scrubland; (7) mangrove; (8) water features. Training areas for each land use and, or land cover class are marked on the NMAP print out and their boundaries read and input to the computer. A STAT program computes the means and standard deviations of the pixel values for each band in each training data set of land use and, or land cover class (Fig. 6.19). A variance-covariance matrix is also generated. A histogram per band is produced at the same time for each class of land use and, or land cover. This permits one to examine how good the training statistics are.

A normal distribution with a narrow spread of pixel values for the training set is acceptable as the signature for that class. This process can be repeated until the best training statistics are obtained. There is also a program called USTAT which can clean the training statistics by removing pixel values which deviate too greatly from the mean.

Fig. 6.18
Flow chart of ORSER image analysis programs

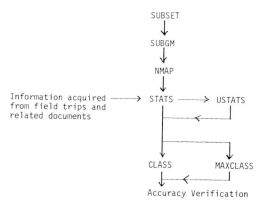

Functions of the programs:

SUBSET--extracts the study area from the CCT

SUBGM--performs cartographic correction and scaling

NMAP--produces brightness line-printer maps

STATS--computes training statistics

USTATS--cleans training statistics

CLASS--classifies by Euclidean distance method

MAXCLASS--classifies by maximum likelihood method

```
        STATISTICAL INFORMATION FOR CATEGORY  1, /CLEAR WATER   TRAINING AREA
        ===========================================================================

     NUMBER OF-OBSERVATIONS:         513

     UNNORMALIZED DATA USED

     CHANNELS USED :   1  2  3  4

    MEANS AND STANDARD DEVIATIONS FOR GIVEN CHANNELS
    ------------------------------------------------

  CHANNEL        1        2        3        4
  MEAN         14.28     8.98     3.89     0.07

  ST. DEV.      1.13     0.81     0.82     0.26


    VARIANCE-COVARIANCE MATRIX
    --------------------------

               1        2        3        4
     1        1.27
     2        0.48     0.66
     3        0.04    -0.02     0.68
     4       -0.01     0.00     0.03     0.07


    CORRELATION MATRIX FOR GIVEN CHANNELS
    -------------------------------------

               1        2        3        4
     1        1.00
     2        0.52     1.00
     3        0.04    -0.03     1.00
     4       -0.04     0.02     0.16     1.00
```

```
        HISTOGRAM FOR CHANNEL  1      0.50 - 0.60  MICRONS.
        ----------------------------------------------------
            EACH * REPRESENTS    3 OBSERVATION(S).

            GREY
FREQUENCY   LEVEL
---------   -----
      1      11  I1
      7      12  I**
    105      13  I*******************************
    238      14  I*******************************************************************************
     97      15  I********************************
     31      16  I**********
     31      17  I**********
      3      18  I*
```

```
        HISTOGRAM FOR CHANNEL  2      0.60 - 0.70  MICRONS.
        ----------------------------------------------------
            EACH * REPRESENTS    3 OBSERVATION(S).

            GREY
FREQUENCY   LEVEL
---------   -----
     23       7  I*******
     98       8  I*********************************
    268       9  I****************************************************************************************
    116      10  I**************************************
      8      11  I**
```

```
        HISTOGRAM FOR CHANNEL  3      0.70 - 0.80  MICRONS.
        ----------------------------------------------------
            EACH * REPRESENTS    3 OBSERVATION(S).

            GREY
FREQUENCY   LEVEL
---------   -----
     11       2  I***
    161       3  I************************************************
    225       4  I*****************************************************************************
    105       5  I*********************************
     11       6  I***
```

```
        HISTOGRAM FOR CHANNEL  4      0.80 - 1.10  MICRONS.
        ----------------------------------------------------
            EACH * REPRESENTS    5 OBSERVATION(S).

            GREY
FREQUENCY   LEVEL
---------   -----
    480       0  I**********************************************************************************************
     32       1  I******
      1       2  I1
```

Fig. 6.19
An example of the computer output from STAT program. This shows the statistical data for the training area of 'water'. Means and standard deviations for the spectral values for each band, the variance-covariance matrix and the correlation matrix among these bands are all computed. Histograms for each band are displayed. It is clear from this example that the spectral values for 'water' exhibit a narrow spread in all four bands, suggesting that this category of land use is well defined.

Finally, a DCLASS program is run to classify the image of the study area according to the training statistics using a minimum distance to means method. This assigns each pixel to a land use class based on the separation distance between the end points of the mean class vector and the observation (pixel) vector. A limiting distance around the end point of the mean vector of each class is specified by the user. A line printer map of the classification is produced by the computer (Fig. 6.20). It is necessary to experiment with various limiting values for different classes before a final land use and, or land cover map can be obtained. The land use classes on the line printer map can be transferred to the base map by direct tracing or by means of a Bausch and Lomb Zoom Transferscope (Fig. 6.21). The outline given above can be varied by using other programs available in the ORSER system to refine even further the results of classification and hence the accuracy of the land use map produced. There is a DCLUS program which will allow cluster analysis to be carried out in an unsupervised approach without initially resorting to the use of training data. The DCLUS program can be used to generate spectrally distinct natural groups which are then related to 'ground truth' to generate training statistics

Fig. 6.20
An example of the line printer map of the land use of Hong Kong as interpreted from Landsat data using the computer-assisted approach. (see Pl. 1, page 36)

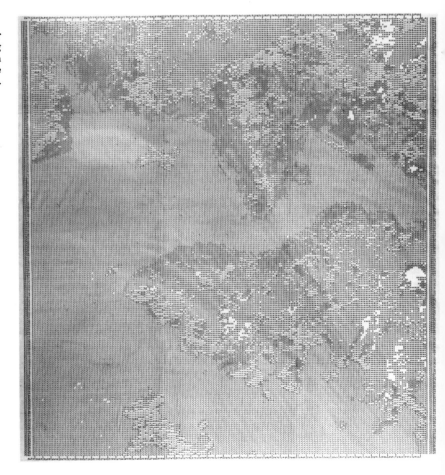

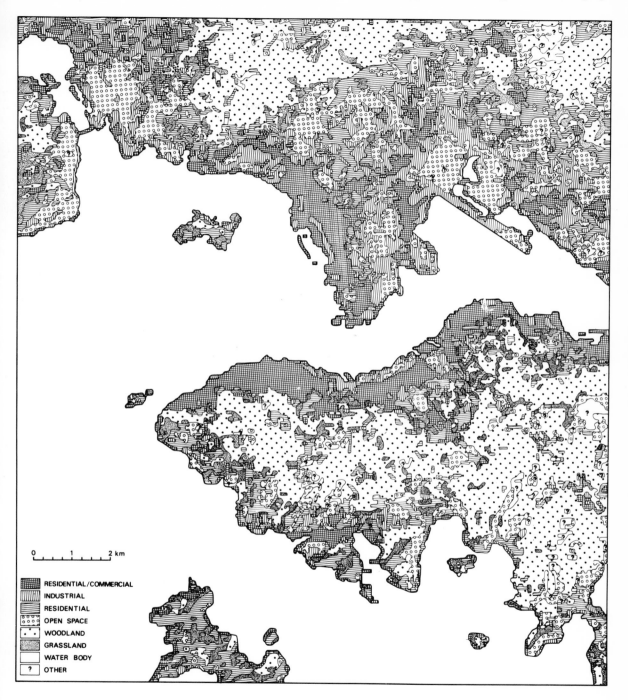

0   1   2 km

	RESIDENTIAL/COMMERCIAL
	INDUSTRIAL
	RESIDENTIAL
	OPEN SPACE
	WOODLAND
	GRASSLAND
	WATER BODY
?	OTHER

Fig. 6.21
A fair-drawn land use and/or
land cover map of Hong Kong
based on the computer print out
map shown in Fig. 6.20

for a supervised classification. This approach was very popular as
witnessed by the works of Welch, Lo and Pannell (1979) for Northeast
China; Gaydos and Newland (1978) for the Puget Sound region in the
USA; Todd, Mausel and Baumgardner (1973) for Milwaukee County,

Wisconsin, USA and Odenyo and Pettry (1977) for the city of Virginia Beach, Virginia, USA. The latter two applications made use of the LARSYS system developed by Purdue University.

There are also other classification methods available for use, notably the Gaussian maximum likelihood method which evaluates both the variance and correlation of the spectral response patterns of different classes when classifying an unknown pixel with the assumption that the distribution of the pixels forming the class training data is normally distributed (Strahler 1980). Another method is the parallelepiped classification which specifies the highest and lowest values in each band of the training set for classification criteria. Geometrically, these appear to draw a rectangular area around a group of pixels in a two-dimensional scatter diagram (Lillesand and Kiefer 1979). In a multi-dimensionsal case, these 'rectangular areas' are known as 'parallelepipeds'.

Many variations of the theme outlined in the case study above are possible. Much more data preprocessing before the actual classification is possible. This aims at reducing the 'noises' in the Landsat MSS data caused by atmospheric and topographic conditions of the imaged area. The method of *band ratioing*, i.e. dividing the reflectance value of a pixel recorded on one band by that of another, is quite usually adopted in land use and, or land cover studies (Sung and Miller 1977). This method has already been discussed in Ch. 5 in connection with the measure of vegetation amount. A ratio of band 7 (near infrared) and band 5 (green) is well correlated with the amount of green biomass within the scene, which might be usefully employed for vegetation classification (Maxwell 1976). Ratioing has also been used in geology for the extraction of lithologic information by band ratioing MSS band 4/band 5 (Ch. 4). Ratioing can also minimize the effects of terrain slopes which cause variations in overall reflected illumination and hence the grey level recorded (Sheffield 1977). In carrying out the mapping of land use and land cover of Taiwan at 1:25,000 scale, Sung and Miller (1977) incorporated 6 band ratios (5/4, 6/4, 7/4, 6/5, 7/5 and 7/6) together with the MSS bands 4, 5, 6 and 7 as input data (altogether 10 channels) for the classification by a supervised approach (stepwise discriminant analysis) into 17 land use and/or land cover classes. However, they found that the band ratios contributed little additional accuracy to the training set classifications, and MSS bands 5 and 7 alone could contribute to an overall training accuracy of 79 per cent as compared to 81 per cent for using all four MSS band ratios. In other words, the improvement in accuracy was incompatible to the time and effort spent.

Another data preprocessing technique used is *filtering*. This is aimed at smoothing the data by means of a digital filter. A commonly used type of filter is the moving average filter in two-dimension which replaces each pixel value with the average value computed within its 3×3 pixel neighbourhood (Maxwell 1976). Itten and Fasler (1979) advocated the use of the filtering technique after a land use classification has been completed to improve the accuracy of land use mapping. Digital filters were designed specifically to remove systematic and non-systematic errors inherent in class of land use. The following examples illustrate some of the filters designed and the purposes intended:

1. Generalization filters:

    (a) smoothing

1	1	1
1	1	1
1	1	1

    (b) to suppress central element

2	2	2
2	1	2
2	2	2

    (c) to enhance central element

1	1	1
1	3	1
1	1	1

2. Special filters:

    (a) to restore dropped lines

1	1	1
0	0	0
1	1	1

    (b) to fill gaps

1	1	1
1	0	1
1	1	1

    (c) to enhance upper left elements

2	2	1
2	1	1
1	1	1

Finally, *linear transformation of variables* is another commonly employed preprocessing technique. The purposes are to maximize the separation of classes along new axes and to satisfy the statistical assumption of independence of variables in multivariate analyses. Both canonical analysis and principal component analysis belong to this group of techniques, although only canonical analysis can maximize the separation between classes. In canonical analysis, the multivariate statistics associated with categories of interest within the data are used to find an orthogonal linear transformation C of the form:

$$y = Cx \qquad [6.1]$$

that defines a relatively small number of statistically independent discriminant axes (Merembeck and Turner 1980). Geometrically, this is equivalent to fitting a set of mutually orthogonal and hence uncorrelated axes to the image data by a rotation, translation and scaling such that the first new axis accounts for the greatest amount of variance with succeeding axes containing lesser and lesser amount (Fig. 6.22). This method has been used with good results in the land use mapping of Famenne area in Belgium at the scale of 1:25,000 using the ORSER system (Ceusters *et al.* 1978). Merembeck and Turner (1980) found that canonical transformation and maximum likelihood method of classification as applied to land cover mapping of a predominantly forested area in Northwestern Pennsylvania, USA using three different image data sets (4-channel Landsat data, 8-channel multi-temporal Landsat data and 22-channel Skylab data)

Fig. 6.22
Canonical transformation involving rotation (θ) and translation for categories A through E. Axes Y_1 and Y_2 respectively depict the first and second transformed axes (*Source*: Podwyscoki *et al.* 1977)

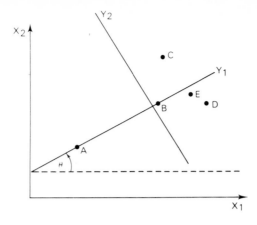

performed better than principal component transformation with maximum likelihood method or any other methods of classification.

Improvements of the classification methods have also been advocated by a number of workers. Strahler (1980) demonstrated the use of prior probabilities in maximum likelihood classification for land use and/or land cover mapping. The prior probabilities depend on the expected distribution of classes in a final classification map. In other words, certain classes of land use and/or land cover have higher probabilities of occurrence than others. The maximum likelihood decision rule can be modified to take into account these prior probabilities (or 'weights'). Strahler (1980) applied this method to map the land cover of the Doggett Creek area in Northern California using Landsat MSS data and digital terrain tape data derived from 1:250,000 USGS map. To obtain prior probability estimates for different cover types, 100 points were randomly selected from a grid covering the study area. At each of these points, the land cover type was determined either from aerial photographs or by actual field visit. From these samples of points, prior probabilities of occurrence of the land covers (mainly forests) related to different elevations and aspects could be obtained. On the whole, this method of classification was found to have improved the classification accuracy from 58 per cent for equal prior probabilities to 71 per cent for unequal prior probabilities. Jensen (1978) also drew attention to the improvement of the parallelepiped classification method by the layered classification method. The parallelepiped

Fig. 6.23
A common single stage classifier. All available channels of information are often used to classify each pixel to its most detailed class (*Source*: Jensen 1978)

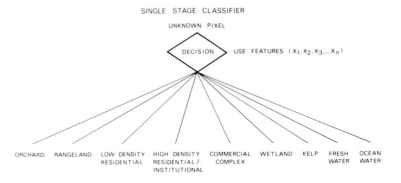

Fig. 6.24
A layered classification decision tree. Each decision uses a subset of the available channels which is optimum for discriminating between classes (*Source*: Jensen 1978)

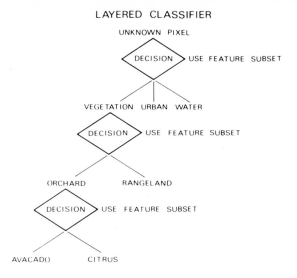

LAYERED CLASSIFIER

UNKNOWN PIXEL

DECISION — USE FEATURE SUBSET

VEGETATION URBAN WATER

DECISION — USE FEATURE SUBSET

ORCHARD   RANGELAND

DECISION — USE FEATURE SUBSET

AVACADO   CITRUS

method defines the upper and lower decision boundaries for each land use and/or land cover class as a single stage classifier (Fig. 6.23). The decision process is applied to a fixed set of features, i.e. usually the spectral measurements, to determine an 'unknown' pixel. On the other hand, the layered classification only makes use of a subset of the available channels which is optimum for discriminating between classes, thus eliminating noises from irrelevant channels (Fig. 6.24). This approach takes into account the fact that a pixel (56×79 m size in Landsat) contains two or more general land cover classes. A recognition algorithm trained to recognize only the homogeneous component classes will not be able to classify this pixel correctly. Jensen (1978) demonstrated the applicability of the layered concept to Goleta, California, a study area to evaluate the algorithms. From field work and/or aerial photographs, layers were developed for the study area as shown in Fig. 6.25. The results of the layered parallelepiped procedures in land use and/or land cover mapping revealed lower combined errors of commission and omission with an absolute overall

Fig. 6.25
Layered classification logic for the Goleta study area in California, USA, using physical composition and associated mixture layers (*Source*: Jensen 1978)

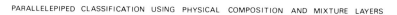

PARALLELEPIPED CLASSIFICATION USING PHYSICAL COMPOSITION AND MIXTURE LAYERS

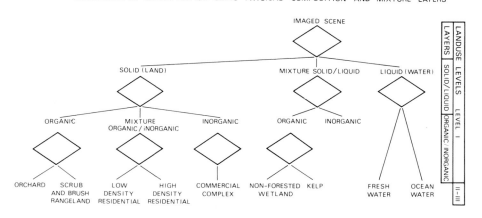

accuracy of 92 per cent for 10 categories as compared with 80 per cent for the traditional approach. It is worthy to note that all the methods of classification discussed above follow a supervised approach. In an evaluation of the performance of five different digital classification methods which included, among others, the maximum likelihood, minimum distance and the layered classifier for crop identification, Hixson *et al.* (1980) observed that the differences in classification accuracy among these methods were small, but a different training method used could significantly affect the accuracy. This confirmed the importance of training data, i.e. ground truth, in land use and land cover mapping.

So far, the supervised classification has been given more attention because of its ease to apply to a relatively smaller area with ground truth data availability. However, the unsupervised classification is useful in a large area with scanty ground truth data and it avoids the difficulty of training sites selection in the supervised approach, which can be very time consuming especially in areas of complex terrain. On the other hand, research undertaken has not been able to confirm the superiority of the unsupervised approach to the supervised one for mapping complex vegetation types using Landsat data (Townshend and Justice 1980).

The digital approach of land use and land cover mapping using Landsat data can surely produce reasonably accurate results. However, great care has to be exercised before meaningful land use and land cover data can be extracted. One should beware of the 'mixed pixel problem' caused by the 56 × 79 m pixel size. In an analysis conducted by Forster (1980) over the residential area of Sydney, Australia, a pixel could contain a mixture of features from houses, buildings, roads, concrete, trees, grass, water and soil. Its reflectance values was further affected by the surrounding pixels. The registration between the pixel and the corresponding ground truth data is not perfect so that the recorded response is not entirely due to the observed land cover characteristics. All these problems can adversely affect the accuracy of the computer-assisted classification based on spectral data alone. Only by reducing the pixel size (i.e. higher spatial resolution) and by improving the pointing accuracy of the satellite can we expect a significant breakthrough in land use and/or land cover mapping accuracy.

Case Study 2:

Land use change detection using Landsat data

One major advantage of the digital approach is the ease in updating the land use map of an area by comparing two Landsat scenes taken on two different dates, pixel by pixel. Three approaches are possible: (1) band ratioing; (2) transformation enhancement of multitemporal data; (3) post-classification comparison change detection.

Band ratioing is the simplest and quickest method of all in detecting land use change. This involves, first of all, accurate registration of images from two different dates. The registration process involves the use of ground control points and the solution of a series of linear equations. It is possible to resample the image data to a 50 m square pixel reformatted to the Universal Transverse Mercator (UTM) grid system. The intensities of reflected energy recorded in one band for the pixels of one Landsat scene are divided by the intensities in the same band for the other scene. In this way, the data are compared on a pixel by pixel basis. If the pixel value is nearly the same in each scene, the result of the division is approximately 1.0 which indicates no change in the land cover. Where change has occurred, the result may be either less than or greater than 1.0. By scaling the ratio values appropriately, one can depict areas of no change with grey and areas of greatest change are indicated by increasing tones towards either the black or the white end of the grey scale. For monitoring vegetation changes, band 7 seems to be appropriate whereas for water-level changes band 5 is normally used. Howarth and Wickware (1981) made use of these two bands to compare changes in vegetation and water-level in the Peace–Athabasca River delta region of Canada between August 1973 and August 1976 using Landsat CCTs (Figs. 6.26(a), (b).) A colour composite can be produced by combining the data from the two ratios to facilitate visual detection of changes. In other words, this approach only identifies the areas of change. Todd (1977) also made use of this approach in mapping the land use change in Atlanta, USA between 1972 and 1974. Only band 5 data of the two dates were used. From this a change map was produced. It was later confirmed that 91.4 per cent of the category of land use and land cover change were correctly identified as change by this method. It is necessary to ensure that the two images compared were obtained under more-or-less identical conditions at two different times before good results be achieved in this approach.

The transformation enhancement approach makes use of the preprocessing techniques mentioned previously. This is best demonstrated by the study of Byrne, Crapper and Mayo (1980). The Landsat data of the same areas recorded on two different dates were superimposed and treated as a single eight-dimension data array. A principal component analysis was carried out. By this method a new set of coordinate axes was fitted to the image data, choosing as the first new axis or component an orientation which would maximize the variance accounted for by that axis (Fig. 6.27). Subsequent axes (components) would account for successively smaller portions of the remaining variance. The principal component analysis differs from the canonical analysis in that it does not maximize the separation between classes. In this study, the principal component analysis decomposed the four-plus-four channels of correlated MSS data into eight orthogonal (uncorrelated) axes. The first- and second-order component images resulted were believed to represent unchanged land cover while the third and later component axes exhibited changes. Changes to be anticipated were of two types: (1) those that would extend over a substantial part of the scene, such as changes in atmospheric transmission and soil water status; (2) those that were restricted to parts of the scene, such as clearing of forests, construction of roads, erection of buildings. This approach was applied to the township of Batemans Bay, New South Wales in Australia. The image for each component was viewed visually and all images could also be superimposed to

Fig. 6.26
(a) Ratio of band 5, 1973 to band 5, 1976. Brightest tones indicate areas where changes in water-level dominated, especially around the border of the lake (b) Ratio of band 7, 1973 to band 7, 1976. Brightest tones indicate areas where changes in vegetation were dominant (*Courtesy*: P. J. Howarth; *Source*: Howarth and Wickware 1981)

(a)

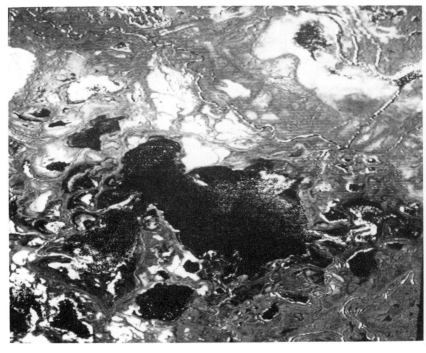

(b)

Fig. 6.27
Elliptical scatter pattern for a hypothetical data set consisting of multichannels X_1 and X_2. The principal component analysis creates a new set of coordinate axes (components) by a rotation and translation such that the first (Y_1) component accounts for most of the variability. The second axis (Y_2) is chosen orthogonal to the first. This concept can be extended to multidimensional space, with each succeeding component axis being oriented orthogonally to the earlier ones and accounting for less and less variation (*Source*: Podwysocki *et al.* 1977)

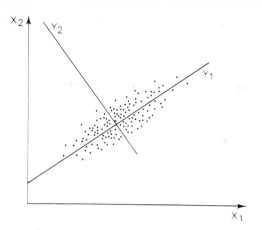

present an enhanced colour composite. It was observed that land cover changes could be identified by means of the black and white areas on the component-3 image which were readily correlated with known changes in land cover. These changes included acacia growth, tidal changes on mudflats, forest clearing, forest regeneration after a fire, urban development, new housing estates and gravel extraction activity.

The post-classification comparison change detection is the best approach of the three in that it can generate detailed statistics of change. The approach is simply to classify independently each Landsat scene by applying the standard procedures of classification described above. The results were overlayed and compared so that the areas and types of change could be identified. To ensure compatibility, categories used in the two classifications should be the same. The following data can be generated: (1) the percentage change in area between the two dates and (2) the change matrix to pinpoint what the changes have been for a specific class from one year to the next. A change map can be produced by comparing the two classifications on a pixel-by-pixel basis. By addition and subtraction of classes a scaled binary theme point can be produced. It is also possible to generate a conflict character assignment map from a line printer, in which the overlap and classification errors between the two Landsat scenes are shown by letters or symbols (Howarth and Wickware 1981). The post-classification comparison change detection has been found to be capable of handling more complex change situations and producing accurate results (Weismiller *et al.* 1977; Milazzo, Ellefsen and Schwarz 1977). A good example of the use of this approach can be found in the detection of coastal changes in Tokyo Bay of Japan for three different years: 1972, 1976 and 1980 (Hong and Iisaka 1982). Care had to be taken to register the three sets of Landsat data as exactly as possible using ground control points read from 1:50,000 scale topographic maps. The LARSY software package was employed for the land use classification. The resultant land use maps were displayed on colour monitor screen and were superimposed two by two to detect changes which were indicated by a distinctive colour. Quantitative estimation of the land use change could also be obtained.

Land use and land cover map accuracy assessment

With the use of non-photographic and satellite remote sensing data for land use and land cover mapping, a major concern is the accuracy of the maps produced. There are a number of sources of errors that can be identified, notably, the quality of the remotely sensed data, the method of interpretation and the area measuring practices. In general, there are two commonly employed methods to check the accuracy of the land use map. One method assesses the 'non-site specific accuracy' by comparing the measured areas of the individual land use categories from the interpretation against the corresponding areas acquired from a more reliable map. This produces a table of comparison as shown in Table 6.8 which is taken from Lins's work (1976) on land use mapping from Skylab S-190B photography discussed previously. The conclusion was that Skylab S-190B photography performed as well as the U-2 high altitude photography in extracting land use data. Another good example of accuracy assessment using this approach is the work by Fitzpatrick-Lins (1978) who compared the land use and land cover maps of the Central Atlantic Regional Ecological Test Site at three scales: 1:24,000, 1:100,000 and 1:250,000, which were produced from high altitude aerial photographs and Landsat multispectral imagery. She was able to specify the accuracy of these Level II maps at 85 per cent, 77 per cent and 73 per cent respectively based on direct observations of selected sample points from a low-flying aircraft.

Another method assesses the 'site-specific' accuracy by comparing two maps point by point. This produces an *error matrix* or *confusion matrix* as shown in Table 6.9, also from Lins's work (1976). The diagonal gives the correct interpretations. If we regard the high altitude

Table 6.8
Comparison of RC-10 and S-190B measurements by Level III land-use categories for the city of Fairfax, Virginia, USA (*Source*: Lins 1976)

Land Use Code Category	RC-10 Hectares/Per cent	S-190B Hectares /Per cent
111	780.4/50.22	657.3/42.30
112	32.2/ 2.07	26.4/ 1.70
120	7.4/ 0.48	– –
121	3.4/ 0.22	– –
122	166.3/10.70	166.4/10.71
123	3.6/ 0.23	1.4/ 0.09
124	0.8/ 0.05	– –
125	60.7/ 3.90	99.8/ 6.42
127	2.8/ 0.18	1.9/ 0.12
129	5.1/ 0.33	1.1/ 0.07
136	32.1/ 2.07	23.7/ 1.53
141	6.9/ 0.44	7.2/ 0.46
146	– –	4.8/ 0.51
150	12.1/ 0.78	4.5/ 0.29
171	54.4/ 3.50	39.6/ 2.55
172	128.2/ 8.25	152.9/ 9.84
411	215.2/13.85	339.5/21.84
412	5.6/ 0.36	– –
431	16.2/ 1.04	– –
760	20.6/ 1.33	27.5/ 1.77
	1554.0/100.00	1554.0/100.00

Table 6.9
Matrix listing the number of occurrences of S-190B interpreted land uses in RC-10 interpreted (field checked) land uses of the city of Fairfax, Virginia, USA sample. Diagonal is the axis of correct S-190B interpretations
(*Source*: Lins 1976)

Land Use Category Code	Skylab S-190B										
	111	112	122	125	126	150	171	172	411	412	760
111	34		1					1	2		
112			1								
122			7								
125					1			1			
136						1					
150											1
171				1				2			
172	1							3			
411								1	9		
412								1			
760											

U-2 photography in Table 6.9 as capable of giving the actual land use on the ground, the upper right half of the error matrix tabulates *errors of omission*, such as land use type 111 on the ground classified as 122, 172, and 411 in the Skylab photography. The lower left half of the error matrix, on the other hand, displays *errors of commission*, such as land use type 172 on the ground classified as 111 in the Skylab photography. In obtaining the error matrix, normally a sample of points is selected from the interpreted land use map for comparison with ground truth data in order to save the need to check every point. A stratified systematic unaligned sample design is normally adopted because it is believed to be the most bias-free method of sampling for areal data (Berry and Baker 1968; Fitzpatrick-Lins 1981; Rosenfield and Melley, 1980). A square grid is placed over the land use map. The side of dimension D of each square is decided by the following equation:

$$D = [(X_{max}-X_{min})\ (Y_{max}-Y_{min})/n]^{\frac{1}{2}} \qquad [6.2]$$

where *n* is the desired sample size, and X_{max}, X_{min}, Y_{max}, Y_{min} are the maximal and minimal X, Y coordinates of the area. The number of squares covering the map should exceed the minimum desired sample size. The origin for the unaligned pattern is selected by using a pair of random numbers to fix the coordinates of the upper left square (point A in Fig. 6.28). The *x* coordinate of A is then used with a new random *y* coordinate to locate B, a second *y* coordinate to locate E and so on across the top row of squares. By a similar process, the *y* coordinate of A is used in combination with random *x* coordinates to locate C and all successive points in the first column of squares. The random *x* coordinate of C and *y* coordinate of B are then used to locate D; the random coordinate of E and *y* coordinate of F to locate G, and so on until each square contains a sample point. To determine the sample size (*n*), a number of studies have been carried out. Hord and Brooner (1976) introduced a method of using the binomial distribution for sample size based on the confidence interval for the mean ($\bar{x}$). They also pointed out that for a classification to be correct it should be correct at all levels reported, i.e. Levels I, II, III, etc.

Fig. 6.28
Design of a stratified systematic
unaligned sample (*Source*: Berry
and Baker 1968)

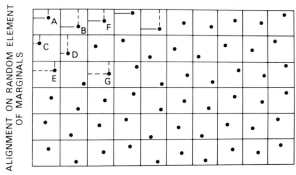

ALIGNMENT ON RANDOM ELEMENT OF MARGINALS

Van Genderen and Lock (1977) suggested the need to determine the optimum sample size, i.e. the minimum number of points to be checked in the field, for each category of land use. They chose the minimum sample size according to the probability of making incorrect interpretations at specified accuracy levels. They computed a table by the binomial expansion: $(p + q)^x$ where $q = 1 - p$, p is the probability of making no interpretation errors and x is the sample size. The ground truth samples should be collected using stratified random sampling. Thus, they observed that to satisfy the 85 per cent of accuracy stipulated in the USGS classification system (Anderson *et al.* 1976), the sample size for each land use category should be at least 20. This method has been adopted by Fitzpatrick-Lins (1981) in checking the accuracy of the land use and land cover map of Tampa, Florida, USA produced by the USGS at a scale of 1:250,000. Her study also experimented with the use of a computer-based automatic sample selection where digitized land use map data were available. A computer program was written to carry out a stratified systematic unaligned sampling followed by a random selection stratified by category to ensure that all categories were adequately sampled. This was found to be far superior to the manual approach.

Research continues in the development of statistical methods to evaluate the accuracy of the land use and land cover maps. Rosenfield (1982) drew attention to the analysis of variance method which could simultaneously compare the classification accuracy of different categories for two or more scales of land use and land cover mapping. The nonparametric statistical techniques were also demonstrated as applicable to the analysis of areal data in land use mapping where the probabilistic distributions for the data were uncertain (Rosenfield 1982). These parametric tests can help to decide whether there is any association between two maps and if so, how strong the association is.

Apart from the statistical methods developed to assess the accuracy of the land use and land cover maps produced from remotely sensed data, there were attempts to examine some specific causes of the errors. Henderson (1980) experimented with the effects of different manual interpretation techniques on land use mapping accuracy, using 1:30,000 colour infrared photography over a study area in southern USA. Nine different interpretation methods based on a grid cell matrix were used. It is interesting to note that best results were not always obtained with the smallest grid cell and less complex interpretation techniques often produced more accurate data. The type of environment studied seemed to have an impact on the accuracy. Crapper (1980) drew attention to the importance of shape and size of the polygons in affecting the accuracy of the area measurement by dot grid in a land use map prepared from Landsat data. Errors of commission or omission occur at the boundary region. These are accentuated if the areas measured are small and the boundaries highly contorted.

Conclusions

From the applications described above, the important role played by remote sensing in land use and land cover mapping and change detection has been firmly established. It is clearly the quickest and the best means to obtain reliable land use and land cover data. A great deal of research has focused on the use of computer-assisted techniques to extract spatial information of land use and land cover from satellite imagery while aerial photography has become the major source of ground-truth information for the analysis. More recently, the computer-assisted approach has been used even with aerial photography which is traditionally interpreted manually. Quirk and Scarpace (1982) demonstrated the use of digitized aerial photography for computer land cover mapping. A colour infrared aerial photograph (1:24,000 scale) was scanned by a microdensitometer through narrow band interference filters centered at 0.45, 0.55 and 0.65 μm, three separate times for conversion into a digital format. These digitized colour infrared data were then analysed by the normal supervised classification procedure with the computer. The result was compared with that from manual interpretation. It was found that the computer analysis of colour infrared photography could be used as a substitute for a manual photo-interpretation. The major advantage was the prospect of quick, accurate interpretation over large areas. On the other hand, it is important to determine how accurate these land use and land cover maps are. The development of the various statistical techniques to measure accuracy is a healthy sign and is conducive to a realistic appraisal of the impact of the various error sources, notably the interpretation technique, the image quality and the measurement practice, on the resultant product. It seems to be a paradox that, while on the one hand improvement in the spatial, spectral and temporal resolutions for the remotely sensed data, especially from the resources satellites, is being advocated, on the other there is already research to compress data for fear of the excessive storage requirements (Harris 1983). Surely, one has to decide in the near future the best balance to strike between the amount of input data and the desired accuracy of the output.

As for the non-photographic sensors, radar continues to play its reconnaissance role in both the developed and developing countries although some attempt to extract detailed information for the urban area has been made with some success using synthetic aperture radar data. The role of thermal infrared scanners in land use and land cover mapping is rather limited, with focus mainly on heat loss detection.

To conclude, land use and land cover mapping has shown an obvious trend towards an increasing degree of automation and reliance on satellite data. This has the advantage of the development of a computer-based geographic information system which permits the spatial integration of resources (land use and land cover) data with environmental (climatic and hydrologic) data.

References

Adeniyi, P. O. (1980) Land-use change analysis using sequential aerial photography and computer techniques, *Photogrammetric Engineering and Remote Sensing* **46**: 1447–64.

Aldrich, F. T. (1981) Land use data and their acquisition: In Lounsbury, J. L., Sommers, L. M. and Fernald, E. A. (eds) *Land Use: A Spatial Approach*. Kendall-Hunt Publishing Company: Dubuque, Iowa, USA and Toronto, Ontario, Canada, pp. 79–95.

Allan, J. A. (1980) Remote sensing in land and land use studies, *Geography* **65**: 35–43.

Anderson, J. R. (1970) Major land uses: In Gerlach, A. C. (ed) *National Atlas of the United States of America*. US Department of the Interior: Geological Survey, Washington, DC, pp. 157–159.

Anderson, J. R. (1971) Land use classification schemes, *Photogrammetric Engineering* **37**: 379–87.

Anderson, J. R., Hardy, E. E., Roach, J. T. and Witmer, R. E. (1976) *A Land Use and Land Cover Classification System for Use with Remote Sensor Data*, Geological Survey Professional Paper 964, US Government Printing Office: Washington, DC.

Artis, D. A. and Carnahan, W. H. (1982) Survey of emissivity variability in thermography of urban areas, *Remote Sensing of Environment* **12**: 313–29.

Austin, M. E. (1965) *Land Resource Regions and Major Land Resource Areas of the United States*. Agriculture Handbook No. 296: US Department of Agriculture, Washington, DC.

Avery, G. (1965) Measuring land use changes on USDA photographs, *Photogrammetric Engineering* **31**: 620–4.

Baker, R. D., DeSteiguer, J. E., Grant, D. E. and Newton, M. J. (1979) Land-use/land-cover mapping from aerial photographs, *Photogrammetric Engineering and Remote Sensing* **45**: 661–8.

Berry, B. J. L. and Baker, A. M. (1968) Geographic sampling: In Berry, B. J. L. and Marble, D. F. (eds) *Spatial Analysis: A Reader in Statistical Geography*. Prentice-Hall: Englewood Cliffs, New Jersey, pp. 91–100.

Brown, R. E. and Holz, R. K. (1976) Land-use classification utilizing infrared scanning imagery, *Photogrammetric Engineering and Remote Sensing* **42**: 1303–14.

Brown, R. J. (1979) Canadian experience with the use of aerial thermography: *Proceedings of Thermosense I*, American Society of Photogrammetry: Falls Church, Virginia, pp. 121–37.

Bryan, M. L. (1975) Interpretation of an urban scene using multi-channel radar imagery, *Remote Sensing of Environment* **4**: 49–66.

Bryan, M. L. (1983) Urban land use classification using synthetic aperture radar, *International Journal of Remote Sensing* **4**: 215–33.

Budge, T. K. and Inglis, M. H. (1978) Residential heat loss mapping of Farmington, New Mexico using airborne thermal scanning, *Proceedings of Fall Technical Meeting*. American Society of Photogrammetry, pp. 82–91.

Burley, T. M. (1961) Land use or land utilization? *Professional Geographer* **13**: 18–20.

Byrne, G. F., Crapper, P. I. and Mayo, K. K. (1980) Monitoring land-cover change by principal component analysis of multitemporal Landsat data, *Remote Sensing of Environment* **10**: 175–84.

Campbell, J. B. (1983) *Mapping the Land: Aerial Imagery for Land Use Information*. Resource Publications in Geography, Association of American Geographers: Washington, DC.

Ceusters, A., Gomber, R., Gulinck, H., Sougnez, H. and D'Hoore, J. (1978) Application of computer aided analysis of Landsat data to land use studies in Belgium: In Hildebrandt, G. and Boehnel, H. J. (eds) *Proceedings of the International Symposium on Remote Sensing for Observation and Inventory of Earth*

Resources and the Endangered Environment, Vol. III: Freiburg, Federal Republic of Germany, pp. 1497–514.

Clawson, M. (1966) Recent efforts to improve land use information, *Journal of American Statistical Association* **61**: 647–57.

Collins, W. G. and El-Beik, A. H. A. (1971) The acquisition of urban land use information from aerial photographs of the city of Leeds, *Photogrammetria* **27**: 71–92.

Crapper, P. F. (1980) Errors incurred in estimating an area of uniform land cover using Landsat, *Photogrammetric Engineering and Remote Sensing* **46**: 1295–301.

Deane, G. C. (1980) Preliminary evaluation of Seasat-1 SAR data for land-use mapping, *Geographical Journal* **146**: 408–18.

Fitzpatrick-Lins, K. (1978) An evaluation of errors in mapping land use changes for the Central Atlantic Regional Ecological Test Site, *Journal of Research US Geological Survey* **6**: 339–46.

Fitzpatrick-Lins, K. (1981) Comparison of sampling procedures and data analysis for a land-use and land-cover map, *Photogrammetric Engineering and Remote Sensing* **47**: 343–51.

Forster, B. (1980) Urban residential ground cover using Landsat digital data, *Photogrammetric Engineering and Remote Sensing* **46**: 547–58.

Fu, L. L. and Holt, B. (1982) *Seasat Views Oceans and Sea Ice with Synthetic-Aperture Radar*. National Aeronautics and Space Administration and Jet Propulsion Laboratory: California Institute of Technology, Pasadena, California, JPL Publication 81–120.

Gautam, N. C. (1976) Aerial photo-interpretation techniques for classifying urban land use, *Photogrammetric Engineering and Remote Sensing* **42**: 815–22.

Gaydos, L., and Newland, W. L. (1978) Inventory of land use and land cover of the Puget Sound region using Landsat digital data, *Journal of Research US Geological Survey* **6**: 807–14.

Geiser, U., Sommer, J., Haefner, H. and Itten, K. I. (1982) Multistage land use mapping and change monitoring in Sri Lanka: paper presented to *EARSel/ESA Symposium on Satellite Remote Sensing from Developing Countries*, 20–21 April, Austria.

Harris, R. (1983) A comparative study of the effects of data reduction on terrain cover mapping from Landsat, *International Journal of Remote Sensing* **4**: 723–8.

Henderson, F. M. (1975) Radar for small-scale land-use mapping, *Photogrammetric Engineering and Remote Sensing* **41**: 307–19.

Henderson, F. M. (1979) Land-use analysis of radar imagery, *Photogrammetric Engineering and Remote Sensing* **45**: 295–307.

Henderson, F. M. (1980) Effects of interpretation techniques of land-use mapping accuracy, *Photogrammetric Engineering and Remote Sensing* **46**: 359–67.

Henderson, F. M. (1982) An evaluation of Seasat SAR imagery for urban analysis, *Remote Sensing of Environment* **12**: 439–61.

Hixson, M., Scholz, D., Fuchs, N. and Akiyama, T. (1980) Evaluation of several schemes for classification of remotely sensed data, *Photogrammetric Engineering and Remote Sensing* **46**: 1547–53.

Hong, J. K. and Iisaka, J. (1982) Coastal environment change analysis by Landsat MSS data, *Remote Sensing of Environment* **12**: 107–16.

Hord, R. M. and Brooner, W. (1976) Land-use map accuracy criteria, *Photogrammetric Engineering and Remote Sensing* **42**: 671–7.

Howarth, P. J. and Wickware, G. M. (1981) Procedures for change detection using Landsat digital data, *International Journal of Remote Sensing* **2**: 227–91.

Itten, K. I. and Fasler, F. (1979) Thematic adaptive spatial filtering of Landsat landuse classification results, *Proceedings from the 13th International Symposium on Remote Sensing of Environment*. Ann Arbor: Michigan, pp. 1035–42.

Jensen, J. R. (1978) Digital land cover mapping using layered classification logic and physical composition attributes, *The American Cartographer* **5**: 121–32.

Lewis, A. J. (1968) *Evaluation of Multiple Polarized Radar Imagery for the Detection of Selected Cultural Features*. USGS-Interagency Report: NASA-130, Washington, DC, 56 pp. (NTIS #N69-28151).

Lillesand, T. M. and Kiefer, R. W. (1979) *Remote Sensing and Image Interpretation*. John Wiley: New York; Chichester; Brisbane; Toronto; pp. 461–85.

Lins, Jr., H. F. (1976) Land-use mapping from Skylab S-190B photography, *Photogrammetric Engineering and Remote Sensing* **42**: 301–7.

Lo, C. P. (1970) Determining and presenting the third dimension of a city centre: a photogrammetric approach, *Photogrammetric Record* **6**: 625–39.

Lo, C. P. (1971) A typological classification of buildings in the city centre of Glasgow from aerial photographs, *Photogrammetria* **27**: 135–57.

Lo, C. P. (1979) Surveys of squatter settlements with sequential aerial photography – a case study in Hong Kong, *Photogrammetria* **35**: 45–63.

Lo, C. P. (1981) Land use mapping of Hong Kong from Landsat images: an evaluation, *International Journal of Remote Sensing* **2**: 231–52.

McMurtry, G. J., Borden, F. Y., Weeden, H. A. and Petersen, G. W. (1974) The Penn State ORSER system for processing and analyzing ERTS data, Freden, S. C., Mercanti, E. P. and Becker, M. A. (eds), *Third Earth Resources Technology Satellite-1 Symposium: Vol. I: Technical Presentations Section B*, National Aeronautics and Space Administration, Washington, DC, pp. 1805–22.

Maxwell, E. M. (1976) Multivariate system analysis of multispectral imagery, *Photogrammetric Engineering and Remote Sensing* **42**: 1173–86.

Merembeck, B. J. and Turner, B. J. (1980) Directed canonical analysis and the performer of classifier under its associated linear transformation, *IEEE Transactions on Geoscience and Remote Sensing* GE-**18**: 190–6.

Milazzo, V. A., Ellefsen, R. A. and Schwarz, D. W. (1977) Updating land use and land cover maps, *Remote Sensing of the Electromagnetic Spectrum* **4**: 103–16.

Moore, E. G. (1969) *Side-Looking Radar in Urban Research: A Case Study.* USGS-Interagency Report: NASA-138, Washington, DC, 24 pp. (NTIS #N69-16108).

National Aeronautics and Space Administration (1974) *Skylab Earth Resources Data Catalog*: Lyndon B. Johnson Space Center, Houston, Texas.

Odenyo, V. A. O. and Pettry, D. E. (1977) Land-use mapping by machine processing of Landsat-1 data, *Photogrammetric Engineering and Remote Sensing* **43**: 515–23.

Omakupt, M. (1978) Land use inventory of North Thailand using Landsat imagery. *Proceedings of the Twelfth International Symposium on Remote Sensing of Environment*: Environmental Research Institute of Michigan, Ann Arbor, Michigan, pp. 2297–2306.

Paine, D. P. (1981) *Aerial Photography and Image Interpretation for Resource Management*, John Wiley, New York, Chichester, Brisbane, Toronto, p. 95.

Place, J. L. (1977) The land use and land cover map and data program of the US Geological Survey: an overview, *Remote Sensing of the Electromagnetic Spectrum* **4**: 1–9.

Podwysocki, M. H., Gunther, F. J. and Blodget, H. W. (1977) Discrimination of rock and soil types by digital analysis of Landsat data. Goddard Space Flight Center: Greenbelt, Maryland, Preprint No. X-923-77-17.

Pollé, V. F. L. (1974) Landuse surveys in city centres, *ITC Journal*, **No. 4**, 490–505.

Quirk, B. K. and Scarpace, F. L. (1982) A comparison between aerial photography and Landsat for computer land-cover mapping, *Photogrammetric Engineering and Remote Sensing* **48**: 235–40.

Rosenfield, G. H. (1981) Analysis of variance of thematic mapping experiment data, *Photogrammetric Engineering and Remote Sensing* **47**: 1685–92.

Rosenfield, G. H. (1982) The analysis of areal data in thematic mapping experiences, *Photogrammetric Engineering and Remote Sensing* **48**: 1455–62.

Rosenfield, G. H. and Melley, M. L. (1980) Applications of statistics to thematic mapping, *Photogrammetric Engineering and Remote Sensing* **46**: 1287–94.

Schmid, R. (1971) Land use mapping in hill country, Eastern Nepal: interpretation of air photographs in compilation of agricultural statistics: In Boesch, H. (ed.) *Contributions to Land Use Survey Methods*. Geographical Publications Ltd: Berkhamsted, Herts, UK.

Sheffield, C. (1977) Digital enhancement and analysis techniques in processing of earth resources data: In Clough, D. J. and Morley, L. D. (eds) *Earth Observation Systems for Resource Management and Environment Control*. Plenum Press: New York; London, pp. 379–401.

Snyder, D. R. (1981) Using Landsat imagery to study Soviet land use, *Journal of Geography* **80**: 217–23.

Stamp, L. D. (1960) *Applied Geography*. Penguin: Harmondsworth, UK.

Strahler, A. H. (1980) The use of prior probabilities in maximum likelihood classification of remotely sensed data, *Remote Sensing of Environment* **10**: 135–63.

Sung, Q. C. and Miller, L. D. (1977) *Land Use/Land Cover Mapping (1:25,000) of Taiwan, Republic of China by Automated Multispectral Interpretations of Landsat Imagery*. Goddard Space Flight Center: Greenbelt, Maryland, Preprint No. X-923-77-210.

Swain, P. H. (1978) Fundamentals of pattern recognition in remote sensing: In Swain, P. H. and Davis, S. M. (eds) *Remote Sensing: the Quantitative Approach.* McGraw-Hill: New York, pp. 136–187.

Taranik, J. V. (1978) *Characteristics of the Landsat Multispectral Data System.* Open-File Report 78–187: United States Geological Survey, Sioux Falls, South Dakota.

Todd, W. J. (1977) Urban and regional land use change detected by using Landsat data, *Journal of Research US Geological Survey* 5: 529–34.

Todd, W., Mausel, P. and Baumgardner, M. F. (1973) An Analysis of Milwaukee County Land Use of ERTS Data. LARS Information Note 022773: Laboratory for Applications of Remote Sensing, Purdue University, West Lafayette, Indiana.

Townshend, J. R. G. and Justice, C. O. (1980) Unsupervised classification of MSS Landsat data for mapping spatially complex vegetation, *International Journal of Remote Sensing* 1: 105–20.

Van Genderen, J. L. and Lock, B. F. (1977) Testing land-use map accuracy, *Photogrammetric Engineering and Remote Sensing* 43: 1135–7.

Weismiller, R. A., Kristof, S. J., Scholz, D. K., Anuta, P. E. and Momin, S. A. (1977) Change detection in coastal zone environments, *Photogrammetric Engineering and Remote Sensing* 43: 1533–9.

Welch, R. (1974) Skylab-2 photo evaluation, *Photogrammetric Engineering* 40: 1221–4.

Welch, R. (1982) Spatial resolution requirements for urban studies, *International Journal of Remote Sensing* 3: 139–46.

Welch, R., Lo, H. C. and Pannell, C. W. (1979) Mapping China's new agricultural lands, *Photogrammetric Engineering and Remote Sensing* 45: 1211–28.

Welch, R., Pannell, C. W. and Lo, C. P. (1975) *Land Use in Northeast China 1973 – A View from Landsat-1.* Map supplement No. 19, *Annals of the Association of American Geographers*, Vol. 65, No. 4, December.

Chapter 7 The hydrosphere

Characteristics of the hydrosphere

The hydrosphere is a liquid water realm characterized by free flows in response to unequal stresses. The characteristics of *fluids* are the same as gases, exhibiting a layered arrangement according to density in order to maintain an equilibrium at rest. As a result, one expects a close interaction between the atmosphere and the hydrosphere at their interface. On the other hand, in nature water can also occur as a solid (ice) when the temperature is below its freezing point. The ocean is the dominant member of the hydrosphere and in fact occupies some 70 per cent of the earth's surface so that the marine environment has a great importance to our life. The ocean affects our climate and is in turn affected by the atmosphere, acting both as a heat reservoir for storing, distributing and releasing solar energy, and as the major source of atmospheric moisture. It collects the detritus of man and nature and is also an important source of petroleum. Its currents are used to dispense sewage and wastes. It is also an important source of food as well as a delightful area of recreation for people. Yet it has remained the least studied part of our earth due to the difficulty in obtaining accurate horizontal and vertical measurements at frequent intervals. In recent years, developments in space remote sensing techniques have provided a better understanding of the hydrosphere. This chapter focuses most of the discussions on the applications of remote sensing to the ocean although rivers and lakes will also be given some attention.

It is customary to study four major characteristics of the ocean, namely, *physical, biological, geological* and *chemical*. The physical characteristics examine the properties of sea water such as temperature and density, oceanic circulation such as surface currents, deep water circulation, waves and tides and water mass descriptions. Biological characteristics examine the living organisms in the ocean, the fish and plants. Geological characteristics study the properties of different types of sediments deposited in the bottom of the oceans, seas and their coastal areas. Finally, chemical characteristics deal with the composition of the sea water, such as its salinity, acidity, basicity and degree of pollution. Case studies are given below according to such a classification because the role played by each type of remote sensor in each class is less distinctive. In other words, it is common to see the same remote sensor being employed to extract physical, biological geological and chemical characteristics of the marine environment.

Applications to study the physical characteristics

Sea surface temperature (SST)

Water has great capacity for, and conductance of, heat. The temperature of sea water is a vital parameter in understanding the role of the ocean as a heat reservoir. Changes in temperature cause variation in the properties of the sea water and the life it supports. The temperature of the sea water can be measured with a standard mercury thermometer from a water sample collected at the rear of the ship. This method applies only to surface water. With the use of remote sensing method, the 'skin temperature' (about 20 μm) at the sea-air interface where evaporation occurs is measured (Walsh 1976). Normally, this makes use of the thermal infrared radiometer operating within the 8–15 μm range which provides the greatest transparency to radiation emitted from the sea surface. The thermal infrared measurement is adversely affected by the presence of water vapour and clouds in the atmosphere. The use of a low-flying aircraft platform (below 300 m) can minimize these two adverse effects and give results of useful accuracy. However, the area covered by the aircraft is still restricted in comparison with that from a space platform. There are a large number of space satellites equipped with thermal infrared sensors designed for sea surface temperature measurement, such as Nimbus 4 through 7, NOAA 1 through 6, SMS/GOES 1 through 5, Skylab, Tiros-N, DMSP, Seasat and Meteosat (Apel 1983). Much progress has been made towards improving the accuracy of the sea surface temperatures from infrared measurements from space.

The thermal infrared sensor in the 8–14 μm window region is normally used for sea surface temperature measurement. It makes use of Planck's radiation law in the form given in equations [1.7] and [1.8] of Ch. 1 (Singh and Warren 1983). However, the chief absorber of radiation in the 8–14 μm window is water vapour in the atmosphere, which is a highly variable quantity. The spectral radiance emitted by the sea is therefore partially absorbed by the atmosphere before reaching the satellite sensor. One needs to have a knowledge of the atmospheric transmittance τ in order to compensate for this loss. To simplify matters, it is a common practice to assume that the optical properties of the atmosphere vary only in the vertical direction, i.e. the distribution of atmospheric constituents in the horizontal direction is assumed to be homogeneous. This knowledge of τ has to be obtained from simultaneous *in situ* measurement in the field or from temperature and humidity sounding from satellites. The sea surface temperature, Ts, can then be derived from the received signal I_{sat} as follows:

$$I_{sat} = I_{sea} + I_{air} + I_{ref} \qquad [7.1]$$

$$I_{sea} = \tau \, \varepsilon \, \beta_s$$

where I_{sea} and I_{air} are respectively received portions of the upward spectral radiance from the sea surface and the atmosphere; I_{ref} is the downward atmospheric spectral reflectance, reflected upward at the sea surface and attenuated in its path to the sensor; ε is the surface emissivity (about 0.99) and β_s is the spectral radiance of a black body at temperature T_s. The values of τ, I_{air} and I_{ref} can be found ideally from the measured vertical profile of pressure, temperature and relative humidity. Since I_{sat} is known, equation [7.1] can be solved for β_s from which T_s may be found by inverting Planck's radiative transfer equation (see also equation [3.3] in Ch. 3; Sidran 1980). It is important to note that accurate results can only be obtained in cloud-free sky. An example of this approach with some modifications is given by Platt and Troup (1973) who compared the sea-surface temperatures acquired from a satellite, (Nimbus 4 THIR radiometer) operating in the 10.5–12.5 μm region with those from an aircraft using an infrared radiometer operating in 10–12 μm region. This was carried out off-shore from Cairns, North Queensland, Australia. The satellite's altitude was 1,112 km whereas the aircraft flew at heights between 2–3 km. The aircraft also made measurements of atmospheric pressure, temperature and humidity. The amount of water vapour absorption was computed from the following equation:

$$k = k_o + \alpha e^2 \tag{7.2}$$

where k was the total water vapour absorption coefficient, k_o and α were constants for a given wavelength and e was water vapour pressure. After reduction it was found that the Nimbus THIR read an SST lower than the aircraft radiometer. However, by taking into account the effect of slant paths on the Nimbus THIR and the spectral mismatch between the Nimbus THIR filter and the aircraft radiometer filter, the values for the two constants of k and α in equation [7.2] had to be increased, resulting in a discrepancy of 0.3° C between the two sets of SST measurements.

If the pressure, temperature and relative humidity profile is unknown, the accuracy of the corrected SSTs can still be obtained by the multispectral method. This involves making radiance measurements in two infrared wavelengths, one between 7 μm and 9 μm and the other between 10 μm and 12 μm (Anding and Kauth 1970). Since the same physical processes are responsible for absorption in both of these wave bands, the effect in one should be proportional to that in the other. Therefore, one measurement in each wavelength interval could be used to eliminate the atmospheric effect. Simply, this involves in practice the solution of a linear equation in the form:

$$T_s = e_o + e_1 \, T_b \, (v_1) + e_2 \, T_b \, (v_2) \tag{7.3}$$

where $T_b \, (v_1)$ and $T_b \, (v_2)$ are the brightness temperatures of spectral channels v_1 and v_2 respectively, and e_o, e_1 and e_2 are parameters to be determined by the multivariate regression analysis. In this way, the effects of atmospheric absorption and reflection are automatically included. Prabhakara, Dalu and Kunde (1974) made use of such an approach in estimating SST from cloud-free Nimbus 3 and 4 Iris spectral data in the 11–13 μm region over the Arabian Sea by using three spectral channels: 775–831, 831–887 and 887–960 cm^{-1} (expressed in terms of wave numbers). The estimated SST was found to exhibit an accuracy of about 1°C. A further improvement and simplification of the computation procedure of this method was reported by Sidran (1980) using the following equations:

$$T_s = 2.2\, T_b\,(3) - 1.2\, T_b\,(1) \qquad\qquad [7.4]$$

$$T_s = 4.7\, T_b\,(3) - 3.7\, T_b\,(2) \qquad\qquad [7.5]$$

$$T_s = 3.976\, T_b\,(3) - 2.329\, T_b\,(2) - 0.6495\, T_b\,(1) \qquad [7.6]$$

where $T_b\,(1)$, $T_b\,(2)$ and $T_b\,(3)$ are brightness temperature in infrared channels centered at 803, 859, and 923.5 cm^{-1} respectively. In fact, only two infrared channels (i.e. either equation [7.4] or equation [7.5] need be used although the use of all three channels (i.e. equation [7.6]) can bring about a slight improvement in accuracy.

A straightforward application of thermal infrared imagery to map SST patterns without involving atmospheric correction problems was demonstrated by Zheng and Klemas (1982) who combined the remote sensing data with the synchronous or quasi-synchronous measurements *in situ* by vessels in the study area (Yellow Sea and East China

Fig. 7.1
Winter sea surface temperature (SST) pattern in the Yellow Sea and East China Sea derived from Pl. 12. The numbers (in °C) on the isotherms are determined by quasi-synchronous temperature measurements *in situ* or by interpolation (*Source*: Zheng and Klemas 1982)

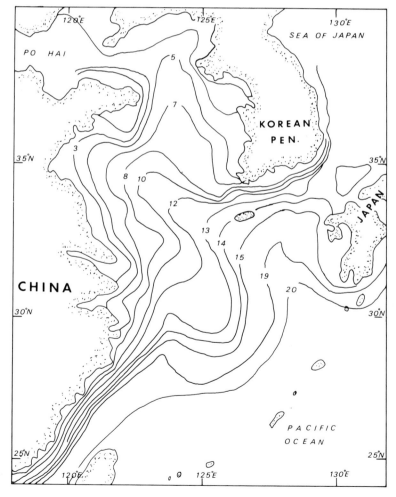

Sea). Thermal infrared imagery taken from a variety of satellites, namely, DMSP, NOAA 5, 6, GMS-1 and TIROS-N over the last 10 years was used. The GMS-1 thermal infrared image of the study area on 7 January 1979 was processed as a false colour density slicing transparency from which the SST isotherms were delineated based on equal colour areas (Pl. 12, page 364, and Fig. 7.1). Temperature measurements taken at the same time in the study area and their interpolation values were used to calibrate these SST isotherms. The SST isotherm pattern map was then compared with similar maps produced from other thermal infrared images. This permitted the detection of a 'sandwich' structure, i.e. two warm tongues sandwiching one cold tongue, and the existence of three strong coastal fronts in winter. The pattern was found to persist from the end of November to the end of April of the next year. From the SST isotherm pattern, the winter surface current pattern could also be deduced (Fig. 7.2). For application of this type, it is essential to rectify the thermal infrared image properly in order to relate to *in situ* measurements.

Fig. 7.2
Winter surface current pattern in the Yellow Sea and East China Sea determined from the SST patterns in Fig. 7.1 (*Source:* Zheng and Klemas 1982)

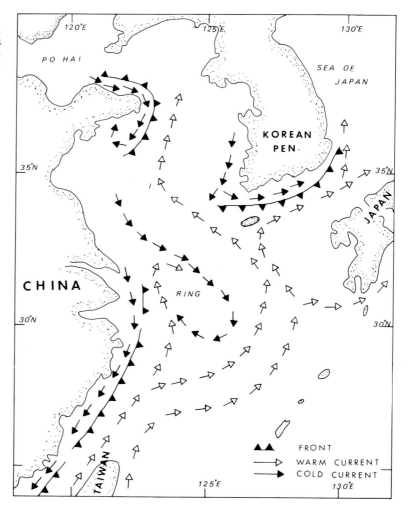

Another approach to measure SST by a remote sensor from an air or space platform is the microwave radiometer which is less accurate than the thermal infrared instruments. This is because microwave radiation is much weaker than naturally emitted radiation in the visible or infrared portions of the electromagnetic spectrum (Krishen 1975). The microwave sensor also has a poorer spatial resolution. However, the microwave radiometer can be used even in cloudy conditions to measure brightness temperatures from which sea surface temperatures can be estimated. The brightness temperature (T_b) measured from a sensor platform at height h can be expressed as:

$$T_b (\theta, h) = [E (\theta) T_g + r(\theta) T_s] \tau(h) + \int_o^h Ta (h) \frac{\delta\tau}{\delta h} \cdot dh \quad [7.7]$$

where T_s is the sky brightness temperature as seen from the surface; T_a (h) is the brightness temperature of the atmosphere at height h; τ is the transmittance of the atmosphere; T_g is the ground temperature; $E(\theta)$ is the emissivity of the surface; $r(\theta)$ is the reflectivity and θ is the viewing angle of the radiometer. This equation applies to a microwave radiometer with a narrow beam-width antenna giving a pencil beam pattern with negligible side lobes. By removing the effects of atmospheric and ground physical temperature, the corrected mean brightness temperatures of the ground scene (sea surface) can be obtained.

The Scanning Multichannel Microwave Radiometer (SMMR) which was carried aboard the Seasat and Nimbus-7 satellites was designed specifically for the determination of global sea surface temperatures and wind speeds from space under different atmospheric conditions (Njoku *et al.* 1980). The SMMR measured earth radiation at 6.6, 10.7, 18.0, 21.0 and 37.0 GHz with vertical and horizontal polarizations. The SMMR data have to be corrected for radiometric calibration and antenna pattern errors first, before the retrieval of the geophysical parameters. The retrieval technique involves the use of an ensemble of realistic sea surface temperatures, wind speeds, atmospheric temperature profiles, water vapour profiles and cloud models to calculate brightness temperatures by means of the surface emission and radiative transfer model explained above (Wilheit and Chang 1980). Based on the relationship between the brightness temperatures (T_b) and the geophysical parameters, a multiple linear regression model is developed. The geophysical parameters are retrieved from the brightness temperatures on regularly spaced grid cells. The characteristic spatial resolutions are approximately 150 km for sea surface temperature, 85 km for wind speed and 55 km for the atmospheric water parameters. An accuracy test of the Seasat SMMR data for sea surface temperature and wind speed was carried out by Njoku and Hofer (1981) in the North Pacific. By comparing the SMMR-derived sea surface temperatures and winds with surface truth point measurements with expendable bathythermographs deployed from ships and aircraft (about 0.2° accuracy) and from buoys, they found that root means square errors of about 1.5°C for SST and 2.5 m/s for wind speed were obtained. However, in this type of comparison, one should note the difficulty of comparing point surface measurements with 'footprint' (i.e. instantaneous field of view) averaged SMMR measurements.

Detection of sea ice

The detection of sea ice in the polar regions has great significance in climatic research and navigation activity. Sea ice occupies vast areas of the world's oceans, about 10 per cent in the Northern Hemisphere and 13 per cent in the Southern Hemisphere. The sea ice floats and is subject to large seasonal and annual variations in extent and composition. It insulates the ocean from the much colder air overlying it, thereby reducing the heat flux from the ocean. Its high albedo insulates the ocean from solar heating and hence the sea ice cover acts as a heat sink to cool the air and to reduce the turbulence in the atmosphere. The movement of the ice towards the open ocean (ice-floes) helps to cool down the ocean water in the high latitudes (Gudmandsen 1983). Winds and currents play a major role in the motion of the ice as well as the melting and forming of the ice. The ice concentration changes all the time, thus causing change in the heat flux. From the navigation point of view, the dynamic nature of the ice floes represents a major hazard and an understanding of their movement patterns is essential.

The early use of remote sensing to detect sea ice was confined to aerial photography in the 1940s and 1950s as reported by Teleki (1958). Time-lapse aerial photography was also used to measure ice movement (Gerson 1958). But, although aerial photographs permit easy distinction between sea ice and open water the manual interpretation of the aerial photography proved to be time-consuming. The launch of the TIROS-1 satellite in 1960 opened up a new age of remote sensing which provided a global view of the earth. The low-resolution satellite images obtained in the visible wavebands were used in conjunction with aerial photography to provide the much needed information on sea ice (Wark *et al.* 1962). This was followed by the development of computer-assisted methods to detect sea ice (Gerson and Rosenfeld 1975). This employed the tonal value of each pixel from the NOAA 2 image data acquired in the visible channel (0.6–0.7 μm) as a basis for discrimination. Because of the low spatial resolution of the data (0.8 km), it attempted to distinguish only four major features: clouds, ice or snow, water and land. The major problem encountered was the difficulty in distinguishing between clouds and ice or snow. With the launch of the earth resources satellite (Landsat 1) in July 1972, multispectral scanner data have also been found useful for sea ice detection. With the greatly improved spatial resolution and the added advantage of spectral resolution of the Landsat data, Barnes and Bowley (1974) have been able to demonstrate the possibility of detecting ice features that are as small as 80–100 m in width and identifying ice types by the combined use of the visible (band 4) and near infrared (band 7) imagery in the Arctic area (Beaufort Sea). The open water which exists as *polynyas* (large cracks) and *leads* (small cracks) within the dynamically active ice canopy could be detected and with the use of sequential Landsat images over the study area, ice movement, ice deformation and significant changes in the leads or polynyas could be mapped. Campbell (1976a, b, c, d and e) supplied further examples of the use of sequential Landsat images to track ice-floes, to study ice lead and polynya dynamics and to examine morphological change of sea ice in the Beaufort Sea area of the Arctic from the Arctic Ice Dynamics Joint Experiment (AIDJEX) project.

The use of the Landsat data for sea ice detection is limited by the prevailing inclement weather conditions of the polar regions to essentially cloud-free days. Microwave remote sensing can overcome this limitation and is a more suitable approach to sea ice detection. Both the passive and active microwave sensors can be used. For passive microwave remote sensing, the radiometer measures the ground's natural thermal emission in the microwave region. The intensity incident on the radiometer antenna is a measure of brightness temperature of the area modified by the atmosphere as expressed in equation [7.7]. Simply, at a given wavelength and viewing angle, the intensity of radiation is directly proportional to emissivity and absolute temperature. Since the relative dielectric constant of sea ice is normally very much lower than that of sea water, an ice-covered ocean has a higher emissivity than an ice-free ocean (Gray 1981). Wilheit *et al.* (1972) who made use of eight passive microwave wavelengths ranging from 0.510–2.81 cm found that the emissivity for ice ranged from 0.8–1.0 while that for sea water varied from 0.4–0.5. Thus, sea ice and open water can be easily distinguished. In addition, the emissivities for sea ice of different age and thickness are also different for microwave frequencies greater than approximately 18 GHz (Gloersen *et al.* 1973; Gray 1981). Thus, differentiating different ice types and conditions is possible with passive microwave sensors (Fig. 7.3). The ice types are generally classified according to

Fig. 7.3
Spectra of the three radiometrically distinct ice types and of open water (*Source*: Gloersen *et al.* 1981)

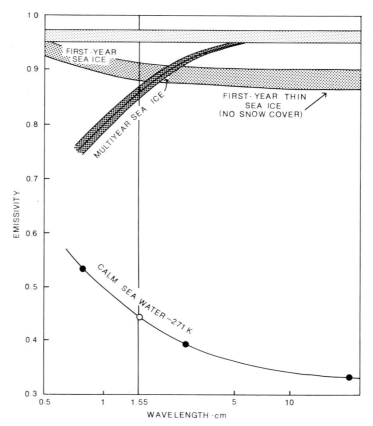

age and, or thickness into: (1) new ice which is composed of weakly frozen ice crystals and has a salinity greater than 5 parts per thousand; (2) thin ice (nilas) which is 10 cm thick; (3) young ice which is 10–30 cm thick; (4) first-year ice which has a thickness ranging from 30 cm–2 m, a smooth surface and a salinity of between 5–10 parts per thousand; (5) old ice including second year and multi–year ice which has a salinity below 1 part per thousand (Ketchum and Lohanick 1980). In general, the multi-year ice was found to exhibit a lower emissivity (brightness temperature) but higher backscatter than the first-year ice. The lowest brightness temperatures in the sea-ice environment occurred in areas of open water and in very recently formed thin ice which still had a wet surface. As the ice becomes older and thicker, it becomes radiometrically cooler (i.e. lower brightness temperatures). However, as the physical properties of the ice change by erosion, deformation and other ageing processes, the radiometric temperatures become more complex as revealed by many older multi-year ice-floes. All these observations have been confirmed by Ketchum and Lohanick (1980) in their study of the Chuckchi and Beaufort Seas in the Arctic Ocean, using 33.6 GHz (Ka band) passive microwave imagery (Fig. 7.4) acquired from altitudes ranging from 300 to 2,700 m. They concluded that the stresses in ice might have caused the lowering of the brightness temperatures because greater volume scattering had occurred. Passive microwave remote sensing has been employed in satellites, notably the Nimbus-7 Scanning Multichannel Microwave Radiometer (SMMR) to prepare global maps of sea ice concentration, age and surface temperature of the polar regions (Gloersen *et al.* 1981a). However, the results from the first SMMR observations were acceptable only after further development algorithms and calibration procedures (Gloersen *et al.* 1981b).

Fig. 7.4
Photographic (top) and passive microwave image (33 GHz) (bottom) of sea ice in the Arctic Ocean. Different ice types and features are displayed. My = Multi-year ice; FY = first-year ice (*Courtesy*: R. D. Ketchum, Jr., *Source*: Ketchum and Lohanick 1980)

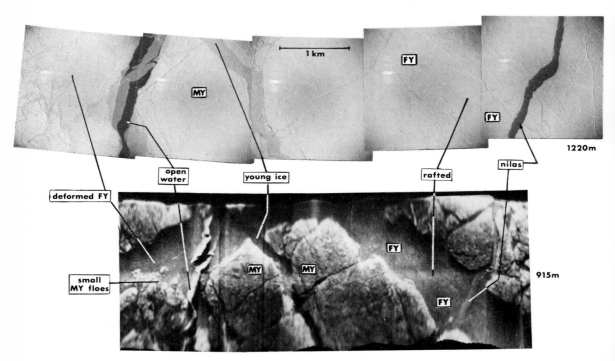

Another approach to detect sea ice is to use active microwave remote sensing in the form of side-looking radars (SLAR). Both real and synthetic aperture radar have been used. Johnson and Farmer (1971) have made use of SLAR to identify the various ice conditions in the North American Arctic region, using a synthetic aperture radar operating at the frequency of 16.5 GHz (Ku band) flown at 2,438 m. They visually interpreted the radar imagery and the results were compared with ground truth data collected during the *SS Manhattan* cruise of the Northwest Passage. They have been able to develop some general guidelines to relate ice type and radar image characteristics as keys to interpretation based on the degree of roughness of surface characteristics (Table 7.1). They concluded that SLAR was a promising tool for

Table 7.1
Key to the interpretation of ice conditions from SLAR imagery (*Source*: Johnson and Farmer 1971)

Ice Type	Image Characteristics
1. Young ice	Even tone dark grey. May have bright straight lines indicating ridging.
2. First-year ice	Dark grey to black and smooth. May have light straight lines indicating ridging.
3. Second-year ice	Even tone grey. May have ridging. Ridges more jagged than in first-year ice. Relatively smooth topography yields even tone.
4. Multi-year ice	Mottled tones of grey probably caused by high weathered ridges and interconnecting melt holes. Old multi-year drainage channels can sometimes be traced.
5. Fast ice	Depend on age. Contact line of shore and shore fast ice delineation very good by evident rise and change of topography.
6. Open water	Dark area of no radar return.
7. Fracture	Linear no-show area that may be parallelled by bright linear return of fractured edge of ice-floe. If radar beam is perpendicular to long axis of crack, far edge will yield a bright return.
8. Crack	If one side is higher than the other, long linear return from cracked edge of floe occurs. Linear return may appear dark if not cracked completely.
9. Polynya	Irregular shaped no-show water area. If quality of imagery is good, floe edges will be delineated.
10. Lead	No-show water area. May contain brash ice and nilas or young ice.
11. Pressure ridges	Bright striations criss-crossing haphazardly on darker-tone background. May be new ridges with sharp bright returns or weathered ridges of subdued bright returns.
12. Hummocks	Give nearly the same radar returns as pressure ridges but possess less linearity.
13. Concentration	The ratio of sea surface actually covered by ice to the entire area of sea surface imaged can be easily estimated due to large area coverage of SLAR.

identifying the following characteristics of sea ice: ice concentration, ice-floe size and number, and water openings. Dunbar (1975) made use of a real aperture SLAR system operating at X-band from the altitude of 610 m over the Nares Strait and the Arctic Ocean. The ranges used were 25 and 50 km which correspond to scales of 1:250,000 and 1:500,000 respectively. Visual interpretation suggested that all ice of a given age category did not look alike either to the eye or to the SLAR. This is particularly true of multi-year ice which shows very different surface characteristics according to its age and history of development. But even first-year ice which is generally either very smooth or very rough can vary considerably from place to place. The guidelines given in Table 7.1 should be used with great care because of the over-simplification. The study of Dunbar (1975) pointed out the possibility of distinguishing three types of ice in the Nares Strait: (1) old ice-floes of a great variety of sizes; (2) very roughly deformed first-year ice; (3) first-year or young smooth ice (Fig. 7.5). Old ice (marked A, B and C on Fig. 7.5) exhibited a wide range of grey tones and appeared brighter or darker than the surrounding younger ice. First-year ice was less varied in topography than old ice being either very smooth (undeformed) or broken by sharply angular ridges which appear bright on the SLAR image (D, E, and F on Fig. 7.5). Younger forms of ice, however, were hard to interpret without the aid of ground truth. The discrimination between smooth ice and water also presented problems because they all exhibited similar black tones. This handicapped the estimation of ice concentration. Dynamic features such as lead orientation, shear zones and drift patterns could be easily detected. A shear zone running as a straight line from Cape Lieber to Cape Beechey could be seen on Fig. 7.5. Drift patterns appeared in the form of eddies in the radar image. Ketchum and Tooma (1973) drew attention to the use of multifrequency side-looking radar to extract sea-ice data over the sea ice fields north of Alaska. The multifrequency radar system had four independent transmitters operating at X band (8.910 GHz), C band (4.455 GHz), L band (1.228 GHz) and P band (0.428 GHz) and a four-channel receiver. This research indicated that the shorter wavelength X band radar appeared to have the greatest potential for sea ice study when more definitive information, such as mapping, distributions of stages of ice development and fracture pattern analysis, was required. The L band radar was best used to map the extent of surface topography while the P band radar could only detect the most prominent features, such as large fractures and floes only. The combined use of all these multifrequency radar images helped to detect the different sea ice conditions better. All these research projects have shown that the parameters used by the radar system during data acquisition, notably frequency, depression angle, resolution, polarization and the output dynamic range, have a great impact on the accuracy of interpretation. In this connection, one should recall the variations in tone and resolution according to range in radar images (see Ch. 6). Because the same ice type can have different degrees of surface roughness, the interpretation of ice type using radar image brightness alone is extremely difficult. Only with the experience of the interpreter who makes use of shape, texture and other image characteristics can the ice types and conditions be correctly interpreted in many cases.

Fig. 7.5
SLAR image of sea ice in Hall Basin, Arctic Ocean taken on 13 January 1973. Altitude 610 m and range 25 km (*Source:* Dunbar 1975)

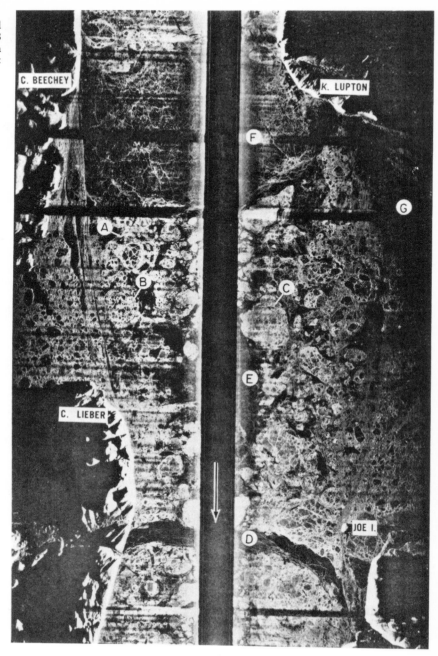

An active microwave remote sensing system – the synthetic aperture radar (SAR) – has been carried on board the short-lived Seasat which has acquired quite a lot of imagery over the Beaufort Sea area. From the optically or digitally processed SAR images, the main morphological features of sea ice can be clearly detected, such as leads, polynyas, pressure ridges, shorefast ice, multi-year ice-floes and mixtures of the multi-year and first-year floes (Fig. 7.6; Fu and Holt

Fig. 7.6
Seasat SAR image of Banks Island in the Beaufort Sea area, Arctic Ocean, taken on 3 October 1978 showing structural variations within pack ice. Multiple polynyas widened up from narrow leads between the cemented first-year and multi–year floes. Their dark appearance indicates either open water or recently frozen ice (*Courtesy*: Jet Propulsion Laboratory; *Source*: Fu and Holt 1982)

1982). From sequential images, it is possible to measure ice motion with an accuracy of ±150 m using ground control points located in each end of an image 1000 km long (Gudmandsen 1983). On the whole, Seasat SAR has proved to be capable of giving a wealth of information useful for sea ice studies at micro-, meso- and macro-scales under all weather conditions and during both day and night. However, there are still a lot of problems to be solved in the interpretation of the different sea ice types and conditions, especially in the summer season. Research conducted with the aid of aircraft and field work is required to resolve many of these problems. Finally, it should be pointed out that the combined use of passive and active microwave systems can help to uniquely classify sea ice conditions and offers much promise (Campbell *et al.* 1978; Livingstone *et al.* 1981).

Oceanic circulation

The circulation of waters in the oceans involves both horizontal movements of water across the surface and vertical-horizontal movements within the body of ocean water itself. These movements depend on the difference in the density of water which is in turn influenced by its temperature and salinity. The horizontal movements of water at the surface also depend on the wind flows which are controlled by atmospheric pressure distributions. These surface movements are in the form of currents and drifts which can be easily detected by a variety of remote sensing techniques. The deep ocean circulation, on the other hand, is not directly evident from the sea surface. In this section, the focus is on remote sensing applications to sea surface circulation only.

One approach to detect currents is to interpret patterns of water turbidity from aerial photographs as demonstrated by Hunter and Hill (1980). This was carried out in near shore waters of the northwestern Gulf of Mexico. Colour aerial photography taken vertically from an

Fig. 7.7
Map of south Texas coast showing regional patterns of water turbidity, surface waves and bands of turbid water on 23 December 1968 (*Source*: Hunter and Hill 1980)

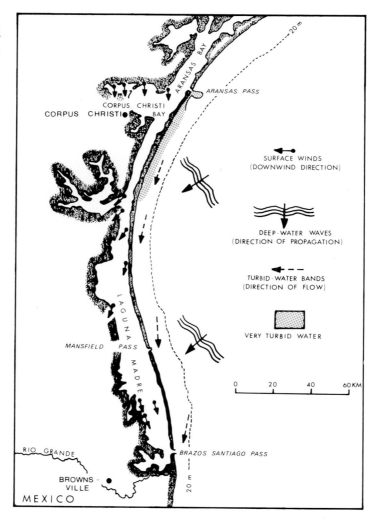

295

altitude of 3,700 m was used. The date of photography was 23 December 1968. The time interval between successive passes over a point varied from 6 minutes at Arkansas Pass to 80 minutes at the mouth of the Rio Grande (Fig. 7.7). The scale of the photography was about 1:24,000. Colour infrared and thermal infrared imagery were also obtained at the same time for comparison. The tonal or colour patterns were used to distinguish water-turbidity features from bottom features in the shallow water. It was also necessary to distinguish the degree of concentration of suspended matter in the water. High concentration in the water areas were indicated by rich yellowish-brown colours. By noting the displacement of distinctive points of the turbidity-pattern during a known time interval between successive passes of the aircraft over the area, the current velocities can be determined. This was done using the turbidity features off a part of northern Padre Island during 25 minute intervals in the study area (Fig. 7.8). The time interval used should not be too long or too short. If it is too long, the specific turbidity features can no longer be recognizable. If it is too short, the movements of the features are not great enough and the measurements become inaccurate. Distinctive points of the turbidity features marked on the photographs were also transferred to a base map (1:24,000 scale) with the aid of a vertical sketch master and the directions of flow of the current could be plotted (Fig. 7.9). It could be seen that the currents moved southwards and were nearly parallel to the shoreline, increasing from about 17 cm/sec in an offshore zone to about 40 cm/sec at the line of breaking waves. The shoreward increase in velocity probably revealed them as wave-

Fig. 7.8
Aerial photographs showing turbidity features off a part of northern Padre Island. Photograph B was taken 25 minutes later than A. One longshore-migrating rip-current plume is labelled X on both photographs. Note its movement relative to features onshore. The separation between the two bars across the beach is 2.5 km (*Source*: Hunter and Hill 1980)

Fig. 7.9
Map showing measured drift of turbidity features off a part of northern Padre Island during a 25 minute period. Turbidity features are shown as they were at end of the period (*Source*: Hunter and Hill 1980)

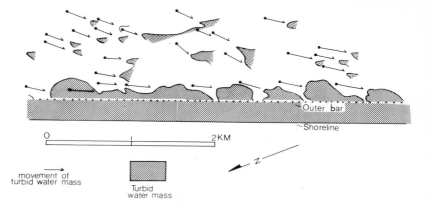

driven longshore currents. The near shore turbidity features were rip-current plumes which moved normally to and away from the shore. Further to the southwest the rip currents became weaker than the longshore current and were drifting with it. They were therefore subjected to the horizontal shear defined by the seaward decrease in longshore velocity. This deformed the plumes into bandlike form and made them to rotate in an anti-clockwise direction.

From the aerial photographs, the source of the suspended fine-grained sediments which caused the turbidity pattern could be located at Arkansas Pass, a tidal inlet from which a large ebb-tidal plume of turbid water was issued (Fig. 7.9). Further to the south of the study area, a banded pattern of turbid and less turbid waters was observed (Fig. 7.10). The bands were nearly parallel to shore at the seaward edges but became more inclined to shore with decreasing distance from shore. They exhibited strong parallelism to the current directions. The

Fig. 7.10
Aerial photograph showing turbidity features off southern Padre Island. North arrow is 0.5 km long (*Source*: Hunter and Hill 1980)

spacing of the bands also varied in the directions along the shore and perpendicular to the shore. The spacing decreased southwards with decreasing water depth. The bands also became straighter and more regular in spacing in the south. All these seemed to suggest that a secondary circulation consisting of counter-rotating pairs of convection cells whose axes were aligned with the primary current had been developed under the water. This type of wind-induced convection pattern is known as Langmuir circulation (Langmuir, 1938). The turbid bands probably marked the lines of upwelling and surface divergence of the secondary circulation. Thus, one sees that much information on currents and circulation can be interpreted from aerial photography alone. It is worthy to note that the thermal infrared imagery did not show many of the turbid water masses but helped to reveal water masses of contrasting temperature not distinguishable from the aerial photographs.

The use of a space platform permits a better vantage point to carry out similar oceanographic interpretations from photography. Mairs (1970) made use of Apollo IX colour photographs to detect distinctive water masses, large-scale suspended sediment patterns and to infer the coastal processes at work for the North Carolina coast of the USA. The suspended sediments stood out as discoloured water masses. The plume-like patterns of suspended sediments were noted and, in view of the small amount of sediments being carried down by the rivers, the tidal origin of the sediment plumes was inferred. The horizontal distance travelled by the plume as it flowed out of the inlet could be determined by multiplying the average velocity during the ebb cycle by the duration. Since the current-velocity curve approximates a sine or cosine curve and for a cosine curve the ratio of the mean ordinate to the maximum ordinate is $2/\pi$ or 0.637, the average current velocity will be the strength of the maximum current multiplied by 0.637. Using a tidal current table for the study area, the average current velocity was determined to be 1.6 knots (0.82 m/s). By combining the space photograph data with climatological, meteorological and tidal current data of the study area, it proved possible to develop a model to explain the origin and dissipation of the suspended sediments (discoloured water masses) depicted in the Apollo photograph. Thus, one can extract useful information on surface current movements from space photography. On the other hand, one needs to exercise care in the interpretations of these observations which ideally should be used in conjunction with concurrent 'sea truth' information. As an example, Stevenson *et al.* (1977) observed the frequent occurrence of circular ocean eddies from Skylab photography; they were defined in circular cloudless areas and the hypothesis that the circular cloud features were indicators of mesoscale turbulence in ocean currents was put forward. Maul (1978) who made use of hand-held photographs from the Apollo aircraft taken over the eastern Gulf of Mexico in July 1975 and concurrently obtained ship data rejected this hypothesis based on observations of sea surface temperatures, streaks on water surface, cloud features and sea surface reflectance patterns. He concluded that the circular cloud features could be better explained as meteorological in nature and were not associated with cold core ocean eddies. His study indicated the usefulness of the sunglint patterns as indicators of current boundaries. A major problem noted was to locate the space photographs in the oceanic area under study. In this case, the cloud patterns on the photographs were matched with reference to those observed in the near-simultaneous GOES 1 imagery over the same area as a means to location determination.

The high-resolution Landsat imagery is even better suited to detect surface water circulation, using the same principle of turbidity analysis mentioned above. As a typical example, Finley and Baumgardner (1980) made use of 22 Landsat images in the red band (band 5, 0.6–0.7 μm) to study the development of plumes of turbid surface water in the vicinity of Arkansas Pass, Texas (the northern part of the same study area mentioned before as shown in Fig. 7.7). The images were optically enlarged to a scale of 1:125,000 and visual interpretation of the near-shore suspended sediment distribution was carried out. The water was classified according to the levels of turbidity observed. Environmental data at the time of satellite passage were also studied in relation to the observed turbidity patterns. These data included wind velocity and direction as well as tidal current conditions. The low astronomical tidal range and shallow bay depths reacted with the wind stress to set up a wind-drift current here. The circulation patterns observed revealed this primary influence by wind. It was possible to categorize the different circulation patterns into five groups: (1) onshore wind group; (2) offshore wind group; (3) persistent offshore wind group; (4) southwest wind group; (5) calm wind group; based on wind, tide and circulation patterns revealed by turbidity. It was also possible to map the surface water circulation as shown in Fig. 7.11 which was based on an interpretation of the Landsat images in Fig. 7.12. From the shape and position of the plume observed, the

Fig. 7.11
Turbidity and surface water circulation as interpreted from the Landsat image of the Arkansas Pass near shore area (Fig. 7.10) (*Source*: Finley and Baumgardner 1980)

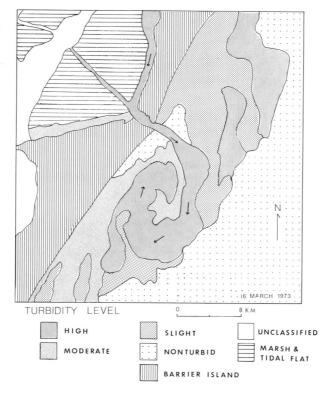

16 MARCH 1973

TURBIDITY LEVEL

0   8 KM

- HIGH
- MODERATE
- SLIGHT
- NONTURBID
- UNCLASSIFIED
- MARSH & TIDAL FLAT
- BARRIER ISLAND

Fig. 7.12
Enlarged Landsat band 5 image
of the Arkansas Pass near shore
area (scene 1236-16323-5, 16
March 1973) (*Source*: Finley
and Baumgardner 1980)

role played by the jetty and the wind regime in affecting sea surface
circulation was noted. It was further pointed out that water penetration in band 5 was limited to approximately 1 m and the information
provided could only supplement site-specific observations of coastal
currents and suspended sediment concentrations undertaken in the
field. There are numerous applications of this type not only in the sea
but also in bays, inlets (Klemas *et al*. 1974; Gatto 1982) and lakes
(Sydor 1976; Pluhowski 1976).

Finally, one should note that thermal infrared and radar imagery
can be used to detect surface currents. As already observed, the
thermal infrared imagery can detect water masses of contrasting
temperatures. In fact, this is closely related to sea surface temperature
determination mentioned before. A good example is the Gulf Stream
off the eastern North Atlantic. From satellite imagery, notably the
Very High Resolution Radiometer (VHRR) on the polar orbiting
satellite NOAA-5 and the Visible and Infrared Spin-Scan Radiometer
(VISSR) on the Geostationary Operational Environmental Satellite
(GOES), the false colour displays revealed the Gulf Stream in pink
against a background of blue (Pl. 13, page 364) (Legeckis, Legg and
Limeburner 1980). The synthetic aperture radar onboard Seasat has

also been able to pick out the boundaries of the Gulf Stream clearly (Fu and Holt 1982). The difference in water temperature on the two sides of the current gives rise to a change in wind stress, with smaller wind stress occurring on the colder side (Fig. 7.13). This produces a difference in backscattering. The Seasat SAR which has a wide swath (100 km) is therefore suitable to monitor the evolution of large-scale current systems in the ocean. Along the coastal areas, the Seaset SAR also recorded vortices generated by highly variable coastal currents as well as coastal eddies produced by mixing of waters of different temperatures and salinity characteristics.

Fig. 7.13
The Gulf Stream off Cape Hatteras as imaged by Seasat SAR on 31 August 1978. Note the wave-like fluctuations of the Gulf Stream (*Courtesy*: Jet Propulsion Laboratory; *Source* Fu and Holt 1982)

Sea state

Sea state refers to the degree of surface roughness in the sea. This is closely related to wind stress at the interface between the hydrosphere and the atmosphere expressed in the form of waves. It is usual to distinguish between two types of waves: the *capillary* and *gravity* waves. The capillary waves are small ripples with less than 1.73 cm in wavelength. Their shape is controlled by surface tension of the water and approximates a sinusoidal form but with slightly V-shaped troughs. Gravity waves have much longer wavelengths and are controlled by gravitational forces. The strength of the waves is determined by wind velocity, the fetch or the distance of water surface over which the wind blows and the duration of the wind blowing. Aerial photography can be used to determine wind directions and gravity wave heights in the sea. The previous case study of the interpretation of near shore current pattern along the South Texas coast from aerial photographs carried out by Hunter and Hill (1980) has also illustrated how these could be done. The wind streaklines visible on water surfaces in areas of sun glitter were used to determine wind directions at a number of places. Wave heights were estimated from the fact that the waves were breaking on the outermost of the three near shore bars in the study area. A probable water depth at the crest of the outer bar was about 2 m. The breaker height was therefore about 1.6 m. More accurate determination of wave heights using photography requires the application of photogrammetry. Stereoscopic pairs of aerial or terrestrial photography have to be obtained preferably with a metric camera and control points have to be established for the absolute orientation of the stereomodel. An analogue or analytical approach can be followed to determine the coordinates of the points defining the waves accurately. More detailed discussion of the photogrammetric method will be found in Ch. 8. This method permits wavelength, wave direction, wave type and wave height to be determined. In an application reported by Mitchell (1983) the height coordinates exhibited standard errors varying between ± 30 and ± 60 mm. One major problem of this approach is the difficulty in establishing control points and shore features have to be used.

Satellite remote sensors have been developed for the observation of ocean waves. These are the precision radar altimeter and the synthetic aperture radar (Tucker 1983). The radar sensors can penetrate cloud cover and are therefore more superior to the photographic method. The radar altimeter in space sends out a pulse to the earth surface and records the shape of the return pulse. This measures the distance of the satellite above the earth and the mean roughness of the earth's surface. Over the oceans the surface height changes show variations in the earth's gravitational field while the roughness changes are due to waves. As the satellite carrying the radar altimeter orbits around the earth, the sea surface is sampled at a specific rate and the radar returns from the individual pulses are highly variable as shown in Fig. 7.14a. To extract the wave heights, the mean shape of all the pulses used has to be obtained (Fig. 7.14b). Then, the difference between successive sample values from the mean curve is computed, thus producing a 'noisy' Gaussian (normal) shape to which a perfect Gaussian curve can be fitted by the method of least squares (Fig. 7.15). The wave height is derived from the following equation:

Fig. 7.14
(a) Typical leading edge shapes of individual pulses plotted from the 16 samples of received signal power. (b) An average of 320 pulses similar to those shown in (a) plotted with the same vertical scale. The small vertical arrows indicate the sample number corresponding to the mean arrival time of the signal (*Source*: Gower 1979)

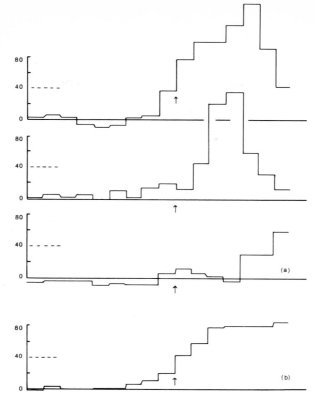

Fig. 7.15
Top: The average leading edge shape of 320 pulses (3.2 sec. of data) is averaged to form the top trace. Bottom: A Gaussian curve (dashed) is fitted to the differential of this trace (*Source*: Gower 1979)

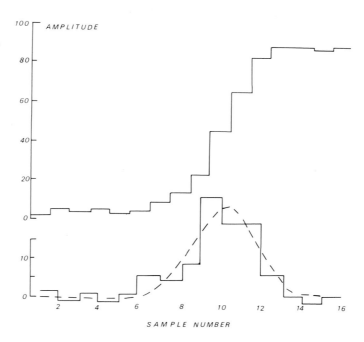

$$H_s = 0.6 \sqrt{\sigma^2 - \sigma^2{}_p - \sigma^2{}_j} \qquad [7.8]$$

where H_s denotes the significant wave height in metres, σ is the standard deviation of the shape of the radar return pulse in nanoseconds, σ_p is the standard deviation of the transmitted pulse shape in nanoseconds and σ_j is the timing loop jitter (i.e. the tracking loop adjusting the delay between the time of transmission and the time of sampling by the gates) in nanoseconds. The Gaussian curve fit techniques provide a simple method to estimate the value of σ.

The above approach of computation was followed by Gower (1979) in obtaining ocean wave height using the short-pulse radar altimeter data from the GEOS-3 satellite launched in April 1975. The data consisted of 100 pulses recorded per second in segments of 3.2 sec, or a total of 320 pulses. The shape of an individual pulse was sampled by 16 gates spaced at 6.25 ns intervals. The result of the computation gave wave heights from each data segment which showed a scatter of ± 1.6 m about a 16 sec (5 data segments) running means. Gower (1979) has shown that by refining the error curve, fitting and correcting for time variations this scatter could be reduced to ± 0.5 m. The radar altimeter was also carried on board Seasat. The carrier frequency used was 13.5 GHz and the effective pulse length was just over 3 ns in time or about 10 cm in space. The height accuracy achievable was about ± 0.1 m (Cartwright and Alcock 1983). The precision of the radar altimeter is therefore quite amazing. This is only achievable with a high precision in computed orbital elevation, which requires a detailed knowledge of the gravity field and non-gravitational forces.

The synthetic aperture radar (SAR) is the second wave sensor which produces an image of the sea surface. It is capable of imaging waves through the physical process producing backscattering known as 'Bragg resonance' (Tucker 1983). This simply means that the incident radar waves are backscattered by the short-wave component of the sea surface roughness whose wavelength matches that of the radar waves. The 'Bragg resonant wave' whose crest is at right anlges to the radar range direction has a wavelength λ_B given by:

$$\lambda_B = \lambda_R/(2 \sin \theta) \qquad [7.9]$$

where λ_R is the radar wavelength and θ is the angle of incidence (Fig. 7.16). However, this only applies when there are no long waves present in the sea. Once there are longer waves, the backscatter will

Fig. 7.16
Bragg resonance. Coherent backscatter is given by waves whose crests are perpendicular to the radar incidence and whose wavelength λ_B is given by $\lambda_B = \lambda_R/(2 \sin \theta_i)$ *(Source:* Tucker 1983)

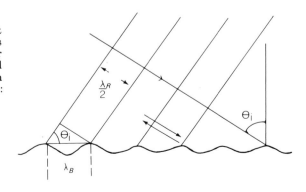

be modulated by them. For the Seasat SAR which used a radar wavelength (λ_R) of about 23.5 cm, the Bragg resonant waves recorded were short gravity waves with a wavelength (λ_B) of about 34 cm. Figure 7.17 is an example of the Seasat SAR image revealing clearly wave refraction patterns around the Shetland Islands in the north of Scotland. Thus, Seasat SAR imagery is most suited for use to study on a synoptic scale the generation and propogation of long ocean waves in the open ocean. The distribution of the short gravity waves recorded by the Seasat SAR shows correlation with a number of significant larger scale phenomena, such as local wind structure, long gravity waves, current shear boundaries and occasionally the local bathymetry. The usefulness of Seasat SAR depends on how well these various phenomena can be separated (Beal 1981). Previous experiments conducted to evaluate the capabilities of Seasat SAR confirmed that waves with a wavelength greater than 100 m could be detected under the condition that the significant wave height exceeded 1 m and the surface wind speed exceeded 2 m/s (Fu and Holt 1982).

Apart from the surface waves, trains of internal waves can also be detected by Seasat SAR as alternating bands of rough and smooth sea surface. These waves have longer wavelengths (which vary from 800 m to over 1,000 m) and are dispersive in nature. They appear to be generated mainly through the interaction of tidal currents and

Fig. 7.17
Seasat SAR image of the Shetland Islands, Scotland (15 September 1978). Note the refraction and diffraction patterns of surface waves around the islands to the top (Foula) and to the bottom (Fair Isle) of the image (*Courtesy*: Jet Propulsion Laboratory; *Source*: Pravdo *et al.* 1983)

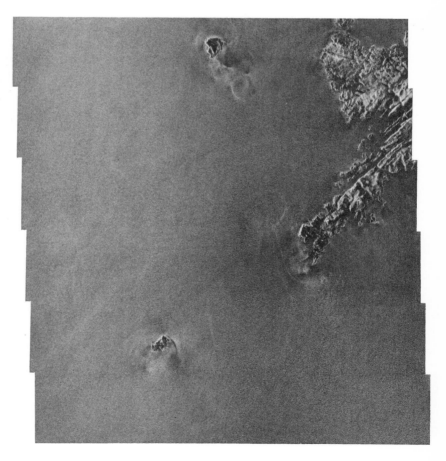

Fig. 7.18
Seasat SAR image of Gulf of
California (29 September 1978)
showing the internal wave
patterns. See Fig. 7.19 for
explanations (*Courtesy*: Jet
Propulsion Laboratory; *Source*:
Fu and Holt 1982)

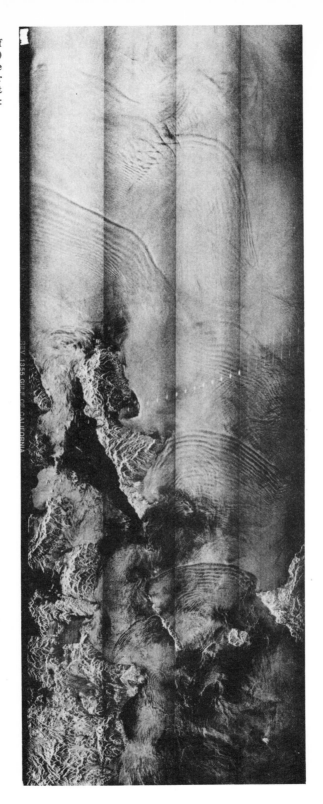

Fig. 7.19
Bathymetric chart of the Seasat SAR image of the Gulf of California shown in Fig. 7.18. The different internal wave patterns detectable on the image are indicated by letters A through H on the chart. These waves were probably generated as a result of the interaction of strong tidal currents and bottom topography. Note that within a particular wave group, the wavelength and amplitude decrease with the distance from the leading crest. The maximum wavelength is about 2 km except for groups A and B which have wavelengths less than 1 km (*Source*: Fu and Holt 1982)

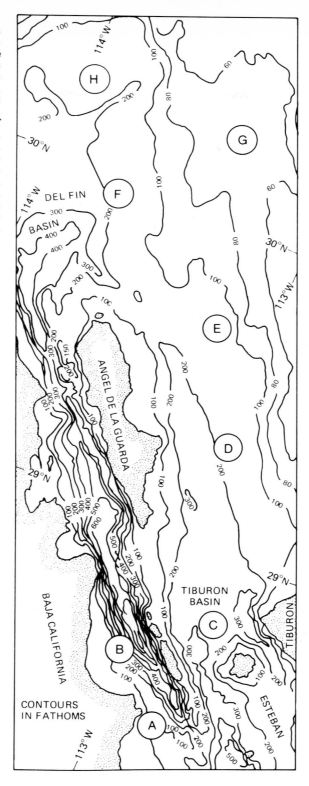

abrupt topographic features in the coastal areas. Figures 7.18 and 7.19 illustrate very well these internal wave patterns of different wave-lengths in the Gulf of California.

The change in sea-state has also been studied with Landsat MSS images. Apel *et al.* (1975) have interpreted internal and surface waves from Landsat MSS band 7 (0.8–1.1 μm) data. Wald and Monget (1983) also demonstrated the use of the same MSS band 7 to detect sea-state local variations. This was based on the assertion that the observed change in reflectance in this infrared channel was mainly induced by sea-state. For a perfectly smooth water surface, the radiation reflected from the surface should be concentrated in the direction of the spec-ular reflection. When the sea is rough, the reflected radiation is scat-tered around this direction, due to the capillary waves acting like small mirrors. Part of this reflected radiation appears as glitter reflec-tance on satellite imagery. A model developed by Cox and Munk (1954) described the glitter as a function of the sea-state for wind speeds less than 14 m/s. When the sea becomes rough, foam appears and the amount of reflectance is increased. Hence, reflectance is greater for a rough sea than for a calm sea, except when the sun and the observer are in opposite direction to each other (i.e. the difference of the azimuthal angles between the sun (ϕ_0) and the observer (ϕ) equals 180° (Fig. 7.20). Figure 7.20 depicts the horizon system with azimuthal angles for the sun and the observer represented by ϕ_0 and ϕ respectively, and the zenith angles for the sun and the observer represented by θ_0 and θ respectively. The reflectance angle is w. Glitter reflectance decreases when the azimuthal angle difference between the observer and the sun (i.e. $|\phi-\phi_0|$) decreases. An increase in the sun zenith angle θ_0 leads to a decrease in the glitter reflectance. Wald and Monget (1983) have shown that the glitter reflectance induced by sea-state variations and observed from a near-polar orbiting satellite such as Landsat depends on the period of the year. This is minimum during late autumn and winter and maximum during June. The reflectance contrast between a calm sea (wind speed of 5 m/s) and a rough sea (wind speed 14 m/s) observable with Landsat for different latitudes throughout the year is shown in Fig. 7.21. It is clear that a rough sea will always appear brighter than a calm sea in Landsat MSS band 7 imagery. Figure 7.21 also permits us to deter-

Fig. 7.20
Reflection geometry. The solar radiation reflects against the wave slope, defined by the normal vector n. The reflected radiation is observed in the (θ, ϕ) direction (*source*: Wald and Monget 1983)

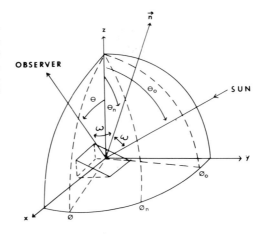

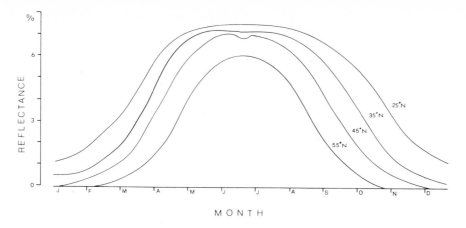

Fig. 7.21
Reflectance contrast observed
on Landsat imagery for a
non-absorbing atmosphere
throughout the year for various
latitudes: 25°N, 35°N, 45°N, and
55°N (*Source*: Wald and Monget
1983)

mine the best months to discriminate between calm and rough seas in different latitudes.

Applications to study the biological characteristics

Sea-water colour

Sea-water colour refers to the spectrum of upwelling radiance just beneath the sea surface. The visible radiant energy (light) coming directly from the sun is partly reflected at the sea surface and the remaining is transmitted into the sea. The transmitted energy is absorbed or scattered by the water molecules and other substances in it. Part of the light (3–5 per cent) is scattered backward into the atmosphere again. Passive remote sensing attempts to pick up the backscattered light (upwelling energy) in the hope to unravel the constituents of the water just beneath the sea surface. Apart from waters in close proximity to coastlines and at the confluences of rivers and the sea, biological constituents, notably the *phytoplankton*, are the most important constituents. Phytoplankton are mostly microscopic and unicellular organisms restricted to the photic zone of the sea where the highest productivity is found. Photosynthesis by the phytoplankton takes place here. The phytoplankton contain *chlorophyll a*, the dominant photosynthetic pigment which strongly absorbs the energy in the blue and red regions of the visible spectrum. As the chlorophyll *a* increases in concentration in the sea water, the colour of the water is changed from the deep blue of its purest state to green. There is also a detrital pigment of chlorophyll *a* – phaeopigments *a* – which possess almost the same absorption spectrum as chlorophyll *a*, and hence cannot be separated from chlorophyll *a* easily. The phaeopigments *a* concentration is usually lower than the chlorophyll *a* concentration by a factor of 5–10. Apart from these, the sea water also contains sediments brought down by rivers and dissolved organic matter (yellow ultra-violet absorbing material or 'Gelbstoff'; Yentsch 1983). Sea water can therefore be divided into two classes: Case 1 water and Case 2 water (Sathyendranath and Morel 1983). Case 1 water contains phytoplankton and their by-products which largely determine the optical properties of the water body (Fig. 7.22a). The by-products include both detrital particles and yellow substances liberated by algae and their debris. It can be seen that the reflectance tends to decrease below the clear water spectrum at wavelengths below around 540 nm and slightly increase at higher wavelengths (Fig. 7.22a). Case 2 water contains resuspended sediments, dissolved organic matter (terrigenous yellow substance) and terrigenous particles from rivers and glaciers. The sediment-dominated Case 2 water displays an overall increase of reflectance at longer wavelengths (Fig. 7.22b). The water appears bright, discoloured or milky green. The yellow-substance–dominated Case 2 water exhibits a distinct decrease of reflectance towards violet due to increasing absorption (Fig. 7.22c). Thus, different types of water have shown sufficient change in optical properties closely related to the amount of dissolved or suspended material to be distinguished by passive remote sensing, particularly if the water is dominated by a single component. Interpretive algorithms have been developed for this purpose (Robinson 1983).

Apart from the fact that most of the incident energy is back-scattered from particles and molecules in the sea water, a small portion is caused by *fluorescence* from dissolved and particulate organic substances. By fluorescence, one is measuring the wavelengths of light active in photosynthesis. Measurement of the intensity of phytoplankton fluorescence aids in the estimation of phytoplankton biomass (Yentsch 1983). Different types of phytoplankton seem to exhibit different intensities of flourescence emission.

One specific remote sensor designed for the study of living marine resources through the detection of the variations in the concentration of phytoplankton pigments is the Coastal Zone Colour Scanner (CZCS)

Fig. 7.22
Some observed spectra of reflectance: (a) Case 1 waters with different quantities of phytoplankton, (b) suspended–sediment-dominated Case 2 waters and (c) yellow-substance dominated Case 2 waters. The dotted line in these diagrams indicates clear ocean water (*Source*: Sathyendranath and Morel 1983)

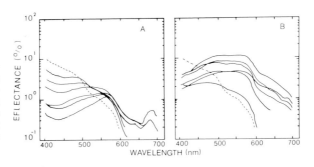

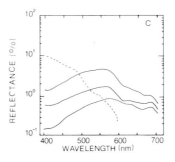

on Nimbus 7 launched in October 1978 (Hovis 1981). The instrument is an imaging multispectral scanner operating in six spectral bands with an instantaneous field of view of 825 μrad (0.047°) for a nadir field of view of about 800 m from a spacecraft altitude of 955 km in a sun-synchronous orbit (Table 7.2). The first five bands belong to the visible spectrum with which phytoplankton pigments detection, estimation and mapping can be carried out. It is worthy to note that the estimation of chlorophyll a is more difficult in coastal waters where the interferences by yellow substances and suspended sediments are greatest. In offshore waters it was found that the pigment concentration could be related to ratios of radiances at various wavelengths rather than to absolute radiances. This had the advantage of partially compensating for the influences of other material in the water and the masking effects of the atmosphere. The ratio of two CZCS channels of 443 nm and 550 mm (L_w443/L_w550) was found to correlate well with ground truth data better than 0.85 (Gordon *et al.* 1980). However, the CZCS measured radiances have to be corrected first for the effects of atmospheric scattering before use. The total radiance (L^λ) observed by the sensor at wavelength λ consists of three parts:

$$L^\lambda = L_R^\lambda + L_A^\lambda + t^\lambda L_W^\lambda \qquad [7.10]$$

where L_R^λ is the radiance due to Rayleigh scattering, L_A^λ is that due to aerosol scattering and L_w^λ is that due to backscattered radiance from the ocean transmitted through the atmosphere and t^λ is the atmospheric transmittance, all at wavelength λ. The aerosol scattering (L_A^λ) is highly variable in concentration, composition and size distribution and its effect is difficult to remove. A correction algorithm developed by Gordon (1978) made use of two wavelengths (λ_1 and λ_2) in equation [7.10] above so that one obtained the following equation:

$$t^{\lambda 2} L_W^{\lambda 2} = L^{\lambda 2} - L_R^{\lambda 2} - \alpha(\lambda_1, \lambda_2)[L^{\lambda 1} - L_R^{\lambda 1} - t^{\lambda 1} L_W^{\lambda 1}] \qquad [7.11]$$

where $\alpha(\lambda_1, \lambda_2)$ is a constant which is related to the optical properties of the aerosol. By choosing λ_1 so that $L_w^{\lambda 1} \approx 0$, then $L_w^{\lambda 2}$ can be determined for any λ_2, provided $\alpha(\lambda_1, \lambda_2)$ is known. The 670 nm band of the CZCS is the most suitable for this purpose except in regions of very high turbidity. *In situ* measured values of L_w^λ at one location are then substituted in equation [7.11] to obtain $\alpha(\lambda_1, \lambda_2)$ for the whole image.

Table 7.2
Coastal Zone Colour Scanner characteristics
(*Source*: Hovis 1981)

Spectral Bands	443 $\pm$ 10 nanometres
	502 $\pm$ 10 nanometres
	550 $\pm$ 10 nanometres
	670 $\pm$ 10 nanometres
	750 $\pm$ 50 nanometres
	10.5 to 12.5 micrometres
Spatial Resolution	0.825 milliradians, 800 metres at nadir
Operation	Day – All channels
	Night – 10.5 to 12.5 micrometres only
Glint Avoidance	Tilt of scan mirror $\pm$10° along track for $\pm$20° pointing
Swath Width	1636 kilometres

This approach has been applied by Gordon *et al.* (1980) to determine the phytoplankton pigment concentrations from the corrected radiances for eastern Gulf of Mexico as derived from Nimbus-7 CZCS imagery (Pl. 14, page 364). Two ratios were used: L_{443}/L_{550} and L_{520}/L_{550}. Simultaneous shipboard measurement for L_W^λ and pigment concentrations were made. The corrected CZCS imagery revealed eddy-like structures in ocean circulation which were not observed before. The pigment concentrations measured by the CZCS were consistently lower than the ship measurements (Fig. 7.23), but the overall agreement with the ship measured data was good.

Although the same approach can also be applied to Landsat MSS data over sea areas, the sensor response which is designed for land use mapping is incompatible with accurate sea observations, particularly because of the poorer spectral and radiometric resolutions of the MSS. Landsat has the advantage of a longer wavelength band (band 7: centred at 950 nm), for which L_W is always zero, but L_A for this band is too small and is not resolved by the sensor with sufficient accuracy to determine L_A for the other channels in each pixel. Therefore, no attempt is normally made to apply a differential atmospheric correction across a scene if Landsat data over sea areas are used for mapping phytoplankton pigment concentrations. A typical approach is illustrated by Gower, Denman and Holyer (1980) who made use of Landsat MSS data to map the phytoplankton distribution in the oceanic area south of Iceland. They first reduced the effects of clouds by substracting the band 7 image from the band 4 image. The pixel values were then averaged over groups of 8 × 6 pixels to reduce instrument noise and to give a roughly square grid of about 474 m spacing for further processing. Fourier transformation was then carried out over three overlapping areas of 256 × 256 averaged pixels each about 120 km square. The Fourier transforms of the pixel arrays were then coverted to power or signal-variance spectra. The patterns of the enhanced image of band 4 were interpreted as being due to phytoplankton and their distribution appeared to be controlled by advection in variable ocean currents (or mesoscale 10–100 km water motions) – hence the occurrence of eddies noted earlier.

The distribution of chlorophyll *a* can also be carried out with

Fig. 7.23
Values of C_2 = chlorophyll *a* + phaeopigments *a* (in milligrams per cubic metre) from CZCS data compared with a track line of concentrations measured aboard the R.V. *Athena II* on 13 and 14 November 1978 (*Source:* Gordon *et al.* 1980)

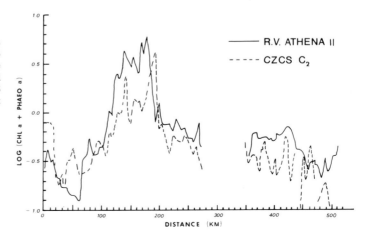

multispectral scanners carried aboard aircraft. Chlorophyll *a* is regarded as an important environmental parameter indicative of the water quality, nutrient contents and pollution effects in coastal zones, not simply as an indicator of water productivity. Johnson (1978) reported of experiments in which quantitative maps of chlorophyll *a* distributions were made for the James River, between Norfolk and Hopewell, Virginia, and New York Bight USA, from a Modular Multispectral Scanner (M2S) and the Ocean Colour Scanner (OCS) respectively. The M2S was flown at 2.4 km altitude operating in 11 bands in the visible, near infrared and thermal spectral ranges (380–1060 nm). The OCS was flown at an altitude of 19.7 km with an U-2 aircraft operating in 10 bands in the visible and near infrared spectral ranges (423–782 nm). A stepwise regression analysis was carried out to relate *in situ* chlorophyll *a* measurements from the sea or river with remotely sensed multispectral data. For M2S the regression equation resulted was:

$$\text{Chlorophyll } a \text{ in mg/m}^3 = 25.62 - 18.06 M2SR2 + 7.23\ M2SR6 + 9.46\ M2SR8 \qquad [7.12]$$

where *M2SR2*, *M2SR6* and *M2SR8* were bands 2 (440–490 nm), 6 (620–660 nm) and 8 (700–740 nm) respectively. In other words, these were the bands that contributed most significantly to the estimation of chlorophyll *a*. The correlation coefficient was 0.96 with a standard error of estimate of ±1.75 mg/m³. For OCS, the regression equation became:

$$\text{Chlorophyll } a \text{ in mg/m}^3 = -4.51 - 4.96\ OCSR3 + 11.44\ OCSR6 \qquad [7.13]$$

where *OCSR3* and *OCSR6* were bands 3 (499–519 nm) and 6 (610–630 nm) respectively. The correlation coefficient was 0.83 with the standard error of estimate of ±3.87 mg/m³. The chlorophyll *a* concentration in each pixel was determined from equations [7.12] and [7.13] as the case might be. A smoothing algorithm was executed to remove spatial and spectral noises and a computer-drawn contour map was plotted for each study area (Fig. 7.24a and b). From these maps, the degree of polluton in the coastal area was revealed.

Finally, the fluorescence of the green algae can be measured directly with a laser induced fluorometer (LIDAR) flying at low altitudes by aircraft. The instrument indices fluorescence by sending out pulses of blue or green light into the water column and then measures the fluorescence emitted. In this way, one can measure the excitation spectra for chlorophyll *a* fluorescence.

Detection of fish stock and aquatic plants

Apart from the plankton, the marine environment is full of other living organisms such as fish, seaweed and algae which contribute significantly to the food chain. The detection of fish stock can be carried out visually by using low-flying aircraft to direct the fisherman to the most productive site. A large number of fish types found near the surface of the sea are detectable and aerial photog-

Fig. 7.24
Quantitative distribution of chlorophyll *a* (a) in the James River near Hopewell, Virginia, USA on 28 May 1974 and (b) in the New York Bight apex on 13 April 1975 (*Source*: Johnson 1978)

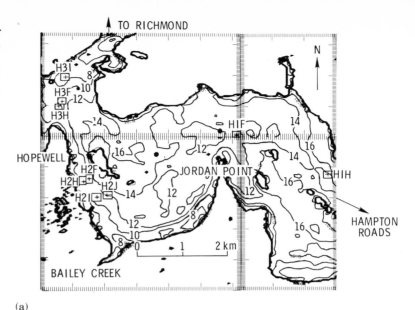

(a)

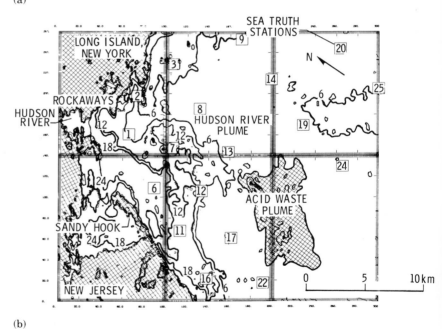

(b)

raphy has been used quite successfully this purpose. The anchovy and sardine fleets of Chile and parts of Peru, the US menhaden fleets and the California sardine industry all employed aircraft to guide their vessels to productive fishing grounds (Hughes 1979). For optimum results aerial photography should only be taken in clear, cloudless skies at solar angles between 20° and 50° from the horizontal to enhance contrast and to avoid glare – the direct method in fish stock detection by remote sensing. More recent interest tends to focus on the indirect method which makes use of such information as sea temperature, wind direction, water salinity and water colour all collectable from remote sensors to locate fish (Hughes 1979). The occurrence of phytoplankton revealed by the chlorophyll *a* concentrations is a good indicator of feeding grounds for fish. All these parameters, as already explained above, are measurable with some degree of confidence from satellite platforms using appropriate sensors. Fiuza (1979) gave an example of the importance of sea surface temperature (SST) and chlorophyll content in finding and predicting the location of sardine stocks on the western coast of Portugal. The fact that the largest catches of sardines were in cold water of about 14 °C seemed to suggest that sardine abundance was associated with thermal gradients (fronts) or a transition from warm to cold water of about 16 °C to 14 °C (Fig. 7.25). Kemmerer (1979) employed Landsat images to map the distribution of menhaden in the Mississippi Sound and off Louisiana in the northern Gulf of Mexico, using ocean colour as the main criterion. Correlation coefficients between menhaden distribution and radiance values in each spectral band were calculated and were found to be highly significant. A supervised classification approach was adopted to classify the sea into high and low probability menhaden areas. The accuracy of classification was high. The menhaden seemed to respond to light intensity because most fish were caught near shore in relatively clear water when light intensity was low. As light intensity

Fig. 7.25
Frequency of sardine catch and number of fishing attempts relative to surface temperatures (*Source*: Fiuza 1979)

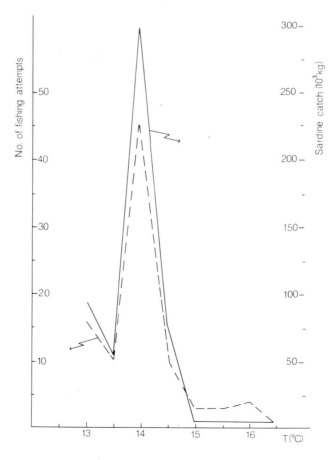

increased, the fish moved offshore into deeper and more turbid waters. In the late afternoon, catches were made further offshore and in clearer waters. However, the use of satellites for fish detection is restricted by cloud cover except with the use of microwave sensor such as Seasat SAR. Such an instrument can detect changes in the pattern of capillary waves caused by slicks of oil associated with certain species of fish.

Algae which are aquatic plants living in the bottom of the sea are not easily detected by remote sensors. One type of algae called *cyanophyta* (blue-green algae) was detectable in the Baltic Sea using Landsat images especially after digital enhancement (Schmidt 1979). The blue-green algae occurred in dense masses at the sea surface during warm late summer periods. They tended to form bands of several kilometres long at right angles to the wind direction. They served as indicators (tracers) for wind-generated surface currents or internal waves (Pl. 15, page 380). These algae depended on phosphate for their mass growth and can be indicative of a possible entrophication of the Baltic. They are capable of nitrogen fixation from the air and their occurrence is connected with the nitrogen budget of the Baltic.

Applications to study the geological characteristics

Coastal water turbidity

Turbidity refers to the churning up of sediments by water to form a dense, heavy flow especially in the sea. As the water becomes more turbid, the transmittance of light through the water becomes more difficult and the reflectance becomes stronger, rendering the suspended sediments highly visible. The colour of turbid water is also distinctly different from that of clear water. Therefore, the study of turbidity is also closely linked to the detection of sea-water colour mentioned in connection with the biological characteristics of the hydrosphere. Both low-altitude aerial photography and space satellite imagery have been used to measure turbidity and to map the pattern of sediment concentration. In a way, this is closely related to the study of water quality, to be discussed in the next section, associated with the chemical characteristics of the hydrosphere. The coastal environment is particularly favourable for water turbidity study because it is here that the rivers deposit their contents into the sea. The heaviest sediments tend to be deposited first, nearest the river delta. Thus, remote sensing of the coastal zones can provide data on sediment volume, characteristics and depositional areas. Of course, the study of turbidity can also be carried out in rivers, lakes and reservoirs.

An important basis for this kind of application is the linear relationship between film density and scene reflectance. Klooster and Scherz (1974) demonstrated in their study of a well-defined paper-mill plume discharge into a Wisconsin river in the USA, that turbidity was strongly correlated to image brightness. Thus, by using a microdensitometer to measure image brightness, turbidity could be quantitatively determined. Photography was taken in mid-afternoon in order to eliminate sun glint. A colour infrared film (Kodak film type 8443) was used with a No. 12 Wratten filter. The infrared energy can only penetrate a few inches into the water thus completely eliminating the effects of the river bed. Water samples were also collected at the same time of the overflight. A 50 μm spot was employed in the colour microdensitometer which measured an area equivalent to 0.3 m² on the water surface for the flying height used. It was found that there is a good positive correlation between turbidity and reflectance (Fig. 7.26). Turbidity was found to depend on the wavelength used, the size and shape of the particles present and their reflectance qualities. From Fig. 7.26 it can be seen that at A the curve flattens. This is the point where water is so saturated with sediments, i.e. water becoming opaque and mud-like, that increases in solid content cause no significant increase in reflectance. It is worthy to note that the field curve derived from the film is always higher than the laboratory curve. This is because the aerial photographs, apart from showing volume reflectance, also show reflection of skylight from the water surface.

In another study by Ritchie, Schiebe and McHenry (1976), suspended sediment concentration of surface water in six northern Mississippi reservoirs was found to be linearly related to reflected solar radiation. From Fig. 7.27, one can see that reflected solar radiation between 450 –900 nm increased as the concentration of suspended sediment increased. The peak of reflected solar radiation shifted from about 550 nm at low sediment concentration to above 600 nm at higher sediment concentrations. However, between 700–800 nm, the change exhibited a more regular pattern than that between 400–700 nm. Therefore, the best region of the spectra for predicting suspended sediment from linear regression equations, using reflected solar radiation or reflectance, would be between 700–800 nm (Fig. 7.27). One should, however, note the influence of the sun angle. If the data were separated according to sun angles, those with a sun angle of less than 40° were found to exhibit greatly improved correlation coefficients in the linear regressions by 5–15 per cent than those with a sun angle of greater than 40°. This study indicates the feasibility of using Landsat data for similar works.

Landsat images provide a synoptic view of the coastal region and are ideal for mapping turbidity or suspended sediment distribution pattern. Three approaches are possible: (1) a qualitative visual interpretation of the Landsat scenes in isolation or as a composite;

Fig. 7.26
Curve of reflectance versus turbidity and suspended solids for paper-mill waste (*Source*: Klooster and Scherz 1974)

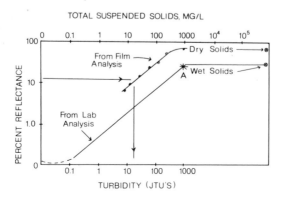

Fig. 7.27
Relationship of reflected solar radiation with wavelength for different concentrations of suspended sediments in surface water (*Source*: Ritchie *et al.* 1976)

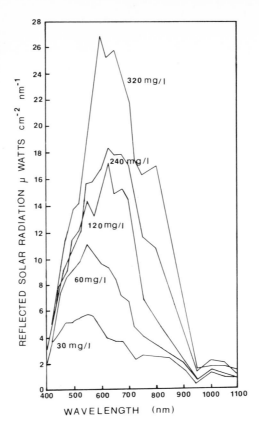

(2) enhancement of the image by optical density slicing techniques to facilitate delineation of turbidity zones; (3) quantitative analysis using digital Landsat data. The work of Klemas *et al.* (1973; 1974) over Delaware Bay, an estuary on the east coast of the US flanked by the states of New Jersey and Delaware, illustrates these three approaches clearly. Imagery and digital tapes from ten Landsat-1 passes and one Skylab pass over Delaware Bay during different portions of the tidal cycle was analyzed with special emphasis on suspended sediment concentration, current circulation, waste disposal plumes and coastal frontal systems. The total length of Delaware Bay is over 200 km, but a less extensive body of water can be defined if a lower limit of the estuary is used. Thus, the bay is 75.2 km long encompassing an area of 1,864.8 km^2 and a volume of about 1.78×10^{13} litres. The Delaware River is the major source of fresh water input at an average rate of 316 m^3/sec, which is considered to be a large volume of fresh water. The river also discharges a heavy load of suspended and dissolved material as a result of the river's large water catchment area characterized by intensive agricultural and industrial land use. The Bay exhibits complex circulation patterns and tidal flows because of its shape and bathymetry. The Landsat images examined covered the period from October 1972 to July 1973 and each time they were taken at about 1514 GMT or 10.14 a.m. local time, giving rise to low solar angles from 23° to 39°, a condition favouring visibility of water features. For mapping suspended sediment concentrations, it was

Fig. 7.28
Landsat MSS band 5 image of Delware Bay area, an estuary on the east coast of the US, acquired on 7 July 1973 (*Courtesy*: V. Klemas; *Source*: Klemas *et al.* 1974)

found that the pattern appeared most distinctly in band 5 (0.6–0.7 μm) (Fig. 7.28). The highest sediment concentration was observed to occur in the Delaware River with a decreasing concentration further down into the upper Bay. The sediment patterns formed boundaries and striations throughout the Bay with the sediment load decreasing towards the mouth of the Bay. The relatively higher concentrations of sediments along the New Jersey side of the Bay were seen and might be due to the shallowness of the water there. Near the mouth of the Bay at Cape May a complex sediment pattern was also seen, which might be related to the currents and intricate bottom topography. At the time of Landsat overpasses, Secchi disk* measurements and sediment concentrations were also obtained in the Bay. There was a strong correlation between sediment concentration, inverse Secchi depth and Landsat image radiance for the 26 January 1973 scene (Fig. 7.29). For the upper one metre of the water column, band 5 correlates closely with turbidity and sediment load measured from boats. On the other hand, if one wishes to distinguish between different sediment types, the reliance on a single spectral band is not enough. This is because different wavelengths penetrate different depths and the use of multispectral data helps to separate different sediment

* Secchi disk measures the amount of light penetration using a white procelain disk which is lowered into the sea by a piece of string until it disappears from sight. The depth measured is known as the depth of visibility.

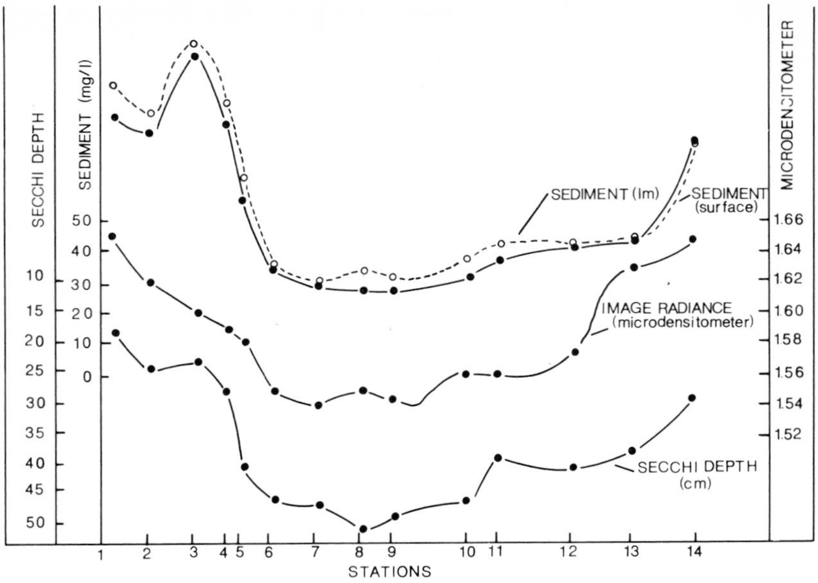

Fig. 7.29
Correlation of Landsat-1 image radiance (from a microdensitometer scan) with suspended concentration and Secchi depth in Delaware Bay, on the east coast of the USA, on 26 January 1973 (*Source*: Klemas *et al.* 1974)

types. For bands 4 and 5 the water penetration depths were estimated to be about 1.8 and 0.16 m respectively. Penetration depths for bands 6 and 7 were even less. Thus, a water mass showing a high radiance in band 7 probably indicates a surface film of oil.

The second approach experimented by Klemas *et al.* (1973) was image enhancement. Colour density slicing was performed for each band. This involved 'slicing' the grey tone into different intervals and with the aid of a colour additive viewer different colours were assigned to the intervals (Pl. 16, page 380). Band 5 was found to exhibit more density steps than bands 4 and 6. In this way, colour density slicing helped to delineate more clearly the suspended sediment patterns and to distinguish turbidity levels.

The third approach was to make use of Landsat digital tapes. This is similar to the approach of supervised classification mentioned in connection with land use mapping. An image brightness map of the Bay was produced from the digital data. Ground truth data were related to the radiance values of the pixels in the selection of training areas. The digital data were classified pixel by pixel with reference to the statistics produced by the training sets. Each class generated was assigned a colour for a maximum of 15 colours and the resultant map for the 7 July 1973 Landsat scene of Delaware Bay is shown in Fig. 7.30. In the above analyses, one assumes that atmospheric effects are uniform throughout the study area. However, in reality, more haze effect was found to occur over the northern portion of the Bay than that over the southern. This probably resulted in a higher radiance gradient over the Bay, with increases more in the north than in the south.

Coastal bathymetry

Bathymetry refers to the measurement of water depth. Unlike turbidity detection, it depends on the degree of penetration of water

Fig. 7.30
Sediment distribution map of Delaware Bay, on the east coast of the USA, as interpreted from Landsat 1 image and field measurement. The highest sediment concentration is shown in black and the lowest is shown in small circles (*Source*: Klemas *et al.* 1974)

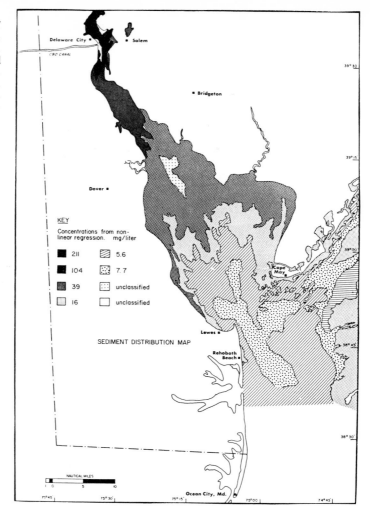

achievable by the radiant energy of a certain wavelength. With remote sensing it is possible to determine the depth of water up to 30 m. Both conventional aircraft and satellites provide platforms to sense the water surface and the ocean bed. Large-scale aerial photography has to be obtained of the water area and the water depth is determined according to tonal variations, especially in the coastal areas. However, from the photogrammetric point of view, it would be difficult to carry out orientation of the stereomodels because no control points can be selected precisely over the sea surface (Bullard 1983a). Water penetration and reflectance from the sea bed depend on the sun's azimuth and elevation angle in relation to the water surface and the sea bed, the nature of the sea bed, the direction of wave propagation and the sea state. Similarly, for satellite remote sensing with the use of Landsat MSS data, band 4 (0.5–0.6 μm) which covers the green to orange range of the visible spectrum is generally preferred for the detection of water depth (Bullard 1983b). The success also depends on reflection on the water surface, sun angle and azimuth, salinity, amount of suspended sediment, nature and reflection of the sea bed, height of platform as well as scatter, absorption and refraction of the atmosphere. Bullard (1983b) illustrated an approach using Landsat band 4 digital data to obtain marine contours for the Red Sea region. The band 4 data were obtained in a high-gain mode, i.e. the radiometric signals being amplified by a factor of three. The band 4 image was density sliced into eight classes each of which was represented by a colour. By tracing out the different colours, form lines were produced for the sea. The marine contours so produced exhibited some discrepancies from the conventional marine chart of the study area. No conclusion could be drawn because one could not establish the accuracy of the marine chart employed for comparison, but the potential of Landsat band 4 images for bathymetry was demonstrated.

Both the conventional aerial photography and Landsat MSS data are passive remote sensing. A more accurate remote sensing system for depth determination has to be an active one. Bullard (1983a) introduced the laser system known as Weapons Research Establishment Laser Depth Sounding (WRELADS) developed by Australia, which has the capability of measuring depths from 2–30 m, an accuracy of depth soundings of 1 m and a position accuracy of 25 m. The system sends off two beams, an infrared and a green, from the laser (Fig. 7.31). The infrared beam which is held vertically provides the surface return signal whereas the green laser beam which sweeps 15° either side of the flight line gives the sea bottom return signal. The infrared beam therefore serves as the datum for depth measurements. Tide gauge measurements in the field are used to correct for water depth variations caused by the tide cycle. The laser system is best employed in deep water areas.

Another active system – the synthetic aperture radar – which

Fig. 7.31
Scanning geometry of Weapons Research Establishment Laser Depth Sounding (WRELADS) (*Source*: Bullard 1983a)

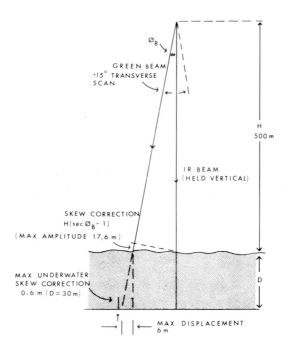

cannot penetrate into deep sea water (less than 1 cm) produces imagery for use in indirectly inferring water depths. Lodge (1983) identified three classes of features observable on Seasat SAR images of the sea that can be correlated with changes in depth. The first category was wave refraction. On encountering shallow water (i.e. the depth of water being less than about half the dominant water wavelength), the gravity waves at sea show changes in velocity, length, height, direction and shape while their period remains unchanged. Therefore, wave refraction patterns can be used to provide quantitative bathymetric information. Figure 7.17, the Seasat SAR image of the Shetland Islands, Scotland, indicates that the waves travelled from west to east and that shallow water, less than 100 m deep extended about 5 km eastward from the island (Foula) at the top of the image. There was also evidence of refraction from an area of shallower water about 10 km north of the island. The second category of features was called shallow-water features or features that could occur at less than a depth of about 50 m. Figure 7.32 revealed linear features in the English Channel, which could be related to sandbanks. Here the depth of the sea seldom exceeded 40 m and might be less than 5 m over the tops of some of the sandbanks. The third category of features was related to deep-water features. Many Seasat SAR images showed wave-like features, with wavelengths of several kilometres, propa-

Fig. 7.32
Seasat SAR image of the English Channel taken on 19 August 1978. The sandbanks are clearly revealed by the surface wave patterns (*Courtesy*: Jet Propulsion Laboratory; *Source*: Fu and Holt 1982)

gating dispersively. These were internal waves which appeared to be associated with changes of depth in water from 100 m to depths of 1000 m or more.

Applications to study the chemical characteristics

Salinity

Salinity and sea-surface temperature are two major factors controlling the circulation of water in the sea. As a chemical characteristic, salinity measures the total amount of dissolved salts in sea water. More formally, it is defined as the total dissolved material in one kilogram of sea water when all carbonate is converted to oxide; all bromine and iodine are replaced by chlorine; and all organic matter is oxidized. It is measured in parts per thousand and is usually between 33‰ and 38‰ with the norm at 35‰. Constancy of composition permits the following standard relationship between salinity (S) and chlorinity (Cl) to be established:

$$S(‰) = 1.80655 \, Cl \, (‰) \qquad [7.14]$$

Salinity also has a great impact on the growth and distribution of phytoplankton as well as the abundance and migration of shrimp and fish population. In the estuarine environment where fresh and sea water mix, the change in salinity and sea water temperature is particularly significant.

Remote sensing of salinity is closely related to that of sea-surface temperature discussed previously. Two approaches are possible. The major approach is to make use of passive microwave sensors. Another is to make use of the spectral reflectance values of pixels in multi-spectral sensor data related to concurrently measured sea truth data. It is not possible to determine sea salinity visually.

The microwave emission of sea water is dependent on temperature and salinity. Experimentally, it can be demonstrated that the apparent temperature of sea water is dependent on salinity. The dielectric constant of sea water therefore seems to depend on salinity. It is known that microwave emission from sea water is more dependent on salinity at lower microwave frequencies and more dependent on temperature at higher frequencies. By selecting the proper frequency, a passive microwave radiometer can be used to measure remotely the salinity of sea water. A suitable frequency that has been used for this purpose is L-band (1.43 GHz; Thomann 1973; Blume *et al.* 1978). The microwave radiometer measures the brightness temperature from the sea surface (T_B), which is given by:

$$T_B \, (\lambda) = e_\lambda \, (T_s, S) \, T_s \qquad [7.15]$$

where e is the emissivity, λ is the wavelength used, T_s is surface temperature and S is salinity. It is possible to produce a graph of brightness temperature plotted against salinity and surface temperature at the frequency employed for example 1.43 GHz (Fig. 7.33). The

Fig. 7.33
Brightness temperature plotted against molecular sea-surface temperature for smooth sea at normal incidence and 1.43 GHz frequency (*Source*: Blume *et al.* 1978)

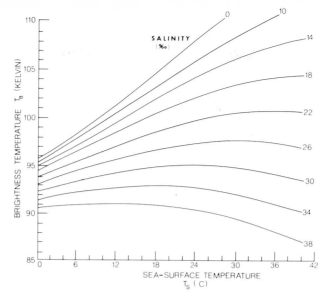

brightness temperature can then be inverted to obtain the corresponding values of T_s and S. An accuracy of $1^o/_{oo}$ salinity and 0.3 °C sea-surface temperature is achievable with this method. Blume *et al.* (1978) made use of a dual-frequency microwave radiometer system at L-(1.43 GHz) and S-(2.65 GHz) bands from an aircraft to measure salinity and temperature of the sea over the lower part of Chesapeake Bay and the adjacent Atlantic Ocean. The radiometers were carefully calibrated before the flight. The sea conditions were fairly calm. Sea truth data (salinity and sea-surface temperature) were also obtained from several locations in the measurement area within 30 minutes of the time when the radiometer data were taken. A comparison of the radiometrically measured values of surface temperature and salinity with those obtained from *in situ* measurements revealed an accuracy $0.91^o/_{oo}$ and 0.59 °C for salinity and surface temperature respectively. The calculated values of salinity and sea-surface temperature could be displayed in map form. Thus, the salinity map (Fig. 7.34) which showed isohalines with 2‰ intervals revealed much higher salinities in the ocean than in the Bay region where the mixing with fresh water outflow was considerable. The lower-salinity Bay water flowed out of the southern part of the Bay entrance and towards the south into the ocean as a result of the effect of the Coriolis force. Thomann's (1973) mapping of salinity over the estuarine environment of the Mississippi Sound in the USA using an L-band radiometer from an aircraft, similarly produced accurate results of salinity within accuracies of 3–5‰. We have seen that the improved performance of modern radiometers has significantly improved the accuracy of the measurement of salinity. One should, however, note that there are major error sources to avoid, namely, cloud reflections, land interference, sun glint and sea state changes if good, accurate salinity measurements are to be achieved.

Satellite platforms can be used to make salinity measurements. Lerner and Hollinger (1977) experimented with S-194 data

Fig. 7.34
Isohalines of the lower Chesapeake Bay in 2‰ increments on 24 August 1976 (*Source*: Blume *et al.* 1978)

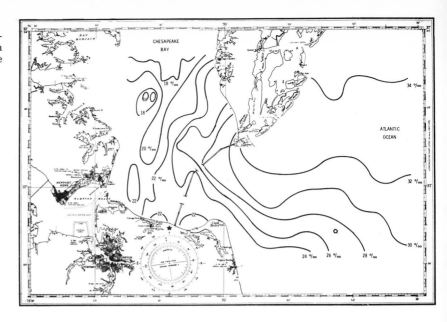

(1.400–1.427 GHz) from Skylab and found that they were relatively insensitive to sea-surface temperature but sensitive to changes in surface salinity. The analysis of data taken over the open indicated that salinity could be determined to an accuracy of ±2‰.

A second approach is to make use of Landsat MSS digital data. In his study of the San Francisco Bay delta, Khorram (1982) found that the mean radiance values of Landsat data could be used to determine salinity. Surface salinity values were measured by a refractometer at 29 predetermined sites in the San Francisco Bay delta region at the time of the Landsat overpass. High-altitude colour and colour infrared photography was used at the same time (altitude 19,800 m; 70 mm format). Landsat digital data were acquired and the sample sites were located in the Landsat coordinate system by applying a coordinate transformation equation between the Landsat data and the United States Geological Survey (USGS) topographic maps. Each sample site was represented by a nine-pixel block. The mean radiance values for 29 sample sites were then computed. It was found that these mean radiance values could be related to the ground measurements to produce a best-fit regression model based on bands four, six and seven of Landsat data only:

$$S = a + bx_4 + cx_6 + dx_7 \qquad [7.16]$$

where S is salinity expressed in parts per thousand (‰), x_4, x_6 and x_7 are respectively mean radiance values in bands four, six and seven of Landsat data. The coefficients a, b, c and d were determined empirically to be 59.96, −1.228, −3.004 and 8,981 respectively. This regression equation gave rise to a coefficient of determination of 0.75. A major source of error of the model was the effect of sea-bed reflection of the shallow areas on the radiance values of sample sites. The relationship established between the surface measurements and Landsat digital data was extended to the entire study area using a simple linear

Fig. 7.35
Salinity distribution of the San Francisco Bay delta as derived from Landsat digital data (*Source*: Khorram 1982)

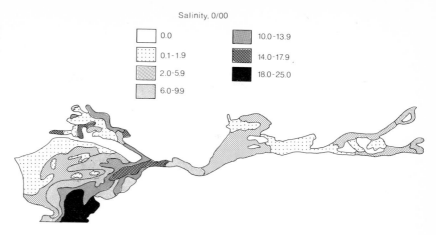

discriminant function. This gave rise to a map to show salinity distribution in the study area (Fig. 7.35), which seemed to conform quite well with salinity distribution patterns obtained by conventional measurements. Of course, the accuracy of the salinity values so determined is not as high as that from microwave sensors.

Water quality

Water quality refers not only to the chemical but also physical and biological characteristics of water in the hydrosphere. The chemical characteristics include the organic and inorganic substances such as heavy metals, pesticides, detergents and petroleum. The physical characteristics of turbidity, colour and temperature as well as the biological characteristics of plankton and chlorophyll pigment have already been discussed previously. Water quality has attracted a great deal of attention to environmental scientists and represents an area where remote sensing can be profitably employed. In this section, our discussion will focus on one major aspect of water quality – chemical pollutants such as oil slicks and sewage sludge plumes in the hydrosphere. All types of remote sensors have been employed for this purpose. One important parameter indicative of the pollution strength of wastes in water is the biochemical oxygen demand. Highly polluted water has a large biochemical oxygen demand or a high content of dissolved oxygen. Bhargava's (1983) study of the Ganga and Yuma rivers in India using a very low-altitude aerial photography suggested that the material causing this was turbid and greyish coloured, both of which correlated strongly with the photographic optical density of the film. Ocean outfall plumes can be visually interpreted in this way. Psuty and Allen (1975) applied trend-surface analysis to the digitized aerial photographic data in order to map the area patterns of two ocean outfall plumes of Bradley Beach and Asbury Park in New Jersey USA. The trend surface analysis is equivalent to a least-squares regression model incorporating the spatial dimension. The analysis separates a systematic trend from the nonsystematic component. A trend surface generates a plane that minimizes the sum of squares of the residual values. The surface can be in different degrees of complexity. It was found that by mapping the residuals (i.e. deviations) from the model for each degree of solution the uneven dispersion characteristics of the two effluent plumes were revealed. Water samples collected in the field also exhibited positive correlations between dissolved-oxygen content and residual values, indicating an increase in dissolved-oxygen content values in the direction of effluent dispersion. The drift and dispersion of ocean-dumped waste have also been studied with the use of Landsat data (Klemas and Philpot 1981). Industrial wastes were dumped at a site 64 km off the coast of Delaware, east USA. The wastes came mainly from the manufacture of titanium dioxide pigment. Sixteen Landsat images were analysed by visual and digital techniques to obtain information on acid waste plume drift, spreading (dispersion) and discrimination from other substances (Fig. 7.36). Drift distances and directions were measured directly from enlarged Landsat films. Plume drift speeds were found by dividing the distance between the plume centroid and the centre of the waste dump site by the time between dump and satellite overpass. Plume spreading rates were obtained by measuring plume width in the Landsat images. To

Fig. 7.36
Acid waste plume visible in Landsat MSS band 4 image taken on 28 August 1975 (during dump) (*Courtesy*: V. Klemas; *Source*: Klemas and Philpot 1981)

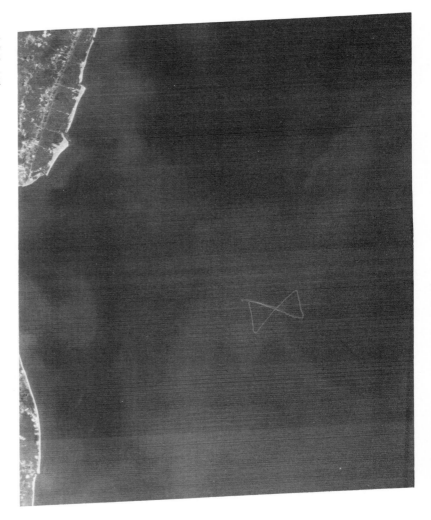

differentiate acid waste plumes from other pollutants and clouds, principal component analysis was performed using all four Landsat MSS bands. One may recall from Ch. 4 that the technique reduces the dimensionality of the input variables with the minimal loss of information. This analysis helped to separate the open ocean water into different ocean colours which could be identified as clear water, sediment-laden water, ice, clouds and an iron-acid industrial waste. The whole scene could then be classified based on the statistics so generated (Fig. 7.37). By suppressing irrelevant background information, the iron-acid waste plume could be clearly mapped. On the whole, the project revealed that the 16 waste plumes drifted at average rates of 0.59 km/hr–3.39 km/hr towards the southwest. The plume width increased at a rate of about 1.5 cm/sec during calm sea conditions and attained average spread rates of 4 cm/sec on days when winds reached speeds of 25 km/hr–38 km/hr. During storms, rapid waste movement towards the shore occurred as the plume became dispersed and diluted.

This leads us to a discussion of an important pollutant – oil spills at sea. Estes and Senger (1972) have already spelled out the importance of detecting and measuring oil pollution in the marine environment. Most of the oil spills were associated with vessels. The

Fig. 7.37
(a) Original computer printout of iron-acid waste plume of 19 January 1976, against a cloud background off the Delaware Coast, east USA; (b) enhancement of the same plume with cloud background removed (*Source*: Klemas and Philpot 1981)

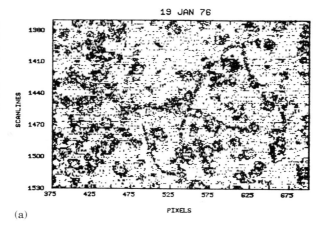

(a)

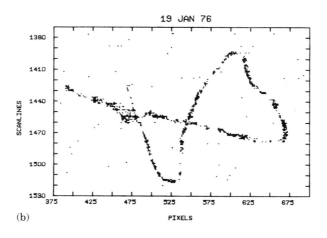

(b)

329

behaviour of oil on water is complex and an understanding of this is necessary before effective containment and clean-up operations can be undertaken. To be useful, the remote sensing system developed should be capable of not only determining accurately the location, areal extent and degree of surface coverage of an oil slick but also identifying its type and thickness. It was found that ultraviolet, thermal infrared and microwave sensors were potentially suited to oil detection. For the ultraviolet in the spectral band 0.32–0.38 μm a nonlinear inverse relationship was observed to exist between oil radiance and oil film thickness. A high contrast between oil (light tone) and water (dark tone) was apparent in the near ultraviolet image. This is unmatched by the panchromatic photography which exhibits very low contrast ratio between the two. Thermal infrared imagery could not be an effective tool under certain conditions. Because a thermal infrared sensor records thermal emission from the top 0.02 mm of the water surface only through such atmospheric windows as 3.5–5.5 μm or 8–14 μm, an oil film covering the surface of the sea produces a different emittance from that of oil-free water. At 10 μm the emissivity of oil-free water is higher (0.993) than that of oil-covered water (0.972). However, in reality, areas of oil-covered water usually appear to be both warmer and colder than the background of clean water. This may be due to reduction of heat transfer across the air-water interface by the altered surface tension caused by the oil slick; or, the result of greater heat conduction of the oil film; or the infrared energy from the background sea water being imaged through the oil slick. If a relationship between oil film thickness and apparent radiometric temperature could be established, it would facilitate volumetric measurement of an oil spill. Unlike infrared radiation, microwave radiation (0.1–100 μm) is more linearly dependent on temperature, as already explained above in connection with sea-surface temperature measurement. Water exhibits a low brightness temperature in the entire microwave region. The presence of an oil film tends to reduce the surface roughness of the ocean giving more specular reflection so that the object-to-background contrast ratios are enhanced. Passive microwave sensors appear to have the potential to identify oil type and thickness at sea. Radar, an active microwave sensor, is the most suitable tool to map oil slicks because of its large areal coverage on a single image. It has an all-weather capability and has been successfully used to detect oil slicks.

Guinard (1971) studied the oil slicks at Chedabucto Bay, Nova Scotia and Santa Barbara Channel, Southern California using a dual polarized synthetic aperture system which obtained imagery in the P, X, L and C bands. The ability of radar to detect oil was found to increase with increased sea state. The oil film on the sea helped to damp out the capillary and small gravity waves normally generated by the local winds, thus rendering the oil surface appear darker on the radar image. It was further observed that image contrast was a function of the polarization of the radar signal and that vertically polarized signals produced maximum contrast. There appeared to be a strong functional relationship between the incident radar wavelength, the sea state and the thickness of the oil film. The minimum thickness of oil detectable was 1.0 μm in low sea states. It might be possible to determine oil thickness, but not oil type with the radar. It is important to maintain the radar viewing angle within 45° of the horizontal to avoid specular returns. The performance of the real- and synthetic-aperture radars was also evaluated with regard to the detection of surface oil slicks in the Santa Barbara Channel in Southern California (Kraus *et al.* 1977). An X-band horizontally polarized AN/APS-94D real-aperture radar and an X-band vertically polarized coherent-on-receive synthetic aperture radar were flown over the study area concurrent with sea-truth support activities. It was concluded that the synthetic-aperture system detected surface concentrations of oil from all directions significantly better than the real-aperture system. The real-aperture system could only detect oil when it was looking approximately into the direction of wind and swell. The Seasat SAR images over the same study area of Estes *et al.* (1980) who observed that the numerous natural seeps of oil, tar and gas in the coastal zone exhibited themselves as oil slicks in dark bands (Figs. 7.38, 7.39). These should not be confused with other dark areas on the image which were basically due to low winds. They had masked any oil slicks signatures that might have been produced by the seeps southeast of Santa Barbara. It appeared therefore that meteorological and oceanic conditions prevailing at the time of radar imaging greatly affected the detectability of oil slicks in sea water. In view of the complexity of oil slicks, a multispectral approach in remote sensing is more appropriate if early detection and monitoring of oil in the marine environment are to be achieved.

Fig. 7.38
Seasat SAR image of the Santa Barbara Channel, Southern California where numerous natural seeps of oil, tar and gas exist in the coastal zone (taken on 18 July 1978). Note the slick patterns (dark bands) along the coast. See also Fig. 7.39 for explanations (*Courtesy*: Jet Propulsion Laboratory; *Source*: Fu and Holt 1982)

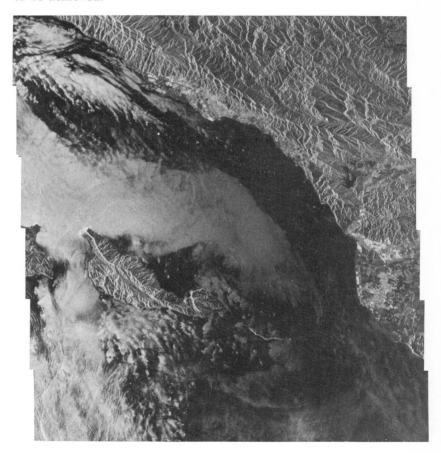

Fig. 7.39
Map of the Seasat SAR image of Santa Barbara Channel, Southern California (Photograph 7.38) indicating the occurrence of oil slicks, offshore platforms and internal waves (*Source*: Fu and Holt 1982)

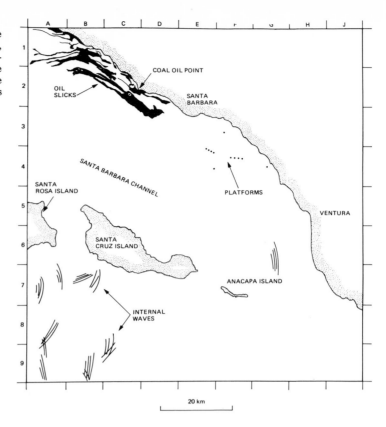

Conclusions

The application of remote sensing to the hydrosphere is perhaps the most varied of all other applications so far examined. The diversity is witnessed by the nature of data and the techniques of analysis employed. The importance of non-photographic remote sensing systems in collecting data concerning the physical, biological, geological and chemical characteristics of the sea, the major component of the hydrosphere, is obvious. Quantitative reduction of data analogous to that discussed under Ch. 3 (The Atmosphere) is necessary in extracting sea-surface temperature, sea-water colour and sea-water salinity from remotely sensed data. The microwave sensors, both the passive and active types, have been found to be extremely useful in understanding the hydrosphere. Passive microwave radiometers have been used in detecting sea-surface temperature, sea ice and salinity. Radar, the active microwave sensor, has been employed in mapping sea ice, measuring wave heights, detecting aquatic plants, determining water depths and sensing oil slicks. The availability of satellite-borne sensors makes a synoptic coverage of the ocean possible, denied by the *in situ* surface platform. The oceans of the world can be sampled many times in the same day by a satellite sensor. In other words, the satellite-borne sensors relieve the problem of collecting sea truth data using research vessels. However, there are doubts on the accuracy of the remotely sensed data from the satellite platform. In fact, Allan (1983) argued that far from diminishing the need for research vessels satellite missions actually required more surface data to help in interpreting and validating much of the data from space-borne sensors. On the other hand, there are specific types of satellite-borne sensors designed for oceanographic applications, such as the Coastal Zone Colour Scanner (CZCS) launched in 1978 on Nimbus-7 designed specifically to provide information to marine biologists and pollution detection agencies, and the Seasat experiment launched on 28 June 1978 aimed to evaluate the performance of a suite of microwave sensors to study the ocean surface. It was anticipated that accurate data on wind speed, wind direction, significant wave height and high-resolution images of the ocean surface could be obtained. Despite the premature failure of Seasat, the data gathered within the 100 days of its performance showed such encouraging results that future satellite systems for remote sensing of the ocean surface have been planned, such as MOS-1 (Marine Observation Satellite) of Japan, SPOT-2 (Poseidon) of France and the ERS-1 (European Remote Sensing) mission of the European Space Agency (Allan 1983).

Apart from the environmental satellite systems, resources satellite systems, notably Landsat, have also contributed significantly to our understanding of the hydrosphere. Both the manual and digital techniques have been employed with a view to map the ocean outfall plumes and the levels of turbidity.

To conclude, remote sensing of the hydrosphere has placed an increasing emphasis on the design of the satellite-borne sensors and the development of more sophisticated techniques of analysis in order to accurately extract physical, biological, geological and chemical data for the environmental scientists.

References

Allan, T. D. (1983) Oceanography from space: In Cracknell, A. P. (ed.) *Remote Sensing Applications in Marine Science and Technology*. D. Reidel: Dordrecht; Boston; Lancaster; pp. 409–33.

Anding, D. and Kauth, R. (1970) Estimation of sea surface temperature from space, *Remote Sensing of Environment* 1: 217–20.

Apel, J. R. (1983) Remote measurement of the ocean: an overview: In Cracknell, A. P. (ed.) *Remote Sensing Applications in Marine Science and Technology*. D. Reidel: Dordrecht; Boston; Lancaster; pp. 1–16.

Apel, J., Byrne, M. H., Proni, J. R. and Charnell, R. L. (1975) Observations of oceanic internal and surface waves from Earth Resources Technology Satellite, *Journal of Geophysical Research* 80: 865–81.

Barnes, J. C. and Bowley, C. J. (1974) *The Application of ERTS Imagery to Monitoring Arctic Sea Ice*. NASA: Goddard Space Flight Center, Greenbelt, Maryland.

Beal, R. C. (1981) The monitoring of large scale synoptic features of the ocean with spaceborne synthetic aperture radar: In Gower, J. F. R. (ed.) *Oceanography from Space*. Plenum Press; New York, London; pp. 505–10.

Bhargava, D. S. (1983) Very low altitude remote sensing of the water quality of rivers, *Photogrammetric Engineering and Remote Sensing* 49: 805–9.

Blume, H. C., Dendall, B. M. and Fedors, J. C. (1978) Measurement of ocean temperature and salinity via microwave radiometry, *Boundary-Layer Meterology* 13: 309–37.

Bullard, R. K. (1983a) Land into sea does not go: In Cracknell, A. P. (ed.) *Remote Sensing Applications in Marine Science and Technology*. D. Reidel: Dordrecht; Boston; Lancaster, pp. 359–72.

Bullard, R. K. (1983b) Detection of marine contours from Landsat film and tape: In Cracknell, A. P. (ed.) *Remote Sensing Applications in Marine Science and Technology*. D. Reidel: Dordrecht; Boston; Lancaster, pp. 373–81.

Campbell, W. J. (1976a) Tracking ice floes by sequential ERTS imagery: In Williams, Jr., R. S. and Carter, W. D. (eds) *ERTS-1: A New Window on Our Planet*. Geological Survey Professional Paper 929: US Government Printing Office, Washington, DC; pp. 337–9.

Campbell, W. J. (1976b) Ice lead and polynya dynamics: In Williams, Jr., R. S. and Carter, W. D. (eds) *ERTS-1: A New Window on Our Planet*. Geological Survey Professional Paper 929: US Government Printing Office, Washington, DC; pp. 340–2.

Campbell, W. J. (1976c) Seasonal metamorphosis of sea ice: In Williams, Jr., R. S. and Carter, W. D. (eds) *ERTS-1: A New Window on Our Planet*. Geological Survey Professional Paper 929: US Government Printing Office, Washington, DC, pp. 343–5.

Campbell, W. J. (1976d) Dynamics of Arctic ice-shear zones: In Williams, Jr., R. S. and Carter, W. D. (eds) *ERTS-1: A New Window on Our Planet*. Geological Survey Professional Paper 929: US Government Printing Office, Washington, DC, pp. 346–9.

Campbell, W. J. (1976e) Morphology of Beaufort Sea ice: In Williams, Jr., R. S. and Carter, W. D. (eds) *ERTS-1: A New Window on Our Planet*. Geological Survey Professional Paper 929: US Government Printing Office, Washington, DC, pp. 350–5.

Campbell, W. J., Wayenberg, J., Ramseyer, J. B., Ramseier, R. O., Vant, M. R., Weaver, R., Redmond, A., Arsenault, L., Gloersen, P., Zwally, H. J., Wilheit, T. T., Chang, T. C., Hall, D., Gray, L., Meeks, D., Bryan, M. L., Barath, F. T., Elachi, C., Leberl, F. and Farr, T. (1978) Microwave remote sensing of sea ice in the AIDJEX main experiment, *Boundary-Layer Meteorology* **13**: 309–37.

Cartwright, D. E. and Alcock, G. A. (1983) Altimeter measurements of ocean topography: In Allan, T. D. (ed.) *Satellite Microwave Remote Sensing*. Ellis Horwood Ltd: Chichester; pp. 308–19.

Cox, C. and Munk, W. (1954) Statistics of the sea surface derived from sun glitter, *Journal of Marine Research* **13**: 198–227.

Dunbar, M. (1975) Interpretation of SLAR imagery of sea ice in Nares Strait and Arctic Ocean, *Journal of Glaciology* **15**: 193–213.

Estes, J. E. and Senger, L. W. (1972) The multispectral concept as applied to marine oil spills, *Remote Sensing of Environment* **2**: 141–63.

Estes, J. E., Wilson, M. and Hajic, E. (1980) *Analysis of Seasat-A SAR Data for the Detection of Oil on the Ocean Surface*. University of California: Santa Barbara, California, USA.

Finley, R. J. and Baumgardner, Jr., R. W. (1980) Interpretation of surface-water circulation, Arkansas Pass, Texas, using Landsat imagery, *Remote Sensing of Environment* **10**: 3–22.

Fiuza, A. F. G. (1979) Airborne SST measurements and the sardine fishery off Portugal, *Proceedings of Applications of Remote Sensing to Fisheries Research*. Valbonne, France; pp. 38–51.

Fu, L. L. and Holt, B. (1982) *Seasat Views Oceans and Sea Ice with Synthetic-Aperture Radar*. JPL Publication 81–120: Jet Propulsion Laboratory: California Institute of Technology, Pasadena, California.

Gatto, L. W. (1982) Ice distribution and winter surface circulation patterns, Kachemak Bay, Alaska, *Remote Sensing of Environment* **12**: 421–35.

Gerson, D. J. (1958) A technique for time-lapse photography of sea ice, *Arctic Sea Ice*. National Academy of Sciences: Washington, DC, pp. 259–64.

Gerson, D. J. and Rosenfeld, A. (1975) Automatic sea ice detection in satellite pictures, *Remote Sensing of Environment* **4**: 187–98.

Gloersen, P., Campbell, W. J. and Cavalieri, D. (1981a) Global maps of sea ice concentration, age, and surface temperature derived from Nimbus-7 Scanning Multichannel Microwave Radiometer data: a case study: In Gower, J. F. R. (ed.) *Oceanography from Space*. Plenum Press: New York; London; pp. 777–83.

Gloersen, P., Cavalieri, D. and Campbell, W. J. (1981b) Derivation of sea ice concentration, age and surface temperature from multispectral microwave radiances obtained with the Nimbus-7 Scanning Multichannel Microwave Radiometer: In Gower, J. F. R. (ed.) *Oceanography from Space*. Plenum Press: New York; London; pp. 823–29.

Gloersen, P., Nordberg, W., Schmugge, T. T., Wilheit, T. T. and Campbell, W. J. (1973) Microwave signatures of first-year and multi-year sea ice, *Journal of Geophysical Research* **78**: 3564–72.

Gordon, H. R. (1978) Removal of atmospheric effects from satellite imagery of the oceans, *Applied Optics* **17**: 1631–6.

Gordon, H. R., Clark, D. K., Mueller, J. L. and Hovis, W. A. (1980) Nimbus-7 Coastal Zone Color Scanner: system description and initial imagery, *Science* **210**: 60–6.

Gower, J. F. R. (1979) The computation of ocean wave heights from GEOS-3 satellite radar altimeter data, *Remote Sensing of Environment* **8**: 97–114.

Gower, J. F. R., Denman, K. L. and Holyer, R. J. (1980) Phytoplankton patchiness indicates the fluctuation spectrum of mesoscale oceanic structure, *Nature* **288**: 157–9.

Gray, A. L. (1981) Microwave remote sensing of sea ice: In Gower, J. F. R. (ed.) *Oceanography from Space*. Plenum Press: New York; London; pp. 785–800.

Gudmandsen, P. E. (1983) Application of microwave remote sensing to studies of sea ice, *Philosophical Transactions of the Royal Society of London* **A 309**: 433–45.

Guinard, N. W. (1971) The remote sensing of oil slicks, *Proceedings of Seventh International Symposium on Remote Sensing of Environment*. Willow Run Laboratories, Institute of Science and Technology, University of Michigan: Ann Arbor, Michigan; pp. 1005–26.

Hovis, W. A. (1981) The Nimbus-7 Coastal Zone Color Scanner (CZCS) program: In Gower, J. F. R. (ed.) *Oceanography from Space*. Plenum Press: New York; London; pp. 213–25.

Hughes, D. G. (1979) Requirements of fisheries research from remote sensing, *Proceedings of Applications of Remote Sensing to Fisheries Research*. Valbonne: France; pp. 14–21.

Hunter, R. E. and Hill. G. W. (1980) Nearshore current pattern off south Texas: an interpretation from aerial photographs, *Remote Sensing of Environment* **10**: 115–34.

Johnson, J. D. and Farmer, L. D. (1971) Use of side-looking air-borne radar for sea ice identification, *Journal of Geophysical Research* **76**: 2138–55.

Johnson, R. W. (1978) Mapping of chlorophyll *a* distributions in coastal zones, *Photogrammetric Engineering and Remote Sensing* **44**: 617–24.

Kemmerer, A. J. (1979) Remote sensing of living marine resources, *Proceedings of Applications of Remote Sensing to Fisheries Research*. Valbonne; France; p. 86.

Ketchum, Jr., R. D. and Lohanick, A. W. (1980) Passive microwave imagery of sea ice at 33 GHz, *Remote Sensing of Environment* **9**: 211–23.

Ketchum, Jr., R. D. and Tooma, Jr., S. G. (1973) Analysis and interpretation of air-borne multifrequency side-looking radar sea ice imagery, *Journal of Geophysical Research* **78**: 520–38.

Khorram, S. (1982) Remote sensing of salinity in the San Francisco Bay delta, *Remote Sensing of Environment* **12**: 15–22.

Klemas, V., Borchardt, J. F. and Treasure, W. M. (1973) Suspended sediment observations from ERTS-1, *Remote Sensing of Environment* **2**: 205–21.

Klemas, V., Otley, M., Philpot, W., Wethe, C., Rogers, R. and Shah, N. (1974) Correlation of coastal water turbidity and current circulation with ERTS-1 and Skylab imagery, *Proceedings of the Ninth International Symposium on Remote Sensing of Environment*. Environmental Research Institute of Michigan: Ann Arbor, Michigan; pp. 1289–1317.

Klemas, V. and Philpot, W. D. (1981) Drift and dispersion studies of ocean-dumped waste using Landsat imagery and current drogues, *Photogrammetric Engineering and Remote Sensing* **47**: 533–42.

Klooster, S. A. and Scherz, J. P. (1974) Water quality by photographic analysis, *Photogrammetric Engineering* **40**: 927–35.

Kraus, S. P., Estes, J. E., Atwater, S. G., Jensen, J. R. and Vollmers, R. R. (1977) Radar detection of surface oil slicks, *Photogrammetric Engineering and Remote Sensing* **43**: 1523–31.

Krishen, K. (1975) Remote sensing of oceans using microwave sensors, in Veziroglu, T. N. (ed.), *Remote Sensing: Energy-Related Studies*, Hemisphere Publishing Corporation, Washington and London, pp. 61–99.

Langmuir, I. (1938) Surface motion of water induced by wind, *Science* **87**: 119–23.

Legeckis, R., Legg, E. and Limeburner, R. (1980) Comparison of polar and geostationary satellite infrared observations of sea surface temperatures in the Gulf of Maine, *Remote Sensing of Environment* **9**: 339–50.

Lerner, R. M. and Hollinger, J. P. (1977) Analysis of 1.4 GHz radiometric measurements from Skylab, *Remote Sensing of Environment* **6**: 251–69.

Livingstone, C. E., Hawkins, R. K., Gray, A. L., Okamoto, K., Wilkinson, T. L., Young, S., Arsenault, L. D. and Pearson, D. (1981) Classification of Beaufort Sea ice using active and passive microwave sensors: In Gower, J. F. R. (ed.) *Oceanography from Space*. Plenum Press: New York; London; pp. 813–21.

Lodge, D. W. S. (1983) Surface expressions of bathymetry on SEASAT synthetic aperture radar images, *International Journal of Remote Sensing* **4**: 639–53.

Mairs, R. T. (1970) Oceanographic interpretation of Apollo photographs *Photogrammetric Engineering* **36**: 1045–58.

Maul, G. A. (1978) Locating and interpreting hand-held photographs over the ocean: a Gulf of Mexico example from the Apollo-Soyuz test project, *Remote Sensing of Environment* **7**: 249–63.

Mitchell, H. L. (1983) Wave heights in the surf zone, *Photogrammetric Record* **11**: 183–93.

Njoku, E. G. and Hofer, R. (1981) SEASAT SMMR observations of ocean surface temperature and wind speed in the North Pacific: In Gower, J. F. R. (ed.) *Oceanography from Space*. Plenum Press; New York; London; pp. 673–81.

Njoku, E. G., Stacey, J. M. and Barath, F. T. (1980) The SEASAT Scanning Multichannel Microwave Radiometer (SMMR): instrument description and performance, *IEEE Journal of Oceanic Engineering* **OE-5**: 100–15.

Platt, C. M. R. and Troup, A. J. (1973) A direct comparison of satellite and aircraft infrared (10 μm–12 μm) remote measurements of surface temperature, *Remote Sensing of Environment* **2**: 243–7.

Pluhowski, E. J. (1976) Dynamics of suspended plumes: In Williams, Jr., R. E. and Carter, W. D. (eds) *ERTS-1: A New Window on Our Planet*. Geological Survey Professional Paper 929: US Government Printing Office: Washington, DC, pp. 157–8.

Prabhakara, C., Dalu, G. and Kunde, V. G. (1974) Estimation of sea surface temperature from remote sensing in the 11- to 13-μm window region, *Journal of Geophysical Research* **79**: 5039–44.

Pravdo, S. H., Huneycutt, B., Holt, B. M. and Held, D. N. (1983) *Seasat Synthetic-Aperture Radar Data User's Manual*. Jet Propulsion Laboratory, California Institute of Technology, Pasadena, California.

Psuty, N. P. and Allen, J. R. (1975) Trend-surface analysis of ocean outfall plumes, *Photogrammetric Engineering and Remote Sensing* **41**: 713–30.

Ritchie, J. C., Schiebe, F. R. and McHenry, J. R. (1976) Remote sensing of suspended sediments in surface waters, *Photogrammetric Engineering and Remote Sensing* **42**: 1539–45.

Robinson, I. S. (1983) Satellite observations of ocean colour, *Philosophical Transactions of the Royal Society of London*, **A309**: 415–32.

Sathyendranath, S. and Morel, A. (1983) Light emerging from the sea—interpretation and uses in remote sensing: In Cracknell, A. P. (ed.) *Remote Sensing Applications in Marine Science and Technology*. D. Reidel: Dordrecht; Boston; Lancaster, pp. 323–57.

Schmidt, D. (1979) Blue-green algae in the Baltic as shown by Landsat imagery, *Proceedings of Applications of Remote Sensing to Fisheries Research*. Valbonne, France, pp. 69–83.

Sidran, M. (1980) Infrared sensing of sea surface temperature from space, *Remote Sensing of Environment* **10**: 101–14.

Singh, S. M. and Warren, D. E. (1983) Sea surface temperatures from infrared measurements: In Cracknell, A. P. (ed.) *Remote Sensing Applications in Marine Science and Technology*. D. Reidel: Dordrecht; Boston; Lancaster, pp. 231–62.

Stevenson, R. E., Carter, L. D., Haar, S. P. V. and Stone, R. O. (1976) Visual observations of the ocean: In *Skylab Explores the Earth*. National Aeronautics and Space Administration: Washington, DC, pp. 287–338.

Sydor, M. (1976) Turbidity in Lake Superior: In Williams, Jr., R. S. and Carter, W. D. (eds) *ERTS-1: A New Window on Our Planet*. Geological Survey Professional Paper 929. US Government Printing Office: Washington, DC, pp. 153–6.

Teleki, G. (1958) The utilization of aerial photographs in sea ice forecasts, *Arctic Sea Ice*. National Academy of Sciences: Washington, DC, pp. 76–9.

Thomann, G. C. (1973) Remote measurement of salinity in an estuarine environment, *Remote Sensing of Environment* **2**: 249–59.

Tucker, M. J. (1983) Observation of ocean waves, *Philosophical Transactions of Royal Society of London* **A309**: 371–80.

Wald, L. and Monget, J. M. (1983) Remote sensing of the sea-state using the 0.8–1.1 µm spectral band, *International Journal of Remote Sensing* **4**: 433–46.

Walsh, D. (1976) Remote sensing in oceanography: In Lintz, Jr., J. and Simonett, D. S. (eds) *Remote Sensing of Environment*. Addison-Wesley: Reading, Mass., pp. 593–636.

Wark, D. Q., Popham, R. W., Dotson, W. A. and Colaw, K. S. (1962) Ice observations by the TIROS II satellite and by aircraft, *Arctic* **15**: 9–26.

Wilheit, T. T. and Chang, A. T. C. (1980) An algorithm for retrieval of ocean and atmospheric parameters from the observations of the Scanning Multichannel Microwave Radiometer, *Radio Science* **15**: 525–44.

Wilheit, T. T., Nordberg, W., Blinn, J., Campbell, W. J. and Edgerton, A. (1972) Aircraft measurements of microwave emission from Arctic Sea ice, *Remote Sensing of Environment* **2**: 129–39.

Yentsch, C. S. (1983) Remote sensing of biological substances: In Cracknell, A. P. (ed.) *Remote Sensing Applications in Marine Science and Technology*. D. Reidel: Dordrecht; Boston; Lancaster, pp. 263–97.

Zheng, Q. A. and Klemas, V. (1982) Determination of winter temperature patterns, fronts, and surface currents in the Yellow Sea and East China from satellite imagery, *Remote Sensing of Environment* **12**: 201–18.

Chapter 8 Cartographic presentation of remote sensing data

Requirements for cartographic presentation of remote sensing data

In previous chapters, remote sensing has been applied to collect qualitative and quantitative data of our terrestrial environment which comprises the atmosphere, the lithosphere and the hydrosphere. The data are more usually spatial in nature and are presented in the form of maps. These maps are thematic maps because they tend to display the spatial variations of a single phenomenon or the relationship between phenomena. The thematic map is more usually small scaled and its accuracy is assessed by the correctness of the spatial pattern displayed. Thematic mapping is possible only when a base map is available. This base map displays topographic features, such as roads, settlements, rivers, coastlines, water bodies and elevations in the form of contours, in their correct positions. A large-scale map of this type is called a topographic map. There are certain requirements to be met by the topographic map in terms of the accuracy of the positions of the points (planimetric or horizontal accuracy) and the accuracy of the elevations (height or vertical accuracy).

These accuracy requirements are normally quite stringent. In the US, the National Map Accuracy Standards (NMAS) specify that for planimetry 90 per cent of all points plotted should be within 1/30 inch (0.85 mm) of their correct position at the scale of publication for all maps published at scales larger than 1:20,000. As for other scales, 90 per cent of all points should lie within 1/50 inch (0.50 mm) of their correct position at the publication scale. In the case of vertical accuracy, 90 per cent of all contours shown should not be in error by more than one-half of the contour interval (Thompson 1979). In Europe, the national mapping standards are even higher. All these standards can be met with the use of land surveying techniques. Is it possible that the remote sensing data can be utilized to produce topographic maps that will meet all these standards? In this chapter, we shall explore in some depth this question of cartographic application of remote sensing.

Aerial photography for topographic mapping

Basic principles of photogrammetry

Conventional aerial photography provides the major source of data for topographic mapping using the theory of photogrammetry which forms a discipline of its own (Wolf 1983; Slama 1980; Moffitt and Mikhail 1980; Kilford 1979; Burnside 1979). Photogrammetry is the science and art of obtaining reliable measurements from photographs, but photogrammetry is always closely related to photo-interpretation (Tait 1970) because one needs to know what one is measuring. In order to meet the stringent standards of height and planimetric accuracies, a metric camera is used to acquire the photography. The camera has been calibrated to determine the position of its principal point (the photo-centre), the position of the projection centre (the lens) and the principal distance (focal length), as shown in Fig. 8.1. This is necessary to ensure that the photographs can reconstruct as exactly as possible the stereomodel produced at the time of photography. The aerial camera makes use of a 23 × 23 cm format stable base film for photogrammetric missions. Normally, vertical aerial photography is preferred for topographic mapping. The flying for aerial photography has been designed to meet the following essential conditions: (1) a straight line course at a constant speed and a predetermined height; (2) a forward overlap of 60 per cent; (3) a lateral lap of 20 per cent between two adjacent flight strips; (4) free from tilt at the time of exposure (Fig. 1.7; Lo 1976). These permit aerial photographs to be joined together into a block. The aerial photography produces stereomodels when viewed under a stereoscope, which form the basis for measurement of heights and positions in topographic mapping. As explained in Ch. 1, the

Fig. 8.1
Geometric relationship of fiducial marks, principal point, principal distance and lens (projection centre) in a vertical aerial photograph

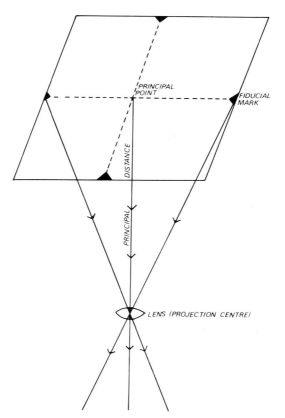

geometry of an aerial photograph is characterized by a central perspective projection with a fixed projection centre for each exposure (Fig. 8.2). This greatly simplifies the formulae developed for the computation of coordinates of points on the photograph. The scale of a vertical aerial photograph (S) at terrain elevation h_A is simply given by:

$$S = \frac{f}{H - h_A} \qquad [8.1]$$

where f is the focal length of the camera lens and H is the flying height (Fig. 8.2). It is clear that the scale is affected by the relief of the terrain and because of the central perspective projection, relief displacements are seen to increase radially from the centre of the aerial photograph. The amount of relief displacement (aa′) can be computed from a single aerial photograph by

$$aa' = \frac{pa' \cdot h_A}{H} \qquad [8.2]$$

$$\text{or } h_A = \frac{aa' \cdot H}{pa'} \qquad [8.3]$$

where *pa′* is the radial distance measured from the centre of the photograph (i.e. principal point) to the object in question, h_A is the elevation of the object, *A*, and *H* is the flying height (Fig. 8.2). For an overlapping pair of aerial photographs, however, heights of objects can be obtained more easily from a measurement of the X-parallax which is defined as the apparent image displacement created along the X-direction of the aerial photographs, i.e. parallel to the line of flight. Figure 8.3 indicates the relationship between X-parallax, the parallactic angle (α) and elevation. The X-parallax can be seen as the algebraic difference in the X-coordinates of a point, i.e. $\alpha = x_1 - x_2$ for

Fig. 8.2
Central perspective projection and the effect of relief displacement

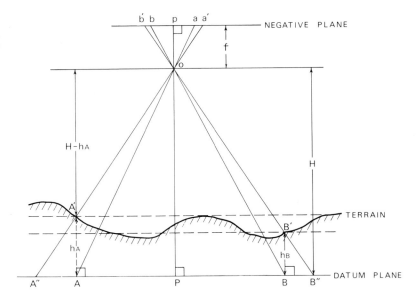

Fig. 8.3
Definition of X-parallax (*Source*:
Lo 1976)

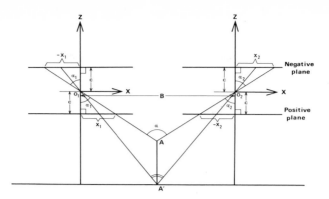

Fig. 8.4
Relationship between the height
of a point and its parallax for
the deriviation of the parallax
equation (*Source*: Lo 1976)

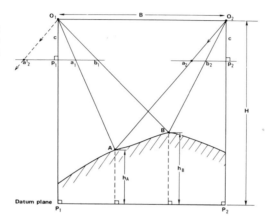

the positive plane or $\alpha = x_2 - x_1$ for the negative plane. From Fig. 8.4, the geometric definition of the absolute stereoscopic parallax or X-parallax at the point A (or P_A) is given by:

$$P_A = a_2^1 p_1 + p_1 a_1 = a_2^1 a_1 \qquad [8.4]$$

since $O_1 a_2^1 // O_2 a_2$. Thus by similar triangles $\Delta a_2^1 O_1 a_1$ and $\Delta O_1 A O_2$, one can obtain:

$$\frac{a_2^1 a_1}{c} = \frac{B}{H - h_A} \qquad [8.5]$$

By substituting equation [8.4] to equation [8.5], we get:

$$P_A = \frac{c \cdot B}{H - h_A} \qquad [8.6]$$

Similarly, for the point B, the X-parallax, P_B, is given by:

$$P_B = \frac{c \cdot B}{H - h_B} \qquad [8.7]$$

341

The difference in parallax between point A and point B (ΔP_{BA}) is given by:

$$\Delta P_{BA} = \frac{c \cdot B}{H - h_B} - \frac{c \cdot B}{H - h_A}$$

$$= \frac{c \cdot B(h_B - h_A)}{(h - h_B)(h - h_A)} = \frac{P_B(h_B - h_A)}{(H - h_A)} \quad [8.8]$$

Let $(h_B - h_A) = \Delta h_{BA}$, then we get

$$\Delta P_{BA} = \frac{P_B \cdot \Delta h_{BA}}{(H - h_A)} \quad [8.9]$$

$$\text{or } \Delta h_{BA} = \frac{\Delta P_{BA}(H - h_A)}{P_B} = \frac{\Delta P_{BA}(H - h_A)}{P_A + \Delta P_{BA}} \quad [8.10]$$

Equation [8.10] is known as the parallax formula and can be used to find the height difference between two points. This difference in parallax can be measured by a special instrument called *parallax bar* or *stereometer* under a mirror stereoscope. The absolute parallax, P_A, can be measured directly from the aerial photographs with an engineer scale. The formula assumes that the aerial photographs are perfectly vertical. This is rarely true. Aircraft tilts usually occur at the time of photography, thus causing deformations of the stereomodel. These tilts can be visualized as rotations about the X (the nose), Y (the wing) and Z (the height) axes of the aircraft, which are respectively known as ω (roll or tilt), ϕ (pitch or tip) and κ (yaw or swing) rotations in photogrammetry. There are also errors caused by shifts in the position of the aircraft in the same three directions, namely, *bx*, *by* and *bz*. These errors give rise to Y-parallaxes which cause difficulty in obtaining a stereomodel visually. To correct for such model deformations, Thompson (1954) proposed the use of a polynomial equation in the following form:

$$dh = a_0 + a_1x + a_2y + a_3xy + a_4x^2 \quad [8.11]$$

where *dh* is the height error, *x* and *y* are coordinate values of the point in question, and a_0 to a_4 are coefficients. This equation essentially assumes the occurrence of two types of errors: the hyperbolic and parabolic errors which are caused by the deformation of the stereomodel due to tilts. In order to calibrate the mathematical model, five ground control points of known heights should be employed so that the five coefficients can be determined. These five control points should be placed in such a way that they evenly cover the whole stereo-overlap (Fig. 8.5). With the general availability of electronic calculators and microcomputers the solution of the five simultaneous equations is easy. Further improvements to Thompson's method can be made by employing more control points distributed in different patterns and using the least squares adjustment with the aid of the computer (Fig. 8.5; Methley 1970). The parallax bar which makes use of the floating dot to measure X-parallaxes and hence heights computed from the parallax equation [8.10] provides the basis for an analogue approach to extract metric data from a stereomodel of the terrain. The more advanced photogrammetric systems are essentially designed on such a principle with more accurate simulation of the stereomodel and

Fig. 8.5
Patterns of control point distribution (*Source:* Lo 1976)

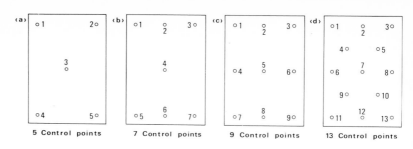

more rigorous solution to the parallax equation. The movement of the floating mark is transferred to the plotting pencil of a pantograph arm so that contours and planimeteric details can be directly plotted from the stereomodel. An analogue photogrammetric plotting machine is equipped with orientation elements (ω, ϕ and κ) and shifts (bx, by and bz) for each photographic carriage, thus permitting a correct reconstruction of the stereomodel for measurement by the floating dot. Three major types of orientation have to be carried out to set up the stereomodel correctly: (1) *inner orientation* or the correct centering and setting of the principal distance of the aerial camera; (2) *relative orientation* or the correct orientation between the two photographs with the elimination of the Y-parallax in the stereomodel; (3) *absolute orientation* or setting the stereomodel at the correct height and scale with reference to ground control points. There are different types of photogrammetric plotting machines according to performance and accuracy. One should, however, bear in mind that good ground controls are absolutely essential if accurate topographic maps are to be produced from aerial photography. Aerial photography can also be used to obtain both horizontal and vertical controls by the photogrammetric method called *aerial triangulation*, based only on a framework of ground surveyed points.

Orthophotography

An important development relating to topographic mapping from aerial photography is *orthophotography*. This is a process which transforms the central projection of an aerial photograph into the orthographic projection required of a map. The orthophotograph produced is essentially a photographic reproduction of the aerial photograph in which the distortions of images due to tilts and relief have all been removed. This is achieved by a method called *differential rectification*. The terrain is broken down into a number of flat areas at different heights above the datum, each of which is then rectified (i.e. removing tilts). This can be carried out mechanically by an analogue photogrammetric instrument. The floating mark is replaced by a narrow slit and a piece of sensitive film material is placed on the plotting table. The whole instrument operates under dark-room conditions. After all orientations have been completed, the slit is automatically scanned across the overlap (the stereomodel). The operator has to set the slit continuously at the correct height as the stereomodel is viewed. As the slit passes over the sensitive film material, an image is exposed. By scanning along the Y-direction and then stepping over in the X-direction, the whole stereomodel can be traversed. Obviously, this method puts a lot of strain on the operator. The accuracy of the orthophotograph depends on the width of the slit and the speed of scanning. A narrower slit and slower scanning speed will produce better orthophotographs because the operator will be able to set the heights more accurately. However, resultant orthophotographs usually show distinct strip effects and feature mismatches, especially over rugged terrains or buildings (Fig. 8.6; Helava 1968).

There are many different types of instruments developed for orthophotograph production. These normally take the form of an attachment to the analogue photogrammetric plotter, such as the Zeiss (Oberkochen) GZ-1 Orthoprojector for the Stereoplanigraph C8 stereoplotter, and Wild PPO-8 orthophoto printer for Wild Autograph A8 stereoplotter. The two, however, differ in their method of producing the orthophotograph. The former transfers the image by geometrically correct reprojection using a third projector while the latter carries out the image transfer by frontal optical projection or tapping the optical train from one of the projectors of the stereoplotter and projecting the image orthogonally onto the film wrapped around a drum.

Because the projection of orthophotographs requires continuous measurement of the stereomodel with the floating mark in a stereoplotter to which the orthophoto printing system is attached, a series of height profiles are in fact produced at the same time during scanning, hence the term profile scanning. It is possible to produce lines of different thicknesses known as 'droplines' according to a pre-set height interval for exposure on the film. By joining the ends of drop lines of the same types contour lines can be produced (Fig. 8.7). Orthophotography has thus speeded up topographic mapping from aerial photographs. An international experiment conducted to check the accuracy of orthophotographs produced by various instruments has confirmed excelled geometric quality of orthophotographs and the related height information (Blachut and Van Wijk 1976). In general, a standard error of ±0.2–0.3 mm in planimetry and ±0.4 per thousand of the flying

Fig. 8.6
A part of the orthophotograph of Kelvingrove area, Glasgow, produced by the Zeiss (Oberkochen) GZ-1 orthoprojector with a scan speed of 2.5 mm/sec. Note the chopping up of the bridge at A and mismatches of the roof of the building at B (*Courtesy*: Department of Geography, University of Glasgow).

Fig. 8.7
Production of droplines: (a) representation of heights by using lines of different thicknesses; (b) a portion of the dropline map produced with a 4 mm slit width and height contours derived from this (*Source*: Lo 1976)

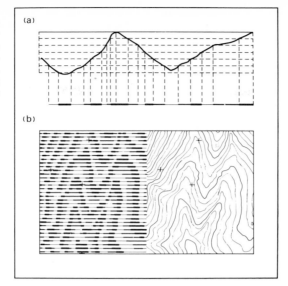

(a)

(b)

height in height accuracy is achievable. As a relatively inexpensive technique, orthophotographs have attracted much attention from cartographers. Many orthophotomaps have been produced as a result. Good examples are the Economic Map of Sweden (1:10,000 and 1:20,000), the West German Grundkarte or Base Map (1:5,000), the Orthophotoplans of Belgium (1:10,000) and the Orthophotoquads of the United States Geological Survey (USGS) (1:24,000) for selective areas as a substitute for or complement to the standard topographic line maps (Petrie 1977; Southard 1978).

Automation in photogrammetry

With the advances in electronics and computer technology, a high degree of automation in photogrammetry becomes possible. The three major approaches are: (1) image correlating; (2) the fully analytical approach; (3) the hybrid system. The image correlation technique makes use of cathode ray tubes (CRT), photomultipliers and an electrical analogue computer. The image densities from two photographs are compared automatically and the photographs are shifted, one relative to the other until a best correlation is found between the two images. In this way, the Y-parallax can be removed and relative and absolution orientation of the stereomodel can be completed automatically. By measuring the X-parallaxes, height information can be obtained. The output from the system consists of an orthophotograph and a drop-line chart. By combining the two together a topographic map can be produced quite automatically. Good examples of such a system are the Wild-Raytheon A2000 Stereomat designed by Hobrough (1959), Bunker-Ramo's Universal Automatic Map Compilation Equipment (UNAMACE) of the US military, and the Gestalt Photomapper (GPM) of Canada, originally designed also by Hobrough (Dowman 1977). The Gestalt Photomapper which is widely used in the commercial field is controlled by a computer which carries out image correlation digitally (Masry *et al.* 1976). A series of closely spaced lines, covering an area of about 9×8 mm and oriented along epipolar lines on one photograph, are correlated with a similar series on the other photograph. The image area is transformed by a high order polynomial to give the best correlation and to ensure a good fit with adjacent areas. The output is an orthophotograph. The X-parallaxes are converted to heights and are used to produce a digital terrain model (DTM) from which contour plots can be obtained. Good accuracy is achieved (Crawley 1975).

The fully analytical approach is to make use of a digital computer to solve mathematically for the unknowns in the *collinearity equations* which relate the model coordinates and image coordinates together as shown below in matrix form:

$$\begin{bmatrix} X_A - X_O \\ Y_A - Y_O \\ Z_A - Z_O \end{bmatrix} = \frac{1}{k} \begin{bmatrix} a_{11} & a_{12} & a_{13} \\ a_{21} & a_{22} & a_{23} \\ a_{31} & a_{32} & a_{33} \end{bmatrix} \begin{bmatrix} x - x_h \\ y - y_h \\ c \end{bmatrix} \qquad [8.12]$$

where X_A, Y_A and Z_A are coordinates for point A in the object space; X_O, Y_O and Z_O are the ground coordinates of the perspective centre (lens); x, y, z are coordinates of point A in the image space; x_h and y_h are coordinates of the principal point of the photograph, c is the principal distance; k is a scale factor; and $a_{11} \ldots a_{33}$ are functions of the orientation elements ω, ϕ and κ. They are called collinearity equations because they express the relationship that the perspective centre, the image point and the object point all lie on a straight line. The operation of the analytical plotter is shown in Fig. 8.8. The photographic plates are roughly oriented and all data concerning the principal distance (focal length), lens distortions, atmospheric refraction and earth curvature corrections as well as calibrated fiducial mark coordinates are entered into the computer storage. The floating mark is set by the operator to measure the coordinates of the fiducial marks on each plate. The computer then computes the correct position of the principal point and the directions of the coordinate axes of each plate.

Fig. 8.8
Components of an analytical plotter (*Source*: Moffitt and Mikhail 1980)

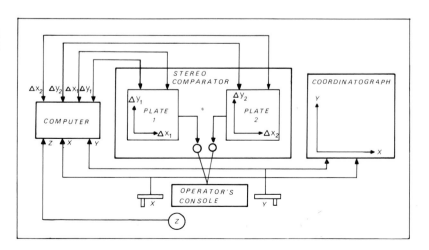

This is inner orientation. A separate photographic coordinate system for each plate is established, which is then related to a 'model' coordinate system by relative and absolute orientation with reference to ground control points. The floating mark is moved around in the model system by the operator to produce model coordinates of points from which the corresponding photographic plate coordinates are computed by the computer using the collinearity equations. These photographic coordinates are used to generate signals to drive the photo carriages to their correct positions for the operator to view conjugate points. The calculation is so fast that the operator cannot sense any delay between his turning of the handwheels and the apparent movement of the floating mark. Absolute orientation requires the operator to enter the coordinates of a sufficient number of ground control points and to record their corresponding model coordinates by means of the floating mark. The model coordinates are then transformed by an iterative process into the corresponding ground coordinates. The operator can trace out planimetric details and contour lines in the same manner as the analogue stereoplotter. The analytical plotter possesses a number of advantages over the analogue plotter. First, it is more accurate because no attempt is made to reconstruct the stereomodel as in the case of the conventional analogue plotter and systematic errors can be easily corrected mathematically. The computer can store redundant information such as extra fiducial marks and more points than the minimum necessary for relative and absolute orientation. Secondly, the analytical plotter is highly flexible and can accommodate all types of images, not restricted to those from conventional aerial photography. Thirdly, the model can be transformed by the computer for output to the plotting table or to a data bank.

One major disadvantage of analytical plotters is their high cost compared with traditional analogue plotters. Early analytical plotters were designed for military applications such as OMI-Bendix AS-11A (1963) and the Bunker-Ramo UNAMACE (1968). Since 1976 commercially oriented analytical plotter systems at more affordable prices have become available, such as the OMI AP/C-4 (1976), equipped with a PDP 11/03 computer, and the Instronics-Gestalt Anaplot, equipped with a PDP 11/45 computer. It is worthy to note that the computers employed have become general purpose ones. More recently, a new generation of commercial analytical plotters of varying degrees of accuracies and prices has emerged. The emphasis has been placed on the ease of operation, data editing and better man–machine interaction such as the Anaplot II System of Canadian Marconi of Montreal (Turner 1982) which featured a PDP 11/34a computer and the APPS-IV of Autometric Incorporated (Greeve 1982). Low-cost systems such as the Wild Aviolyt BC1 (Hasler 1982) and Helava's US-2 analytical stereoplotter (Seymour 1982) have also been introduced to the market. The design of all these commercial analytical plotters has been aimed primarily at a more attractive cost compared with performance ratio. One foresees that the analytical approach will become more and more attractive in topographic mapping as sophisticated computers become less expensive (Case 1981).

In view of the high cost of the analytical plotters, a common approach towards automation is the hybrid system or computer-assisted analogue plotting system (Dorrer, Lander and Torasker 1974). Essentially this makes use of an analogue plotting machine to solve the collinearity condition and a computer to correct for distortions such as instrument errors, lens distortion or refraction. A good example is the Zeiss (Oberkochen) Planitop interfaced with DIREC-1 Display and HP-9810 desktop computer based on Dorrer's (1976) computer-assisted stereoplotting package. This is a software-based system according to Petrie's classification (1981). The computer can be programmed to carry out all kinds of functions for photogrammetric plotting. The DIREC-1 displays the coordinates of the points measured, which can be stored by the computer. The analogue plotter therefore acts both as a measuring and digitizing device. It is possible to make use of much more powerful computers. Petrie and Adam (1980) have designed a software-based system consisting of Officine Galileo Stereosimplex IIc plotter and a more versatile Wang 2200 desktop computer in the University of Glasgow, Scotland, UK. The Wang 2200 makes use of high-level programming language (BASIC) with extensive facilities for error detection and editing. It has a comparatively large memory of 8K, a vast choice of peripheral devices including printers, plotters and mass storage devices, a visual display unit equipped with a full alphanumeric keyboard and a numeric pad. It also has twin cassette drives, each with 500K bytes storage capacity which allows storage and access to large programs and simultaneous recording of large quantities of data. Strong software support is essential for such systems to operate successfully. Programs developed for the Wang 2200 included digitizing, orientation, model connection, strip formation and perspective plotting of ground detail. One can see that the cost involved in software is much more than that for the hardware.

A further development of this software-based system is to have a more powerful computer to serve two or more analogue plotters. This is the multistation time-sharing system based on a large minicomputer. A good example is the digital mapping system developed by the commercial air survey firm of Hunting Surveys (Leatherdale and Keir

Fig. 8.9
Hardware of the Hunting digital mapping system (*Source*: Leatherdale and Keir 1979)

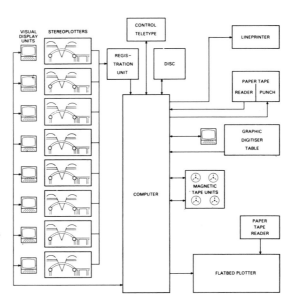

1979). This consists of eight Wild A8 stereoplotters with visual display units, one Ferranti Freescan digitizer and one Ferranti flatbed plotter connected on line to a DEC PDP 11/50 computer with 96K of 16 bit memory and a 40 megabyte disc (Fig. 8.9). This system has almost unlimited capacity and flexibility to handle all types of requirements and applications, such as aerial triangulation, digital map production with a wide range of scales, digital map tapes of topographic detail and volume computation. The high degree of flexibility which cuts down production costs has made such a large mapping system economically viable. The digital approach, however, cannot be used effectively in cases where many features shown on the aerial photographs are obscured and extensive field completion is necessary.

From the above discussion, it can be seen that aerial photography and photogrammetry provide well tested techniques for topographic mapping. The high image quality and the simple geometry of the central perspective projection of aerial photography permit topographic maps of varying scales in high accuracy to be produced economically by an analogue, analytical or a hybrid approach.

Thermal infrared imagery as a cartographic data source

The availability of non-photographic remote sensing systems has produced imagery which detects information hitherto not detectable by conventional aerial photography. They can be employed to acquire imagery under conditions unfavourable to aerial photography. Unlike aerial photography, non-photographic remote systems are dynamic systems because a scanning process is usually involved. The imagery is produced continuously as the aircraft progresses along its flight path. Thus, each point on the film strip is exposed at a different time and consequently the correspondence between image position and ground position is directly affected by the flight trajectory. The image is also obtained indirectly through the photographic copying of the display on a cathode ray tube using a moving film. The fixed perspective and simple geometry of the aerial photograph are destroyed. In the case of the thermal infrared imagery, the scanning in effect produces a cylindrical surface for the image (Fig. 8.10), thus giving rise to a scale compression along the scan direction (Y), which increases away from the nadir line while in the flight direction (X) the image scale remains unchanged, thus producing a quasi-panoramic distortion (see Ch. 1). Apart from this major source of error, there are also distortions caused by the internal geometry of the scanner, such as the axis of rotation not being perpendicular to the axis of the instantaneous field of view (IFOV) and the attitude errors of the aircraft, such as roll (ω), pitch (ϕ) and yaw (κ) (Masry and Gibbons 1973; Taylor 1971). Other errors include weather effects caused by clouds and winds; electronic noise caused by strong interference patterns of aircraft radios; and image processing effects caused by faulty film developing equipment (Sabins 1973). Before thermal infrared imagery can be used as a cartographic data source, elimination of these errors, in particular the scale distortion, is essential.

Fig. 8.10
Line scan geometry of the
thermal infrared line-scan
imaging system (*Source*: Taylor,
1971)

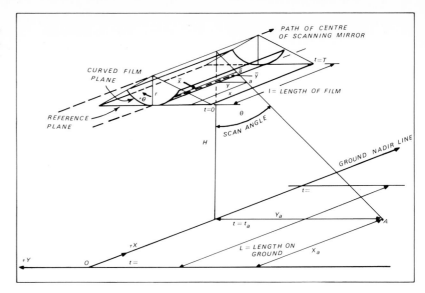

The method employed for this purpose is *rectification* commonly
employed in photogrammetry. Rectification is a process of removing
tilts of the imagery with reference to ground control points. For
thermal infrared imagery, this amounts to changing the curved image
surface into a plane of distortion-free imagery. If the aircraft is tilt-
free (i.e. ϕ, ω and κ values being zero at all times), it is only necessary
to mathematically project the cylindrical image surface to a plane as
shown in Fig. 8.10. The point A on the ground is imaged on the curve
film at $\bar{a}$ at time t_a. The ray from point A to $\bar{a}$ passes through the
reference plane at a, the scan angle being θ. However, orientation
errors caused by tilts are always present. Taylor (1971) derived rectification equations that will take these errors into account. Assuming
roll (ω), pitch (ϕ) and yaw (κ) occur in this order,

$$\bar{x} = x_m + r \left(1 - \frac{h}{H}\right) [\tan \phi + (\tan \theta_m)(\sin \kappa)]$$

$$\bar{y} = r\bar{\theta} = r \{w + \tan^{-1}[(\tan \theta_m)(\sec \phi)(\cos \kappa)] \quad [8.13]$$

where $\bar{x}$, $\bar{y}$ are coordinates of the point on image plane (Fig. 8.10), x_m
and θ_m are the coordinates of ground points after introduction of all
three tilts (ω, ϕ, κ), i.e. the coordinates as measured on the film, r is
the equivalent focal length of the scanner, θ is the scan angle, H is
the flying height of the aircraft and h is terrain elevation (or relief).
  Equations [8.13] can be further transformed from the film plane
into the horizontal reference plane as:

$$x = \bar{x} = x_m + r \left(1 - \frac{h}{H}\right)[\tan \phi + (\tan \theta_m)(\sin \kappa)]$$

$$y = r \tan \bar{\theta} = \bar{r} \tan[\omega + \tan^{-1}(\tan \theta_m)(\sec \phi)(\cos \kappa)] \quad [8.14]$$

If ω, ϕ and κ are all zero, then $x = x_m$ and $y = r(\tan \theta_m)$. For $\omega = \phi = \kappa = 1°$, $r = 28.5$ mm, $\theta_m = 30°$, $h = 0$, then the image displacements computed using equation [8.14] give $\Delta x = 0.787$ mm and $\Delta y =$

0.660 mm. If the image scale is 1:10,000, the displacements are 7.87 m and 6.60 m respectively. However, h is not always zero. Without a knowledge of the terrain elevation, accurate rectification is not possible.

The rectification of a strip of thermal infrared imagery can be carried out instrumentally using an OMI orthophoto printer controlled by the Analytical Plotter AP/2-C (Masry and Gibbons 1973). The strip of imagery to be rectified is placed in the printer carriage and aligned with the centreline of the strip to be parallel to the X-direction of the carriage (Fig. 8.11). As the imagery is scanned, the relationship between the x and y coordinates of an image point is maintained continously by the real-time program in the computer of the analytical plotter. The relationship between the speeds of the drum carrying the sensitive film and the photo-carriage carrying the infrared imagery should also be found and maintained. Another instrument that has been used for rectification and infrared imagery is the Gestalt photomapper (Masry, Derenyi and Crawley 1976). This requires the use of ground coordinates and correction function to control the computer of the analytical plotter. The imagery is placed on one of the photo-carriages and information on the inner orientation of the scanner is set. The correction function is in the form:

$$x = F_1(X,Y)$$

$$\text{and} \quad y = F_2(X,Y) \tag{8.15}$$

where x, y are measured image coordinates and X, Y are the corrected image coordinates. The instruments operate in a patch-by-patch mode and the correction function is used to form the scanning raster within each area. The carriages are moved in the Y-direction with an incremental step over in the X-direction. Rectification proceeds until the whole strip is corrected. It took about 5 minutes to print a rectified strip of 10×23 cm.

It is possible to generate stereo-imagery with the thermal infrared sensor by adopting the following flight configuration: (1) parallel–pass; (2) fore-and-aft stereo systems (Fig. 8.12). The parallel-pass method plans parallel flights separated with a fixed distance between the

Fig. 8.11
Orientation of the centreline of the infrared imagery strip parallel to X-direction of the Orthophoto-Printer carriage (*Source*: Masry and Gibbons 1973)

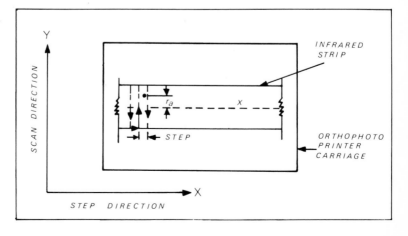

Fig. 8.12
Types of stereo-systems for
linescan infrared imagery: (a)
parallel flights; (b) fore-and-aft
(*Source*: Taylor 1971)

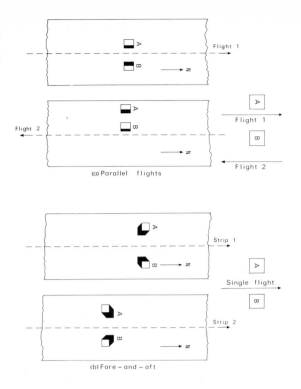

nadir lines of the two passes (Fig. 8.12a). The fore-and-aft method
employs a forward- and backward-looking scanner on the same flight
(Fig. 8.12b). Because of the increasing scale compression away from
the nadir line, the same terrain features appear in different sizes and
shapes on the two imagery strips in the parallel-pass method, making
stereo-vision difficult to achieve. Also, flight lines cannot be maintained precisely parallel. The fore-and-aft method seems to provide a
better stereo image because all terrain features are imaged at essentially the same scale on both strips of imagery. One should, however,
note that the conventional analogue instruments cannot be used to
obtain the stereomodel because of the moving perspective point of the
dynamic scanning system. Leberl (1971) pointed out that the photogrammetric meaning of 'stereo' does not always apply, but the use of
stereo-overlap permits height information to be extracted from the
thermal infrared imagery.

Konecny (1972) has conducted experiments to test the geometric
accuracy of thermal infrared imagery after rectification. Altogether
five parallel strips overlapping by 60 per cent flown at an altitude of
1,520 m above the ground over the Gatineau Park area in Canada
were obtained with an HRB Singer Reconofax with r = 28.65 mm
covering a view of 120° on 70 mm film. The imagery was rectified
numerically first using single strips and then using overlapping strips
to give stereo images (i.e. parallel-pass method). The check points on
the rectified infrared images were compared with the corresponding
points established by analytical aerial triangulation from 1:15,000
scale aerial photographs of the same area. The results are shown in
Table 8.1. Clearly, stereo images gave better planimetric accuracy

Table 8.1
Metric accuracy of rectified infrared imagery
(*Source*: Konecny 1972, p. 30)

Coordinates		$\triangle$x		$\triangle$y		$\triangle$z	
		m	‰h	m	‰h	m	‰h
Single Images	Maximum error	35	23	56	37	–	–
	Standard error (RMS)	±13	±8.4	±12	±8.0	–	–
Stereo Images	Maximum error	35	23	26	17	22	15
	Standard error (RMS)	±13	±8.4	±4.4	±2.9	±7.6	±5.0

h = flying height; RMS = root mean square

than single images especially in the *y*-direction. Such high accuracy can be achieved only with a stable sensing platform in which the rotations and shifts are kept as small as possible. Konecny strongly recommended the use of auxiliary data to help improve the metric accuracy of the thermal infrared imagery. On the whole, thermal infrared imagery is not too suitable for topographic mapping purposes if the stringent standards in planimetry and elevation of the resultant maps are to be met. But its use in thematic mapping is justified.

Side-looking radar imagery as a cartographic data source

Side-looking radar imagery (SLAR) is an active microwave sensing system which possesses the greatest potential for topographic mapping and as a source of valuable cartographic data. Its capability of penetrating clouds is the greatest asset and explains its use in topographic mapping in such countries as Panama, Brazil, Venezuela, Colombia and Peru. However, its use in topographic mapping suffers from poor spatial resolution and a complicated geometry which is best handled with an analytical approach. Since SLAR is also a line scanning system which measures 'echo' time, there are two components in the spatial resolution of the resultant image: the azimuth resolution in the X-direction (flight direction) and the range resolution in the Y-direction (across-track). The former depends on the width of the antenna beam while the latter depends on the length of the emitted pulse (see also Ch. 1). Since the antenna points to the ground at an angle, the resultant image gives the appearance of low sun-angle photography characterized by long shadows. The radar image can be recorded by a slant range sweep or a ground range sweep of the electron beam across the cathode ray tube. The scale distortion suggests compression that increases towards the nadir line (just the reverse of that for thermal infrared imagery). In making use of the SLAR imagery for topographic mapping, it is necessary to correct for a number of errors caused by the geometry of the range projection, changes in exterior orientation (i.e. changes in aircraft speed, height, attitude and time)

353

and incorrect inner orientation (i.e. improper adjustment of the recording unit, lens and film).

An approach commonly used is two-dimensional polynomial transformation relating the imagery to ground control points, best demonstrated by Derenyi's work (1974). He made use of two strips of real-aperture radar imagery of an area covering the south of Washington, DC. Each strip covered an area of approximately 20×100 km with a scale of 1:250,000. The ground relief in the area was rather modest, ranging from sea level in the east end to about 150 m in the west end. Prominent points were selected and marked on the imagery and a total of 130 points were selected on the two strips. The coordinates of these points were measured with a Zeiss PSK comparator under 16 times magnification and in three repetitions. Each strip was subdivided into three sections to accommodate it in the instrument. Ground positions for all these points were obtained from 1:24,000 scale topographic maps. The mean of the three sets of comparator measurements were computed, and for each section of the image a linear conformal transformation using two widely separated points were carried out:

$$X = Ax + By + C_1$$
$$Y = Bx + Ay + C_2 \qquad [8.16]$$

where X, Y are the along-track and across-track ground coordinates respectively and x, y are the along-track and across-track image coordinates respectively. The parameters A and B express a scale change and a rotation while C_1 and C_2 are two translation constants. The root mean square errors of the remaining points were computed after the transformation and were found to be rather large (overall positional error = 261 m). In order to reduce the error, polynomials were established to describe the effects of the changes in the exterior orientation of the sensor on the image position of points as follows:

$$dX = A_1 + A_2X + A_3X^2 + A_4X^3 + B_1Y + B_2XY + B_3X^2Y$$
$$dY = C_1 + C_2X + C_3X^2 + D_1Y + D_2XY + D_3X^2Y + D_4X^3Y$$
$$[8.17]$$

where the terms with A coefficients represent the effect of change in ground speed and y-tilt (pitch); the B terms represent the effect of yaw; the C terms represent the effect of the change in heading; and the D terms represent the effect of change in the flying height. Eight control points were selected in each section of the imagery to determine the coefficients by least-squares adjustment. The most successful polynomial fit produced a root mean square error in position of 142 m. Such a result suggested that SLAR imagery could be used for medium-scale planimetric mapping.

A common approach to produce maps quickly from SLAR imagery is by 'mosaicking'. There are four methods of SLAR mosaicking (Leberl, Jensen and Kaplan 1976). The simplest method is similar to a print-laydown in aerial photography without the use of any ground control points. The individual radar strips are directly laid out on a board so that adjacent images fit together. The most sophisticated and expensive method is to make use of continuous simultaneous SHORAN tracking of the survey aircraft from two geodetically surveyed ground stations. This method was used during the early phase of the Project RADAM (for Radar Amazon) carried out in October 1970 covering an area of about 4.6 million km² (van Roessel and de Godoy 1974). The imagery was obtained with Goodyear Mapping System 1000 (GEMS) flown in North–South direction. The flight spacing provided a side lap of 25 per cent for the adjacent radar strips. Ground range presentation was used and the image scale was about 1:400,000. The aircraft positions were tracked by SHORAN which required 45 ground points to be accurately determined as ground support. The air stations were plotted on a stable base overlay at a scale slightly larger than the final mosaic (1:200,000). Control points were also plotted. Based on the overlay, overlapping pieces of the same radar strip negative were enlarged through an anamorphic lens (capable of differential enlargement in two orthogonal directions). Only the along-track scale was adjusted whereas the across-track scale was left unchanged. The prepared copies of the radar strips were assembled on a piece of Masonite hardboard. The strips were glued down and when the glue had dried they were inspected to ensure that linear features were continuous and the control points matched their plotted positions on the overlay. The planimetric accuracy of the mosaic was found to be within the range of ±200–300 m.

The third method of mosaicking is to make use of two or more 'tielines' as controls. They are radar image strips flown along the perimeter of the mapping area. Ground control points are marked on the tielines or the tielines are tracked by SHORAN. The radar images are laid out on the mosaic board to achieve a fit to the tielines. The planimetric accuracy of this method was found to be about ±500–600 m in the Project RADAM case. This is a frequently used method of mosaicking.

The fourth method was proposed by Leberl (1974) using the photogrammetric block triangulation method. Points in the sidelap between adjacent strips of radar imagery are used to tie the images together into a block. The points in a block are later transformed into a network of ground control points by photogrammetric block adjustment using polynomials. This method was applied to a block of 24 overlapping synthetic aperture side-looking radar images from over an area of about 90,000 km² covering parts of Ohio, West Virginia and Kentucky in the US (Leberl, Jensen, and Kaplan 1976). The resulting radar mosaics gave a root mean square error in planimetry of about ±150 m, using 1:24,000 scale maps as controls. The accuracy was determined by the distribution pattern of ground control points and by the presence of topographic relief. This method, however, was found to be the most cost-effective of all.

The mosaicking method is restricted by the presence of great relief. Another approach is to make use of stereo side-looking radar for topographic mapping. There are two basic stereo flight configurations to obtain: the same-side and opposite-side cases using two separate

Fig. 8.13
Basic stereo-radar configurations (*Source*: Leberl 1979)

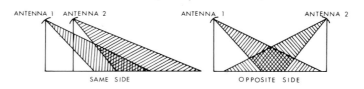

355

flights (Fig. 8.13). There are also other stereo configurations such as cross-wise proposed by Leberl (1979) (Fig. 8.14). The opposite-side configuration will not permit a good visual stereoscopic model to be achieved in mountainous regions because of the varying directions of the shadows. The same-side flight configuration permits a visual stereoscopic model to be achieved with ease and is therefore thought to be the best to use. With stereo-radar images, parallax equations can be developed similar to those in aerial photography. Leberl (1979) has developed the following parallax equations for same-side and opposite-side stereo radar images in both ground range and slant range presentations according to Fig. 8.15:

Ground range presentation:

$$p' = h \cot \theta'$$

$$p'' = h \cot \theta''$$

$$\Delta p = p'' \mp p' = h (\cot \theta'' \mp \cot \theta')$$

$$h = \Delta p/(\cot \theta'' \mp \cot \theta') \qquad [8.18]$$

Slant range presentation:

$$p' = h \cos \theta'$$

$$p'' = h \cos \theta''$$

$$\Delta p = p'' \mp p' = h (\cos \theta'' \mp \cos \theta')$$

$$h = \Delta p/(\cos \theta'' \mp \cos \theta') \qquad [8.19]$$

Fig. 8.14
Stereo-radar flight configurations (*Source*: Leberl 1979)

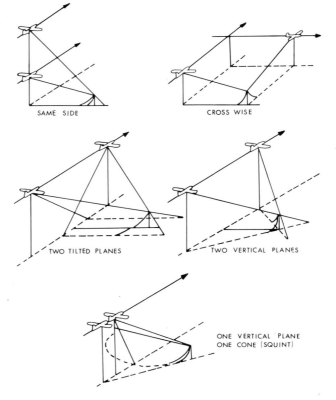

SAME SIDE

CROSS WISE

TWO TILTED PLANES

TWO VERTICAL PLANES

ONE VERTICAL PLANE
ONE CONE (SQUINT)

Fig. 8.15
Definition of radar stereo parallax, Δp and relief displacement, p for (a) same-side and (b) opposite-side stereo (*Source*: Leberl 1979)

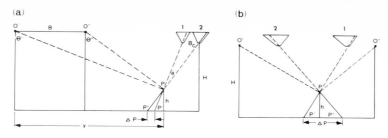

where h is the object height difference, p' and p'' are relief displacements on left and right radar image respectively; Δp is the parallax difference due to height h, and θ' and θ'' are look angles for the left and right radar images respectively. The minus sign applies to same-side geometry and plus sign to opposite-side geometry. It is worthy to note from the equations, that the height difference depends not only on the parallax difference but also on the look angles. In the ground range presentation, it is also possible to obtain the absolute parallax a as follows:

$$a = y'' - y' \qquad [8.20]$$

where y' and y'' are ground distances on the left and right radar images. As for slant ranges, the absolute parallax a can be obtained from:

$$a = (H^2 + y^2)^{\frac{1}{2}} - (H^2 - (y - B)^2)^{\frac{1}{2}} \qquad [8.21]$$

where H is the flying height, y is the ground distance of an object from the left flight line and B is the stereo base or distance between the left and right flight lines. Finally, vertical exaggeration (q) of the stereo-model can be estimated from the base-height ratio as follows:

$$q \approx 5 \cdot B_c/H \qquad [8.22]$$

Since from aerial photography it is known that $B_c/H = \Delta p/h$, therefore with the notation in Fig. 8.15, we find that:

$$q \approx 5HB/(y(y-B)) \qquad [8.23]$$

For airborne radar with 60 per cent overlap and stereo intersection angles ($\Delta\theta$) between 12° and 15°, q ranges from 2.3–1.8 (compare to $q \approx 3$ in aerial photography).

It is therefore possible to carry out topographic mapping using stereo radar images, in particular, synthetic aperture radar because of its much better spatial resolution. They can be used to interpolate coordinates between known points. Reasonably accurate maps can be obtained provided look angles are not less than 40° off-nadir, with intersection angles of about 15°. As for the information content extractable from the synthetic aperture radar images, Dowman and Morris (1982), after comparing details plotted from 1:150,000 scale synthetic radar images of the Ottawa area in Canada with those from 1:50,000 scale aerial photography using a Thompson-Watts MK.2

plotter, concluded that synthetic aperture radar might have a potential for small-scale mapping and map revision only.

A successful attempt to produce small-scale topographic maps from radar images was illustrated by Project RAMP (Radar Mapping in Panama). A topographic map of Darien Province of Panama at a scale of 1:250,000 was produced (Crandall 1969). The radar imagery scale was about 1:172,000 and much auxiliary airborne data were acquired simultaneously as the radar imagery was flown. A complex data reduction procedure had to be carried out with the aid of the computer. Extensive supplemental ground control points were required. The corrected coordinates of the range marks were computed and stored on a tape to drive an automatic plotter to produce a base sheet at the desired scale showing all the supplemental controls. An electronic radar sketching device was used to insert all the details. From the previously determined positions and attitudes of the antenna, the measurements of range and depression angle to the ground were converted to X, Y and Z coordinates and sorted to provide profile values. Those points occurring at the predetermined contour intervals were coded in the computer, from which a magnetic tape was prepared to drive the automatic plotter to scribe the positions of the contour crossings along each profile with the coded value. These were then drawn up to give the contours. Today, with the availability of analytical plotters, such as the Gestalt Photomapper (mentioned earlier in Ch. 8), rectification of the radar imagery is rendered quite easy, using the polynomial equations proposed by Derenyi, as discussed earlier (Masry, Derenyi and Crawley 1976). To conclude, SLAR has a much greater potential to complement aerial photography in small-scale topographic mapping in perenially cloud-covered regions of the world.

Cartographic applications of remote sensing data from space platforms

Methods of data acquisition from space

The advent of space remote sensing of our earth in recent years signifies the end of the ceiling limit to aerial photography from the photogrammetric point of view. It is obviously advantageous to make use of the much wider coverage afforded by the space platforms to the remote sensors. For topographic mapping applications, this would mean fewer photographs to handle and few ground points needed for control. The important question is the extent to which maps produced from the space remote sensing data are comparable in accuracy to those produced from an aircraft platform.

A commonly used space platform is an earth satellite which revolves around the earth in a circular orbit from an altitude of *at least* 180 km. However, in order to ensure a longer life, the altitude of the satellite is often much higher to cut down air drag. The first generation of Landsat (1 through 3) has an altitude of 920 km. Landsat 4 is 705 km from earth while Skylab had an initial altitude of 435 km. This means that an unfavourable base to height ratio is the case which adversely affects the overall accuracy of the resulting maps. There are a number of ways by which imagery can be acquired from the satellite. Three of these are specially relevant to the present discussion: (1) by film cameras with physical recovery of the exposed film; (2) by vidicon cameras with radio transmission of the stored image to ground stations; (3) by optical-mechanical scanning with a similar radio transmission (Petrie 1974).

In method (1), the use of film cameras permits satellite photography suitable for detailed interpretation to be obtained. If topographic mapping is to be carried out, metric aerial cameras should be employed. Essentially, vertical photographs with 60 per cent forward overlap are obtained. However, because of the earth's rotation about its own axis, the earth has already moved further westwards during the time when the satellite has completed one revolution (i.e. the orbital period). The next strip of photography is not adjacent to the first, and it will take perhaps 16 or 18 orbits, and not until later the next day that the satellite will come back over the same area (Petrie 1970). As a result, several days are required to cover a given area. The planning of satellite photography for topographic mapping is therefore rendered more difficult. Also, because the satellite travels very fast in space (e.g. as much as 8 km/sec in low orbit or about 100 times faster than the aircraft), image movement compensation is necessary, which involves moving the film in its focal plane during exposure to prevent blurring of the image.

A good example is the manned Skylab flight which produced orbital photography using two camera systems: the S190-A multispectral film camera and the S190-B earth terrain camera. The S190-A system consists of six Itek high–resolution cameras with each lens having a focal length of 152 mm on an image format of 57×57 mm, giving a negative scale of 1:2,860,000 with an altitude of 435 km. The base to height ratio was 0.19. Four of the cameras carried films covering the spectral bands of 0.1 μm over the range 0.5–0.9 μm on black-and-white film and the other two exposed true- and false-colour films respectively. It used an intralens shutter and forward motion compensation, thus making it more suited to mapping. The S190-B system is the Actron KA-74 long-focus, high-resolution reconnaissance camera with an 114.3×114.3 mm format and a focal length of 457 mm, thus giving a negative scale of 1:948,500. Since it used a focal plane shutter with a long focal length lens, an even poorer base to height ratio (0.10) resulted, making it less satisfactory for mapping purposes. From both systems, stereoscopic coverage with a 60 per cent overlap was possible. There had been research carried out to determine the planimetric and height accuracy of maps produced from these Skylab photographs. Investigations at the University of New Brunswick, Canada using three strips of S190-A and two strips of S190-B photography for small scale mapping have produced results in planimetry compatible with those from more sophisticated aerial triangulation procedures (Derenyi and MacRitchie 1980). They indicated that Root Mean Square Error (RMSE) values of approximately 60 m and 70 m in X and Y for S190-A photography and 20 m in both X and Y for S190-B photography were achievable using the standard aerial triangulation method. As for the heighting accuracy, it was expected to be much poorer in view of the poor base to height ratios of these two types of photography. Mott and Chismon (1975) studied a model of the Himalayas in Nepal acquired by the S190-B system that revealed a maximum height error of 270 m after adjustment. Welch and Lo (1977) found that for S190-A photography of San Francisco, USA a mean RMSE of 148 m in height was obtained. Therefore, one can conclude that the Skylab photography is acceptable only for planimetry up to a scale of 1:250,000, but is unsuitable for use in contouring. It may be possible to produce a topographic map at a scale of 1:500,000 with a contour interval of 250 m for mountainous regions.

In method (2), the use of vidicon cameras with radio transmission produces an electronic image. An optical image of the scene is first focused on the photo-conductive target of the vidicon tube. The conductive property of the target converts the optical image into an electronic image which is then scanned line-by-line by an electron beam. Therefore, the image produced by the vidicon camera should still possess the properties of conventional photography. The big advantage of the vidicon camera is its light weight and its eraseable surface. As soon as one picture is taken and recorded on tape or transmitted to a ground station, another can be taken. In this way, the vidicon camera can produce numerous photographs per day over a considerable period, say, a year or two. The technique was first developed for weather satellites, such as TIROS, NIMBUS, ESSA, ITOS, etc. but later further developed for use in the earth resources satellites (Landsat 1 through 3). In Landsat 1 and 2, three vidicon cameras, known as the Return Bean Vidicon (RBV) cameras, were used, covering the following three spectral bands: 0.475–0.575 μm, 0.580–0.680 μm and 0.690–0.830 μm. The RBV camera gave an image of 4,000 lines over the 25 × 25 mm format which produced a resolution equivalent in photographic terms to about 40 lines per mm. From an altitude of 920 km, the nominal picture scale was 1:7,300,000 and the ground resolution was about 180 m in theory. In practice, other factors (atmospheric and processing) acted to degrade the resolution to 250 m (Welch 1973). For Landsat 3, the resolution of the RBV system was improved by doubling the focal length of the optical system and only two vidicon cameras were used. Also only one broad wavelength band from 0.505–0.750 μm was employed.

In method (3), optical-mechanical scanners with radio transmission of data to ground stations are used to acquire imagery. The scanning is carried out with a rotating mirror whose axis is parallel to the flight direction of the spacecraft. The radiation received is projected onto an array of detectors, each one of which senses the intensity of the radiation in different parts of the electromagnetic spectrum. The scanning speed is high (6,000 scans per minute) and the image is built up sequentially element by element from successive signals as the spacecraft moves forward. The intensity of radiation at each element can be converted to a numerical value which is then recorded on magnetic tape. This record is transmitted to a ground station where it is re-recorded and used to control a cathode ray tube to produce the final image on film. Thus, by scanning, the central perspective projection of the photography is no longer preserved. A good example of this type of scanner is the Multispectral Scanner (MSS) of Landsat whose detectors operate in four spectral bands: 0.5–0.6 μm, 0.6–0.7 μm, 0.7–0.8 μm and 0.8–1.1 μm. The quality of the MSS image is similar to that produced by the RBV although geometrically the two are very different. It is worthy to note that both systems produce imagery in digital form which permits radiometric and geometric corrections of the imagery using a computer.

Cartographic potentialities of Landsat imagery

Although both the RBV and MSS systems onboard Landsat were not specifically designed for topographic mapping applications, much work has been done to evaluate their potentialities in cartographic applications (Kratky 1974; Wong 1975; Mott and Chismon 1975; Welch and Lo 1977; Dowman and Mohamed 1981). In order to correct for any geometric errors inherent in these data, it is necessary to understand the sources of these errors. Basically, these include the altitude, attitude and velocity of the spacecraft; the earth rotation; the errors of the scanning mechanism (for MSS) and the map projection into which the image data are to be mapped (Berstein 1983). Geometric correction of these image data consists of two separate steps: (1) auxiliary data processing which transforms the raw imagery projection into the corrected imagery projection; (2) image correction resampling which consists of obtaining the pixel intensities on a regular grid in the corrected projection (Fig. 8.16) (Friedmann *et al.* 1983).

In auxiliary data processing, a commonly used procedure is to fit the images to ground control by means of different mathematical models. The following types of mathematical models can be used:

1. Similarity transformation in the form:

$$\bar{X} - a = pX + qY$$
$$\bar{Y} - b = -qX + pY \qquad [8.24]$$

where X, Y and $\bar{X}$, $\bar{Y}$ are space image coordinates and corrected image coordinates respectively, *a*, *b*, *p*, *q* are four unknowns representing the amount of rotation and translation required.

Fig. 8.16
The geometric correlation of satellite image data. Auxiliary data are processed to obtain the transformation between the raw and correlated projection. The raw imagery is then resampled (*Source*: Friedmann *et al.* 1983)

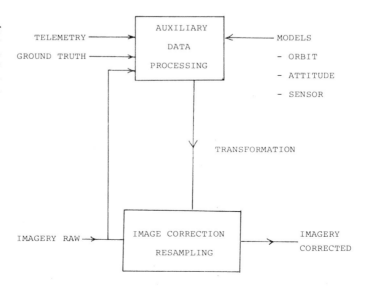

2. Affine transformation in the form:

$$\bar{X} = a_1X + b_1Y + c_1$$
$$\bar{Y} = a_2X + b_2Y + c_2 \qquad\qquad [8.25]$$

where a_1, a_2, b_1, b_2, c_1 and c_2 are six coefficients aimed to correct for first-order distortions such as non-orthogonality, scale difference between along-track and scan directions which may be caused by earth rotation and map projection.

3. High order polynomials according to the number of ground control points used. Wong (1975) made use of the following model using 20 ground control points:

$$\begin{aligned}
(\bar{X} - X) = {}& V_x + b_1 + b_2x + b_3y + b_4xy + b_5x^2 + b_6y^2 + \\
& b_7x^2y + b_8y^2x + b_9x^3 + b_{10}y^3 + b_{11}x^3y + b_{12}y^3x \\
& + b_{13}x^4 + b_{14}y^4 + b_{15}x^2y^2 + b_{16}x^3y^2 + b_{17}y^3x^2 \\
& + b_{18}x^5 + b_{19}y^5 + b_{20}x^3y^3
\end{aligned}$$

$$\begin{aligned}
(\bar{Y} - Y) = {}& V_y + c_1 + c_2x + c_3y + c_4xy + c_5x^2 + c_6y^2 + \\
& c_7x^2y + c_8y^2x + c_9x^3 + \ldots + c_{20}x^3y^3 \qquad [8.26]
\end{aligned}$$

where $\bar{X}$, $\bar{Y}$ and X, Y are corrected and uncorrected coordinates respectively; x and y are measured image coordinates of an image point.

4. Collinearity equations in the form shown in equation [8.12] which take into account variations of the satellite altitude with time.

Wong (1975) has evaluated the RBV and MSS images of ERTS-1 (Landsat 1) using the first three methods mentioned above. By using four RBV scenes (three of which were first-generation 70 mm film transparencies at 1:3,369,000 scale and the other printed on a glass plate at 1:100,000 scale), the similarity and affine transformations between the measured coordinates of the reseau crosses and their calibrated coordinates gave rise to a mean RMSE of planimetry (distortion) of ±58 m and ±50 m respectively. If photo-identified ground control points were used instead of the reseau crosses, the RMSEs were ±131 m and ±105 m respectively. When the 20-term polynomials in equation [8.26] were used to model the electronic distortions at the reseau points and to apply corrections to the image coordinates of the ground control points, the average RMSE was ±90 m. For the MSS system, one frame (70 mm format) was analysed using 189 ground points, the RMSEs were found to be ±290 m, ±187 m and ±57 m respectively using similarity transformation, affine transformation and 20-term polynomials. The geometric fidelity of the MSS system was clearly inferior to that of the RBV system, but both the RBV and MSS images could be corrected to meet the National Mapping Accuracy Standard of the US at 1:500,000 scale. Since the MSS distortions are primarily caused by irregular variations in the attitude of the satellite, accurate calibration of each frame using 25–30 ground control points is required in order to achieve the limiting accuracy of ±55 m for 1:500,000 scale mapping. In recent years research has focused on reducing or even eliminating the number of ground control points required per frame in the rectification of the spaceborne imagery using accurate auxiliary data on the attitude of the satellite (i.e. roll, pitch, yaw and height), the MSS scanning mechanism characteristics and the earth's shape (Friedmann *et al.* 1983, Sawada *et al.* 1981). By so doing, it was claimed that the number of ground control points required could be reduced to two or three.

Although Landsat MSS imagery is not intended to give stereoscopic coverage of the earth, some height measurements can still be obtained in the area of sidelap between two adjacent images. The percentage of sidelap varies from 14 per cent along the equator to 85 per cent at latitudes 80° N and 80° S. Between latitudes 50° and the pole the sidelap is adequate to provide a good stereoscopic model for parallax measurement. Welch and Lo (1977) have made use of a specially designed Zoom Height Finder to measure the parallax of three Landsat MSS models: San Francisco (b/h = 0.14), Vancouver, Washington (b/h = 0.12) and Juneau, Alaska (b/h = 0.10), using both band 5 and band 7 images in 70 mm format. The results revealed a mean RMSE of ±400 m using five control points. In another evaluation carried out by Dowman and Mohamed (1981) using the Wild A-8 plotter a much better height accuracy of ±57 m was obtained using seven control points (b/h = 0.11). This evaluation indicated that Landsat MSS images could be plotted photogrammetrically to produce small-scale topographical maps which can meet the mapping requirements at 1:500,000 and 1:100,000 scales.

The cartographic application of Landsat MSS images clearly focuses on the mapping of poorly mapped areas in the developed countries (Mott and Chismon 1975). However, one major concensus from all these evaluations was the poor spatial resolution which adversely affected the geometric fidelity of the imagery and the completeness of mappable details. The Thematic Mapper data from Landsat 4 (see Ch. 1) have compensated these weaknesses to some extent, notably by a greatly improve pointing accuracy of 0.01° and a higher spatial resolution of 30 m. The spectral resolution has also been improved with the use of seven spectral bands. Some preliminary evaluations of TM imagery confirmed strong discriminatory value of six out of the seven spectral bands of the TM data (Townshend *et al.* 1983). A recent evaluation by Welch and Usery (1984) confirmed the geometric fidelity of the TM data which produced a planimetric accuracy of ±25 m with the use of five to ten ground control points. This meets the standard of topographic mapping at 1:100,000 scale.

It follows from the above discussions that two possible approaches to improve the usefulness of satellite imagery in topographic mapping are: (1) to use a metric camera and physical recovery of the film; (2) to use a vertical line-array camera system designed for stereoscopic coverage. The first approach has always been preferred by the photogrammetrists. This is witnessed by the development of a modified Zeiss (Oberkochen) RMK A 30/23 metric camera with a focal length of 305 mm on 230 × 230 mm format for use in the first European Space Agency (ESA) Spacelab flown in the US Space Shuttle on 5 December 1983 (Dowman 1978). Another example is the large format camera (LFC) built by Itek Optical Systems for US National Aeronautics and Space Administration. The camera has a 30.5 cm focal length lens with a 23 × 46 cm format. It is equipped with automatic exposure sensors and forward motion compensation, permitting the use of high resolution fine grain film. A ground resolution of 10–15 m can be obtained from the nominal Shuttle altitude (259 km). A base-height ratio of 1.2 can be achieved by using a forward overlap of 80 per cent, thus permitting contours of 20 m intervals to be compiled at 1:100,000 scale mapping (Doyle 1982). An expected, heighting accuracy is ±14 m (Dowman 1978). The magazine of the LFC has a capacity of 2,400 frames. To further improve the orientation accuracy, two stellar cameras ($f = 152$ mm or 70 mm film) are to be added, which will be directed horizontally when the LFC is looking vertically. The LFC was flown by the Space Shuttle Challenger on 5 October 1984.

The second approach has attracted a great deal of attention in recent years. A line-array system is an electro-optical sensor consisting of long linear arrays of solid-state detectors (see Ch. 1). These arrays have several thousand elements and can provide more than 10,000 detectors per line. The lens of the camera focuses the image of a line from the ground scene to the line of detectors. In other words, a single strip across the satellite track can be imaged at one time, and this strip preserves the central perspective projection. As the satellite moves ahead, the next line is imaged simultaneously. This method of imaging is called the push-broom technique (Fig. 1.17) (Dowman 1978; Thompson 1979). The spatial resolution and the strip length can be adjusted. Typically, a 30 m resolution push-broom sensor requires 6,300 detectors per spectral band to subtend a Landsat type swath. The advantages of this system are: (1) the elimination of the complex scan mechanisms required for the MSS scanner; (2) greatly improved signal-to-noise ratio; (3) better geometric fidelity of the image. Stereoscopic coverage can also be obtained with this system by using one of the three methods: (1) two cameras pointing fore-and-aft, (2) one camera with a rotating mirror; (3) one camera with the satellite orbit planned to create overlaps. The French have already designed a space linear array system called SPOT (Systeme Probatoire d'Observation de la Terre) which consists of two linear array cameras, known as High Resolution Visible instruments (HRV; Chevrel, Courtois and Weill 1981). A ground resolution of 10 and 20 m over a 60 km swath is achievable with these linear arrays for a satellite flown at an altitude of 832 m in a near-polar (98.7°) orbit. The two sensors can be pointed to obtain nadir viewing by offsetting 0.163° from the nadir so that they overlap by 3 km in a total swath width of 117 km. They can also be programmed to any point target within an off-nadir angle of ±(0.163° + 27° + 2.065°) or observing a region of 950 ±50 km width on earth centered on each satellite track. Another major advantage of SPOT sensors is the possibility of acquiring stereo coverage with base to height ratios equal to or greater than 0.5 on two successive days (Fig. 8.17). The base to height ratio varies according to the latitude because along the equator a given point on the earth's surface can be observed on seven different passes and on eleven along a altitude of 45° (Fig. 8.18). The base is therefore wider for the two satellite positions in two successive days on the equator (108 km) than that on a latitude of 45° (76 km), thus giving base to height ratios of 0.75 and 0.5 respectively. This will permit topographic mapping at the scales of 1:100,000 and 1:50,000 to be carried out.

There have also been investigations on other similar systems in the US, although there are no definite decisions whether they will be implemented. These include the MAPSAT (Colvocoresses 1979) and Stereosat (Welch and Marko 1981). MAPSAT which was designed by the

Plate 12
False colour density slicing transparency of thermal infrared image of Yellow Sea and East China Sea taken from GMS-1 satellite on 7 January 1979 (*Courtesy*: Q.A. Zheng; *Source*: Zheng and Klemas, 1982)

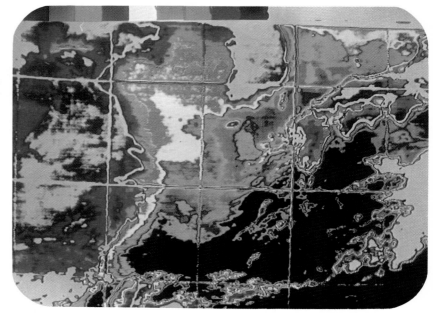

Plate 13
VISSR infrared image of the Gulf of Maine obtained at 1400 GMT on 23 May 1978 at a spatial resolution of about 12 km. The Gulf Stream which appears pink is clearly shown. The waters in the Gulf of Maine are blue and the urban heat islands are black (*Courtesy*: R. Legeckis; *Source*: Legeckis, Legg and Limeburner 1980)

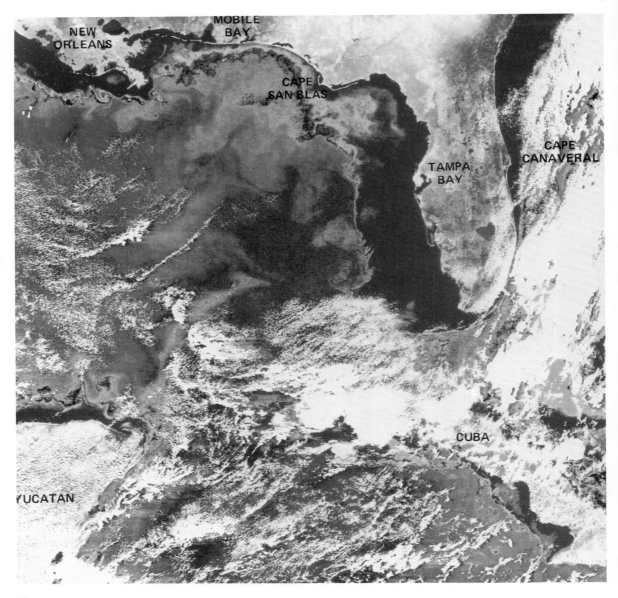

Plate 14
A map of phytoplankton pigments in the Gulf of Mexico on 2 November 1978 derived from the Nimbus 7 CZCS data of orbit 130. Open water pixels are coded according to the colour scale to show estimated chlorophyll *a* + phaeopigments *a*. Cloud and land pixels are represented in quasi-natural colours. Colour scale (value in mg m^{-3}): purple <0.05, dark blue 0.10, blue 0.15–0.20, light blue 0.25, dark green 0.35, green 0.40, light green 0.45, yellow 0.50, light brown 0.60, brown 0.70, dark brown 0.80, black >1.00. (*Courtesy*: NASA, Nimbus Experiment Team for the Coastal Zone Colour Scanner; *Source*: Gordon *et al.* 1980)

Fig. 8.17
Stereoscopic viewing capabilities by SPOT HRV sensors (*Source*: Chevrel *et al.* 1981)

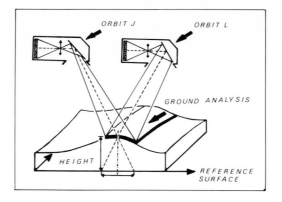

Fig. 8.18
Stereoscopic viewing by the SPOT HRV sensors at latitudes 0° (equator) and 45°. The base to height ratios are approximately 0.75 at the equator and 0.50 at a latitude of 45° (*Source*: Centre national d'etudes spatiales, 1984)

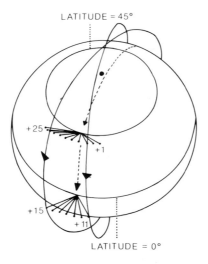

USGS would make use of three linear array cameras looking in the vertical and ±23° fore-and-aft directions to provide stereo data. Each array would have three spectral bands and a minimum 10 m IFOV. The orbit of the satellite would be the same as that for the first generation of Landsat. It was believed that MAPSAT could produce imagery adequate for topographic mapping in 1:50,000 scale with 20 m contour intervals. Another proposed system is the Stereosat which also would employ three linear array cameras mounted in fore, vertical and aft looking directions to obtain base to height ratios of 1.0 for fore-and-aft steropairs and 0.49 for the vertical camera. Stereotriplets for a 6.14 km wide swath would be obtained from a 713 km circular, sun-synchronous, near-polar orbit with an inclination of 98.24° and a repeat cycle of 48 days. The nominal scale of the image would be 1:1,000,000. An IFOV of 15 m and one spectral band would be used. The imagery acquired was thought to be adequate for the production of topographic maps at 1:100,000 scale with contour intervals of 100 m (Welch and Marko 1981). If a coordinate data base can be established for the earth, the availability of Stereosat and MAPSAT will permit the construction of digital terrain models and the integration of satellite derived coordinates with detailed geophysical and environmental surveys conducted by aircraft or field parties.

Conclusions

The present review has clearly indicated the important role played by conventional aerial photography in topographic mapping under normal conditions largely because of the simplicity of the central perspective geometry and the well-established line of photogrammetric instruments available commercially. The essential conditions to satisfy are planimetric and height accuracy of the resulting maps which would meet the stringent requirements of a national mapping agency. There has been increasing interest in applying photogrametry towards a more analytical approach with the aid of computers. Much has been done in hardware and software developments to meet the desire for varying levels of automation. The analytical approach is applicable to imagery from nonphotographic systems, notably thermal infrared and SLAR. The latter has the greatest potential to complement conventional photography as a mapping tool under adverse weather conditions. Unfortunately, the poorer geometric accuracy especially in height restricts the use of SLAR to small-scale mapping projects only. Another restriction is the lower spatial resolution of the radar images which limits the information content of the resulting maps.

One major development in recent years has been space remote sensing and the employment of satellite imagery to produce topographic maps has received much attention by photogrammetrists. Preliminary evaluations have produced hopeful results. Methods of acquiring image data involving the use of conventional aerial cameras, vidicon cameras and multispectral scanners on a space platform have been developed. The Landsat system which was not specifically designed for topographic mapping has provided much valuable experience to the mapping scientists. Advancing solid-state detector technology has made linear-array sensors available for the acquisition of high-resolution image data of the earth from space with a greatly improved geometric fidelity. The designs of SPOT, Stereosat and MAPSAT embody all these concepts. It is anticipated that topographic mapping from space imagery at 1:100,000 scale with contour intervals of 100 m will soon be a reality.

References

Berstein, R. (1983) Image geometry and rectification: In Colwell, R. N. (ed.) *Manual of Remote Sensing*, American Society of Photogrammetry: Falls Church, Virginia, pp. 873–922.

Blachut, T. J. and Van Wijk, M. C. (1976) Results of the International Orthophoto Equipment 1972–76, *Photogrammetric Engineering and Remote Sensing* **42**: 1483–98.

Burnside, C. D. (1979) *Mapping from Aerial Photography.* Granada: St. Albans UK.

Case, J. B. (1981) Automation in photogrammetry, *Photogrammetric Engineering and Remote Sensing* **47**: 335–41.

Chevrel, M., Courtois, M. and Weill, G. (1981) The SPOT satellite remote sensing mission, *Photogrammetric Engineering and Remote Sensing* **47**: 1163–71.

Colvocoresses, A. P. (1979) Proposed parameters for MAPSAT, *Photogrammetric Engineering and Remote Sensing* **45**: 501–6.

Crandall, C. J. (1969) Radar mapping in Panama, *Photogrammetric Engineering* **35**: 641–6.

Crawley, B. G. (1975) Automatic contouring on the Gestalt photomapper: testing and evaluation, *Proceedings of Conference of Commonwealth Survey Officers.* Ministry of Overseas Development: London; Paper **No. E4**, 14 pages.

Derenyi, E. (1974) SLAR geometric test, *Photogrammetric Engineering* **40**: 597–604.

Derenyi, E. E. and MacRitchie, S. C. (1980) Photogrammetric application of Skylab photography, *Canadian Surveyor* **34**: 123–30.

Dorrer, E. (1976) Software aspects in computer-assisted stereoplotting. Presented Paper, Commission II, 13th ISP International Congress of Photogrammetry: Helsinki. Also published as **Dorrer, E.** (1977) Software aspects in desk-top computer-assisted stereoplotting, *Photogrammetria* **33**: 1–18.

Dorrer, E., Lander, E. and Toraskar, K. V. (1974) Analog to hybrid stereoplotter, *Photogrammetric Engineering* **40**: 271–80.

Dowman, I. J. (1977) Developments in on line techniques for photogrammetry and digital mapping, *Photogrammetric Record* **9**: 41–54.

Dowman, I. J. (1978) Topographic mapping from space photography: further developments, *Photogrammetric Record* **9**: 513–22.

Dowman, I. J. and Mohamed, M. A. (1981) Photogrammetric applications of Landsat MSS imagery, *International Journal of Remote Sensing* **2**: 105–13.

Dowman, I. J. and Morris, A. H. (1982) The use of synthetic aperture radar mapping, *Photogrammetric Record* **10**: 687–96.

Doyle, F. J. (1982) Status of satellite remote sensing programs, *USGS Open-file Report* **82–237**, 21 pages.

Friedmann, D. E., Friedel, J. P., Magnussen, K. L., Kwok, R. and Richardson, S. (1983). Multiple scene precision rectification of space borne imagery with very few ground control points, *Photogrammetric Engineering and Remote Sensing* **49**: 1657–67.

Greve, C. W. (1982) APPS-IV, Improving the basic instrument, *Photogrammetric Engineering and Remote Sensing* **48**: 903–6.

Hasler, A. (1982) The analytical stereoplotter Wild Aviolyt BC1, *Photogrammetric Engineering and Remote Sensing* **48**: 907–11.

Helava, U. V. (1968) On different methods of orthophotography, *The Canadian Surveyor* **22**: 5–20.

Hobrough, G. L. (1959) Automatic stereoplotting, *Photogrammetric Engineering* **25**: 763–9.

Kilford, W. K. (1979) *Elementary Air Survey.* Pitman.

Konecny, G. (1972) Geometric aspects of remote sensors. Invited paper for Commission IV, Twelfth Congress of the International Society of Photogrammetry: Ottawa, Canada.

Kratky, V. (1974) Cartographic accuracy of ERTS, *Photogrammetric Engineering* **40**: 203–12.

Leatherdale, J. D. and Keir, K. M. (1979) Digital method of map production, *Photogrammetric Record* **9**: 757–78.

Leberl, F. (1971) Metric properties of imagery produced by Side-Looking Airborne Radar and infrared linescan systems: In Kure, J. (ed.) *Proceedings of the ISP Commission IV Symposium.* Delft: 8–11 September 1970, pp. 125–51.

Leberl, F. (1974) Evaluation of SLAR image quality and geometry in PRORADAM, *ITC Journal* **No. 4**, pp. 518–46

Leberl, F. (1979) Accuracy analysis of stereo side-looking radar, *Photogrammetric Engineering and Remote Sensing* **45**: 1083–96.

Leberl, F., Jensen, H. and Kaplan, J. (1976) Side-looking radar mosaicking experiment, *Photogrammetric Engineering* **42**: 1035–42.

Lo, C. P. (1976) *Geographical Applications of Aerial Photography.* David and Charles: Newton Abbot; London; Vancouver.

Masry, S. E., Derenyi, E. and Crawley, B. G. (1976) Photomaps from nonconventional imagery, *Photogrammetric Engineering and Remote Sensing* **42**: 497–502.

Masry, S. E. and Gibbons, J. G. (1973) Distortion and rectification of IR, *Photogrammetric Engineering* **39**: 845–9.

Methley, B. D. F. (1970) Height from parallax bar and computer, *Photogrammetric Record* **6**: 459–65.

Moffitt, F. H. and Mikhail, E. M. (1980) *Photogrammetry.* Harper and Row: New York.

Mott, P. G. and Chismon, H. J. (1975) The use of satellite imagery for very small scale mapping, *Photogrammetric Record* **8**: 458–75.

Petrie, G. (1970) Some considerations regarding mapping from earth satellites, *Photogrammetric Record* **6**: 590–624.

Petrie, G. (1974) Mapping from earth satellite, *Road design,* **1.** Planning and Transport Research and Computation Co. Ltd: London, UK, pp. 1–20.

Petrie, G. (1977) Orthophotomaps, *Transactions New Series Institute of British Geographers* **2**: 49–70.

Petrie, G. (1981) Hardware aspects of digital mapping, *Photogrammetric Engineering and Remote Sensing* **47**: 307–20.

Petrie, G. and Adam, M. O. (1980) The design and development of a software based photogrammetric digitising system, *Photogrammetric Record* **10**: 39–61.

Roessel, J. W. van and Godoy, R. C. de (1974) SLAR mosaics for project RADAM, *Photogrammatic Engineering* **40**: 583–95.

Sabins, Jr., F. F. (1973) Recording and processing thermal IR imagery, *Photogrammetric Engineering* **39**: 839–44.

Sawada, N., Kidode, M., Shinoda, H., Asada, H., Twanaga, M., Watanabe, S., Moti, K. and Akiyama, M. (1981) An analytic correction method for satellite MSS geometric distortions, *Photogrammetric Engineering and Remote Sensing* **47**: 1195–1203.

Seymour, R. H. (1982) US-2 analytical stereoplotter, *Photogrammetric Engineering and Remote Sensing* **48**: 931–7.

Slama, C. C. (ed.) (1980) *Manual of Photogrammetry.* American Society of Photogrammetry: Falls Church, Virginia.

Southard, R. B. (1978) The USGS orthophoto program, a 1978 update, *Proceedings of the International Symposium Commission IV, New Technology for Mapping*: Ottawa, Canada, 2–6 October 1978, pp. 594–601.

Tait, D. A. (1970) Photo-interpretation and topographic mapping, *Photogrammetric Record* **35**: 466–79.

Taylor, J. I. (1971) Rectification equations for infrared line-scan imagery: In Kure, J. (ed.) *Proceedings of the ISP Commision IV Symposium*; Delft, 8–11 September 1970, pp. 178–94.

Thompson, E. H. (1954) Height from parallax measurements, *Photogrammetric Record* **1**: 38–49.

Thompson, L. L. (1979) Remote sensing using solid-state array technology, *Photogrammetric Engineering and Remote Sensing* **45**: 47–55.

Thompson, M. M. (1979) *Maps for America*, US Geological Survey. Reston, Virginia; US Government Printing Office, Washington, DC.

Townshend, J. R. G., Gayler, J. R., Hardy, J. R., Jackson, M. J. and Baker, J. R. (1983) Preliminary analysis of Thematic Mapper products, *International Journal of Remote Sensing* **4**: 817–28.

Turner, H. (1982) The Canadian Marconi ANAPLOT II system, *Photogrammetric Engineering and Remote Sensing* **48**: 891–901.

Welch, R. (1973) Cartographic quality of ERTS-1 images. Paper presented at the *Symposium on Significant Results from ERTS-1*; sponsored by NASA/Goddard Space Flight Center, March, 1973.

Welch, R. and Lo, C. P. (1977) Height measurements from satellite images, *Photogrammetric Engineering and Remote Sensing* **43**: 1233–41.

Welch, R. and Marko, W. (1981) Cartographic potential of a spacecraft line-array camera system: Stereosat, *Photogrammetric Engineering and Remote Sensing* **47**: 1173–85.

Welch, R. and Usery, E. L. (1984) Cartographic accuracy of Landsat-4 MSS and TM image data, *IEEE Transactions on Geoscience and Remote Sensing*, **GE-22**: 281–8.

Wolf, P. R. (1983) *Elements of Photogrammetry.* McGraw-Hill: New York.

Wong, K. W. (1975) Geometric and cartographic accuracy of ERTS-1 imagery, *Photogrammetric Engineering and Remote Sensing* **41**: 621–35.

Chapter 9 Geographic information systems

Basic concepts of geographic information systems

It has been demonstrated in the preceding chapters that remote sensing applications to the atmosphere, lithosphere and hydrosphere are interrelated. An integrative approach to tackle environmental problems is essential. With the advent of high-resolution multispectral remote sensing from space platforms, the volume of data concerning our terrestrial environment becomes too large to be effectively utilized. This explains the need to carry out geographic data processing with the aid of a computer. It should be noted at this point that the input to the geographic information system (GIS), as this use of the computer is called, is not restricted to remote sensing data. The word 'geographic' implies data with specific locations. In a geographic information system, the data should be referenced in a manner which will allow retrieval, analysis and display on spatial criteria (Tomlinson 1972).

A geographic information system should at least consist of a data processing subsystem, a data analysis subsystem and an information use subsystem as shown in Fig. 9.1. The data processing subsystem includes data acquisition, input and storage. The data analysis subsystem includes retrieval, analysis and output of information in various forms. The information use subsystem permits relevant information to be applied to the problem.

In the design of a geographic information system, specialized graphic input and output components often play a dominant role in shaping the architecture of the remainder of the system. It is necessary to understand in some depth the procedures adopted in dealing with problems of data input or output as well as data organization and processing.

Data input and output

There are three broad categories of data for input to the system: (1) alphanumeric; (2) pictorial or graphic; (3) remotely sensed data in digital form. The entry of alphanumeric data presents no great problem because they are available in computer-readable form. The input of pictorial or graphic data, such as maps and photographs, requires the use of a digitizer which converts the features into strings of coordinate values. A usual approach is to represent polygonal

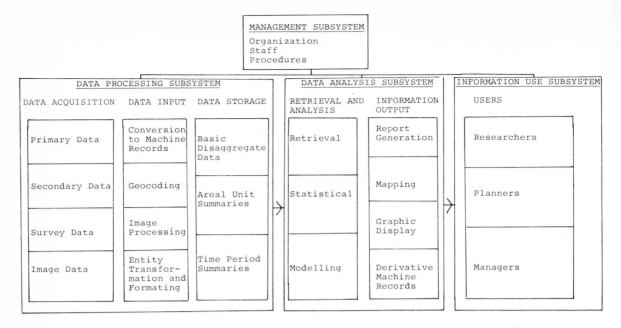

Fig. 9.1
Information system elements
(*Source*: Tomlinson 1972)

boundaries as lines and to represent lines as a sequence of very short, straight line segments which can be represented by an ordered sequence of points defining the line segments (Marble and Peuquet 1983). This results in a *vector format*. Another approach is to make use of an optical scanner or a scanning densitometer to convert the graphic material to computer-readable form automatically. This records spatial data in narrow strips across the data surface, thus producing the *raster format*. Some scanners are particularly designed for use with grey-scale material (such as aerial photographs), which record data in eight-bit bytes for each picture element (pixel). The raster-scanned data are converted into a row or column-ordered matrix of reflectance values. Other scanners are restricted to black-and-white operation, and hence are more suitable for line maps. The raster format includes grids or cells or matrix data-structure. This is not only more compatible with modern input/output hardware but also has the advantage that the order of the data element is determined by their geographical positions.

Remotely sensed data generated by multispectral scanners or high-resolution vidicon cameras from space platforms, are in a raster format. However, these spectral data have to be restored, enhanced, filtered or geometrically transformed by the techniques of image processing before they can be incorporated into the geographic information system. One important problem is to effect an interface between raster information and geographic information system coordinates (i.e. vector format). A common approach is to develop a set of transformation equations (Meyer 1984). The set of equations which converts the vector format of the geographic system into the raster format of the image is:

$$L = f_1(X,Y)$$

$$\text{and} \quad E = f_2(X,Y) \qquad [9.1]$$

where L = scan line;

E = element number within scan line;

X = horizontal coordinates of the geographic information system map projection;

and Y = vertical coordinate of the geographic information system map projection.

Similarly, the set of equations which converts the raster image data set into the vector format of the geographic information system is:

$$X = f_3(L,E)$$
$$\text{and} \quad Y = f_4(L,E) \qquad [9.2]$$

These equations commonly take the form of polynomial functions derived by least squares analysis of control points that can be visually identified on an image.

When multispectral image data sets are used, there is the need to register them together. A common practice is to reconstitute the image data sets in such a way that they are geometrically corrected and registered to a common map base of the geographic information system. This registration is accomplished by resampling techniques which involve defining new pixel positions on the map base and filling them with data determined by nearest neighbour or interpolation algorithms. As an example, the multispectral scanner (MSS) data of Landsat are normally registered to a UTM (Universal Transverse Mercator) grid.

The output devices from the geographic information system are incremental plotters, vector display cathode ray tubes and raster displays and matrix plotters which all show correspondence with the input devices. Line printers are effective low-cost output devices commonly employed.

Data organization and processing

In a geographic information system, algorithms are required to transform the raw data into meaningful information in the form of cartographic products. These data are conventionally divided into *point*-types, *line*-types and *region*-types based on their geometry. Good examples of point data type are mountain peaks, cities and street lights. Line data type includes rivers, roads or topographic contours. Region data type is represented by land use, soil classification, drainage basins and crop type. These geographic data types interact with each other as shown in Fig. 9.2 (Nagy and Wagle 1979). Altogether there are six types of binary relations to consider: point-to-point, point-to-line, point-to-region, line-to-line, line-to-region, and region-to-region. Algorithms have to be written to deal with them. The retrieval of geographic data often involves distance considerations, such as finding the nearest neighbours among a set of points.

Fig. 9.2
Geographic data types (*Source*:
Nagy and Wagle 1979)

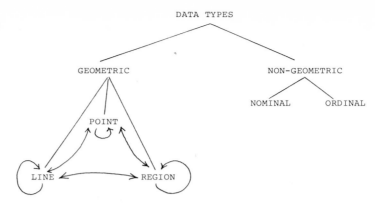

There are two major approaches to internal data organization in a geographic information system: the *cellular* organization and *linked* organization, which correspond roughly to *discrete* and *continuous* data respectively. Cellular organization makes use of a matrix of cells of uniform size, and the data are stored on each cell. The organization corresponds directly to the raster format for input and the matrix format for output. Software development is generally easier for the cellular approach which is therefore most suited to storing surface variables. However, this approach is wasteful in terms of computer storage space. Linked organization makes use of codinate values. Thus, point entities are described directly by their coordinates, line entities by a chain of coordinates, regions with nominal data by their boundaries and regions with numerical data by contour curves. This data organization corresponds directly to the format of data entered through a digitizer and output by an incremental plotter; it is highly storage space. Linked organization makes use of coordinate values. maps and contour maps. Its disadvantages are the complexity of software development in updating and editing data.

External data organization involves examination of the implementation, operation and characteristics of the data. This relates to the performance of the computer system required to carry out all of the operations designed. Should the data be stored as entered and processed as required, or should they be preprocessed to enrich the archived information? These are the kind of questions to be considered when dealing with external data organization. They are intimately related to auxiliary storage organization.

There are two forms of auxiliary data organization: the *data bank* and the *data base*. A data bank is 'separation' oriented with the information segregated either on the basis of the type of entity or on its geographical unit or both. It is simple to implement. Data are generally maintained on disk or tape files. A data base, on the other hand, is interaction oriented. It adds inter-entity relations to the information. The data exhibit a part-whole relationship which is said to be hierarchically organized. Thus, subdivision of a region and tributaries of a river are examples of this approach. On the whole, nongeometrical information may be organized into a data base while the geometric information may be organized as a data bank.

Some examples of geographic information systems

The foregoing discussion provides some criteria to understand and evaluate the performance of geographic information systems. Four specific systems will be examined in this section. They have been selected because of their emphasis on remote sensing data input.

Example 1

The ODYSSEY system

This system was designed by the Laboratory of Computer Graphics and Spatial Analysis at Harvard University, USA (Teicholz 1980). This is based on the series of computer programs such as SYMAP, SYMVU, CALFORM etc. produced previously by the Laboratory, which are made compatible with each other. The purposes of the system are to combine different data sources, notably the DIME files of the US Census Bureau, the Land Use Series of the US Geological Survey, the World Data Banks I and II of the Central Intelligence Agency, the Landsat data of the National Aeronautics and Space Administration and soil surveys of the US Soil Conservation Service, into a common data base and to compare these irregular coverages by means of the analytical process of polygon overlay so that a composite coverage can be created. The ODYSSEY system can create data bases by integrating data from a variety of sources, enabling the manipulation of a data base, performing analytical tasks on the data and displaying the results as coloured or black-and-white thematic maps. Because of this requirement, the ODYSSEY system can accept data in both raster and vector formats. Data management capabilities include: (1) verification and correction; (2) splicing, i.e. merging of multiple data files; (3) overlay or superimposition of two or more networks; (4) geographic aggregation, i.e. the building of small units into successively larger units; (5) line and area generalization; (6) point-in-polygon calculations, i.e. the ability to automatically identify a point that lies within one coverage set with another coverage set to avoid the computationally demanding polygon overlay operation. Attribute management capabilities involve: (1) aggregation required as a result of geographic aggregation; (2) classification or functional transformation of the attributes, usually from larger to smaller groups which could involve the conversion of nominal data into ordinal data; (3) merging, corresponding to splicing, as mentioned in data management, above; (4) correction or editing of data. The analysis and display capabilities include: (1) first-order geometric measurements on data such as length, area and perimeter calculations; (2) generation of base maps – shaded black-and-white or coloured choropleth maps, three dimensional representations of polygon surfaces and line printer maps. Finally, for purposes of archival storage, binary of ASCII files of ODYSSEY-processed data can be output, in both polygon and gridded data formats.

The following is an example illustrating the overlaying capability of the ODYSSEY system which correlates forest classification based on Landsat images with a forest-stand map derived from aerial photographs. Note that the Landsat derived forest classification map (Fig. 9.3a) is in a grid format while the forest-stand map (Fig. 9.3b) is encoded with a line-following digitizer. Fig. 9.3(c) displays the result of the overlay – a composite map. The values associated with each of the polygon files are then compared by computing a cross tabulation between the digitized and the Landsat-derived forest types. Discrepancies can then be automatically calculated. Correspondence between the two maps is shown in Fig. 9.3(d). Agreement between the two maps by forest class (hardwood) can also be depicted (Fig. 9.4a). The results of the cross-tabulation can also be mapped (Fig. 9.4b).

The ODYSSEY currently operates on 32-bit and larger word machines. Versions for VAX II/780 and PDP-10 are maintained at Harvard University. IBM S/370 and CDC 6000 series versions exist. Program modules take from 40K to 60K words. All software is written in ANSI Fortran IV language. A significant contribution of the ODYSSEY system is the technique of local processing incorporated in the programs. This helps to reduce the storage requirement of many programs.

Fig. 9.3
(a) Forest classification derived from Landsat image; (b) Digitized forest-stand map; (c) Composite map resulted from overlaying (a) and (b); (d) degree of agreement (in black) between Landsat and aerial survey data (*Source*: Teicholz 1980)

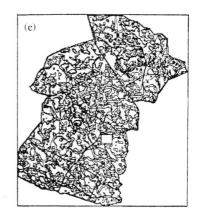

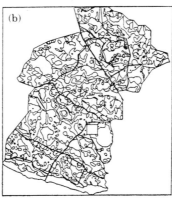

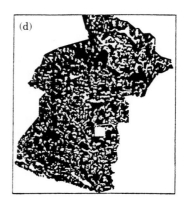

Fig. 9.4
(a) Agreement between Landsat and aerial survey for hardwood; (b) Discrepancies between aerial survey and Landsat classification of forest types (*Source:* Teicholz, 1980)

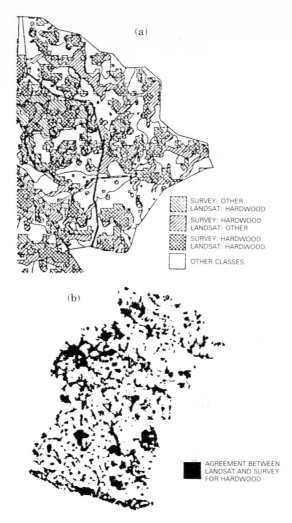

(a)

SURVEY: OTHER
LANDSAT: HARDWOOD

SURVEY: HARDWOOD
LANDSAT: OTHER

SURVEY: HARDWOOD
LANDSAT: HARDWOOD

OTHER CLASSES

(b)

AGREEMENT BETWEEN
LANDSAT AND SURVEY
FOR HARDWOOD

Example 2:

Geographic Information Retrieval and Analysis System (GIRAS)

Unlike ODYSSEY, this geographic information system developed by the US Geological Survey is oriented specifically to land use and land cover maps as the data source. These maps have been produced by the US Geological Survey at scales of 1:250,000 and 1:100,000. The land use and land cover classification system in Level II designed by Anderson *et al.* (1976) provides the basis for the map data. Other source data entering the system include data from political unit maps, hydrologic unit maps, census county subdivision maps, Federal land ownership maps and state ownership maps (Mitchell *et al.* 1977). The system has been designed to input, manipulate, analyse and output digital spatial data developed for the land use and land cover mapping and data

375

circulation programme of the US Geological Survey. The configuration of the system is shown in Fig. 9.5. The first operational version of this system (GIRAS-1) is a batch-oriented sequential system. The initial emphasis was on the editing and correction of digitized land use and associated data bases.

The input of map data into the system depends on cartographic digitizing. In other words, a vector format is used and the linked organization of data is adopted. In order to cut down the volume of the digitizer coordinates, an optimized logical and physical structuring of the data is developed. Since all land use and land cover maps are essentially polygon maps, each polygon map can be defined in terms of such topological elements as node, arc, polygon, island and polygon label (Fig. 9.6). The common boundaries, or arcs, are digitized only once and the arcs are then joined together to form polygons. Arcs are line segments definable by a series of x, y coordinate pairs. A map file with a data structure shown in Fig. 9.7 is produced for each polygon map. The map header contains information about the amount of data in the map, the date of the source material, title and control point

Fig. 9.5
General system flow of GIRAS (*Source*: Mitchell *et al.* 1977)

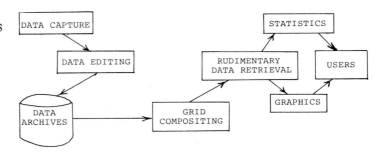

Fig. 9.6
Topological elements of a polygon map (*Source*: Mitchell *et al.* 1977)

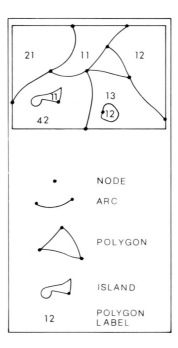

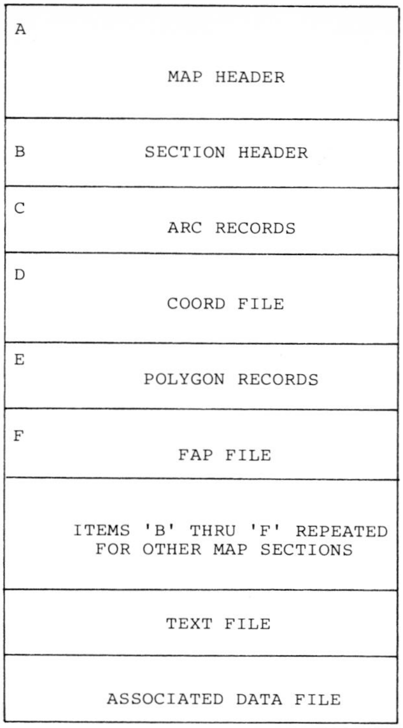

Fig. 9.7
GIRAS file structure (*Source*: Mitchell *et al.* 1977)

```
┌─────────────────────────┐
│ A                       │
│                         │
│       MAP HEADER        │
│                         │
├─────────────────────────┤
│ B     SECTION HEADER    │
├─────────────────────────┤
│ C                       │
│       ARC RECORDS       │
├─────────────────────────┤
│ D                       │
│       COORD FILE        │
├─────────────────────────┤
│ E                       │
│     POLYGON RECORDS     │
├─────────────────────────┤
│ F                       │
│        FAP FILE         │
├─────────────────────────┤
│  ITEMS 'B' THRU 'F' REPEATED │
│   FOR OTHER MAP SECTIONS │
├─────────────────────────┤
│        TEXT FILE        │
├─────────────────────────┤
│   ASSOCIATED DATA FILE  │
└─────────────────────────┘
```

information. A map is then segmented into sections. Within each section there are four subfiles: arc records, coordinate data, polygon records and a subfile assigning arcs to polygons (called FAP). This is then followed by a text file which relates to the entire map and contains definitions of attribute codes to be used in manipulation procedures. The associated file is provided to enable users to carry additional information with the map for their own purposes.

After the necessary data have been captured by the digitizing device to place in the above-mentioned structure, the data go through a sequence of steps, viz., data conversion to standard format, data reduction, data editing, merging and edge matching as shown in the flow diagram (Fig. 9.8). After this, the data can be processed by a number of manipulation and analysis procedures including: (1) rotation, translation and scaling of coordinates; (2) conversion from geographic coordinates to specified map projections; (3) restoration of original digitized map sheet from rectangular coordinate projection; (4) conversion from polygon structure to grid cells of any specified size; (5) production of area summary statistics from polygon or gridded data; (6) geographic interpolation; (7) filtering of nominal spatial data; (8) feature generalization; (9) accuracy estimation of nominal maps. For output, the system can produce colour and black-and-white shading pattern maps, choropleth maps of polygons and gridded polygons, perspective view block diagrams, perspective view contour maps and isometric bivariate histograms.

A new GIRAS-II system makes use of interactive, on-line, time-sharing, random access input processing. Another change is the shift from grid compositing to polygon compositing. The third change

Fig. 9.8
Graphic input procedures for
GIRAS (*Source*: Mitchell *et al.*
1977)

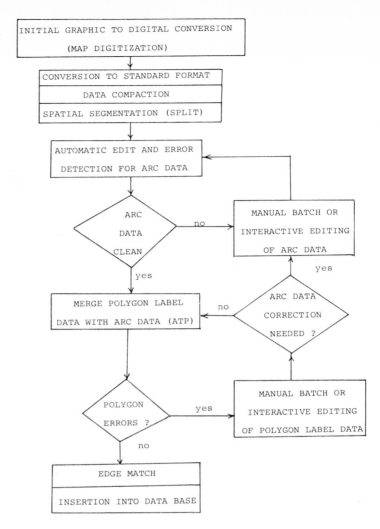

is the interposition of a 'data base management system' between the complex data base and data retrieval. This is required in view of the increasing volumes of spatial digital data added to the US Geological Survey geographic information system as the Landsat images facilitate automatic and semi-automatic mapping of land use and/or land cover. A major problem to resolve in the GIRAS system is the use of digitizers in the data capture stage. A manual approach has been used in digitizing but it becomes obvious that digitizing by an automatic line-following laser scanner can improve the speed and accuracy of the data captured.

Example 3:

IBIS (Image Based Information System)

IBIS is an image-based geographic information system designed to employ digital image processing technology to: (1) interface remotely sensed data with other spatially referenced information; (2) perform interactive data base storage, retrieval and analysis operations (Bryant and Zobrist 1976). All the data are input in a raster format with cellular organization of data. However, other data such as graphical and tabular, can also be used in analysis. This involves transformation of the graphical data, usually obtained in the form of rectangular coordinates, into image space (i.e. raster format). Tabular data are linked to the image data-base through a logical interface. In this system, Landsat imagery provides the planimetric base and a census tract map provides the georeference plane for the data. All other data planes are related to them. The image processing technique carries out geometric rectification procedures to register different sets of image data together. The IBIS system is built upon an existing image processing system called VICAR (Video Image Communication and Retrieval) developed at the Jet Propulsion Laboratory, California Institute of Technology, Pasadena, USA. This carries out all three fundamental elements of digital image processing: image enhancement, multispectral classification and image rectification and, or registration, thus facilitating the updating of data on the image plane in the geographic information system. An important characteristic of data manipulation in the system is the conversion of image data into tabular data and *vice versa*. Maps and tabular reports can be output from the system (Fig. 9.9).

To illustrate the operation of IBIS, an example of application is given below. The parish of St. Tammany near New Orleans, Louisiana, USA is to develop a geographic information system for general purposes of engineering, planning and management. The processing steps required are shown in Fig. 9.10. The study area is divided into 14 transportation districts and subdivided into 44 transportation zones (Fig. 9.11, page 380). The district is digitized on a coordinate digitizer manually with magnetic tape output. The data which are stored by zone in a polygon district file are edited to guarantee topological

Fig. 9.9
IBIS configuration diagram (*Source*: Marble and Peuquet 1983)

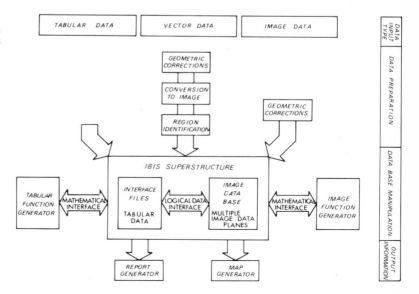

379

Fig. 9.10
Processing steps for the case study using IBIS (*Courtesy*: N. A. Bryant; *Source*: Bryant and Zobrist 1976)

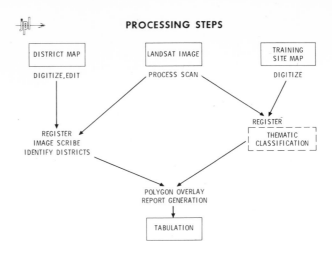

PROCESSING STEPS

DISTRICT MAP

DIGITIZE, EDIT

LANDSAT IMAGE

PROCESS SCAN

TRAINING SITE MAP

DIGITIZE

REGISTER
IMAGE SCRIBE
IDENTIFY DISTRICTS

REGISTER

THEMATIC
CLASSIFICATION

POLYGON OVERLAY
REPORT GENERATION

TABULATION

closure of regions. Landsat digital image data of the region of interest is obtained and converted into a format suitable for classification algorithms of the image processing system. The training area map is digitized on a coordinate digitizer. The polygon district file is then converted into an image (Fig. 9.12). The rubber-sheet stretching method is performed to register the district boundaries to the Landsat image (Fig. 9.13). With the training area data, a mathematical classification of the Landsat image data into a land use map of the study area is carried out (Pl. 17, page 380). A grey scale map (choropleth map) of traffic zones in the district is produced (Fig. 9.14).

Fig. 9.12
An image of St. Tammany Parish, Louisiana, USA transportation zone boundaries (*Courtesy*: N. A. Bryant; *Source*: Bryant and Zobrist 1976)

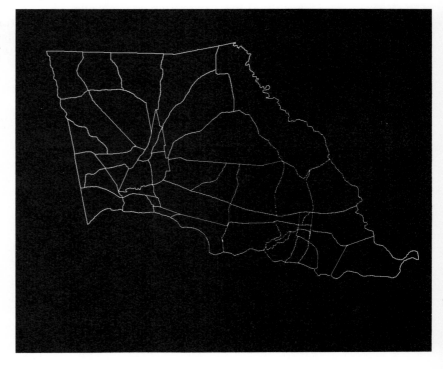

Plate 15
Digitally enhanced Landsat image of the Baltic Sea area using MSS bands 4, 5 and 6 reprojected through blue, green and red filters respectively. This shows the blue-green algae in the Arkona Sea east of Møn Island. They form eddies and surface currents (*Courtesy*: D. Schmidt; *Source*: Schmidt 1979)

Plate 16
Colour density sliced enhancements of Landsat-1 images taken on 10 October 1972 (a) MSS band 4, (b) MSS band 5, (c) MSS band 6 and (d) MSS band 7 (*Courtesy*: V. Klemas; *Source* Klemas *et al.* 1973)

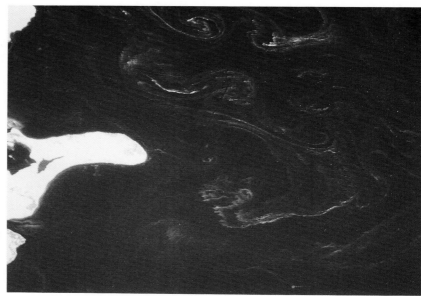

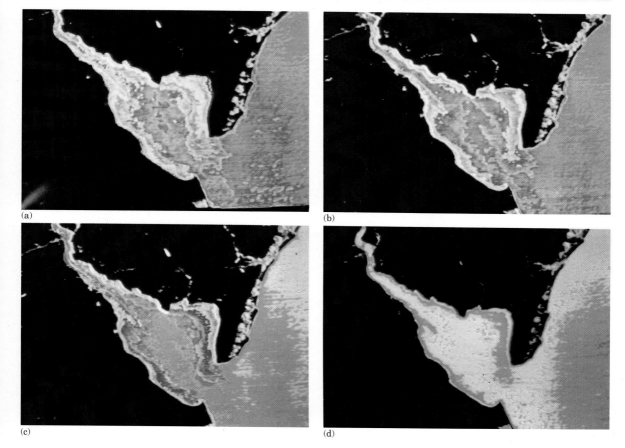

(a)

(b)

(c)

(d)

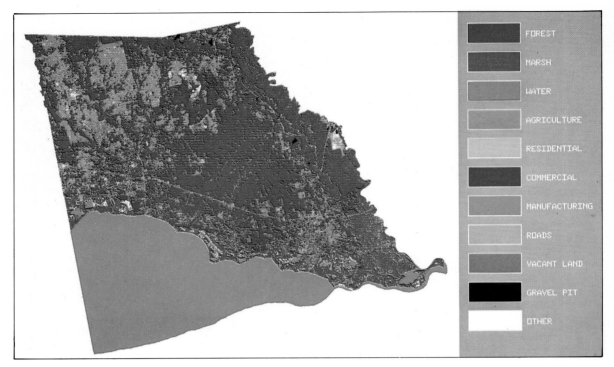

Plate 17
Land use classification map
(*Courtesy*: N. A. Bryant; *Source*:
Bryant and Zobrist 1976)

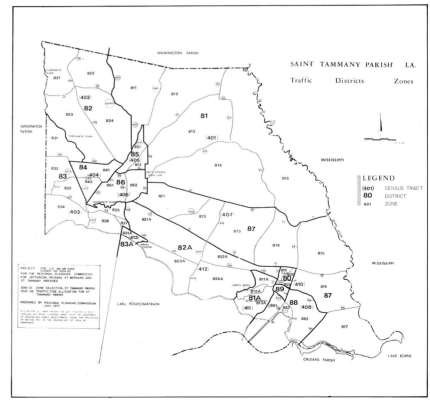

Fig. 9.11
District map for St. Tammany
Parish, Louisiana, USA (*Courtesy*: N. A. Bryant; *Source*:
Bryant and Zobrist, 1976)

Fig. 9.13
Overlay of district boundaries and Landsat image (*Courtesy*: N. A. Bryant; *Source*: Bryant and Zobrist 1976)

Fig. 9.14
Transportation zone chloropleth map (*Courtesy*: N. A. Bryant; *Source*: Bryant and Zobrist 1976)

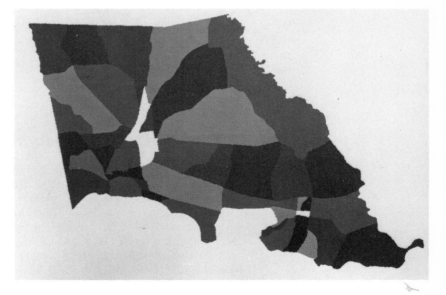

A polygon overlay is performed between the land use map (Pl. 17) and the transportation zone choropleth map (Fig. 9.14). The pixel of the land use map is related to the corresponding pixel of the choropleth map. A tabular report with absolute and percentage values of land use by areas is produced for each traffic zone (Fig. 9.15).

The IBIS system is an evolving system. In recent years it has been

Fig. 9.15
Land use tabulation produced by IBIS (*Courtesy*: N. A. Bryant; *Source*: Bryant and Zobrist 1976)

ST. TAMMANY PARISH
LAND USE ACREAGES BY TRANSPORTATION ZONES

ZONE	AREA	MANUFACTURING		ROADS		VACANT LAND		GRAVEL PIT		OTHER	
		AREA	PCT	AREA	PCT	AREA	PCT	AREA	PCT	AREA	PCT
801	540.	2.	0.4	105.	19.4	93.	17.2	0.	0.0	2.	0.4
802	677.	6.	0.9	103.	15.2	41.	6.0	0.	0.0	6.	0.9
803	453.	7.	1.5	116.	25.6	48.	10.6	0.	0.0	6.	1.3
811	15134.	1.	0.0	794.	5.2	2486.	16.4	0.	0.0	360.	2.4
812	17121.	14.	0.1	959.	5.6	4449.	26.0	225.	1.3	797.	4.7
813	21122.	1.	0.0	1529.	7.2	4494.	21.3	9.	0.0	556.	2.6
814	29398.	4.	0.0	636.	2.3	5422.	19.3	277.	0.9	319.	1.1
815	37136.	66.	0.2	1148.	3.1	3116.	8.3	277.	0.7	2675.	7.1
821	8712.	4.	0.0	726.	8.3	1536.	17.6	0.	0.0	165.	1.9
822	9654.	0.	0.0	575.	6.0	1550.	16.1	0.	0.0	210.	2.2
823	13820.	8.	0.1	1274.	9.2	3835.	27.7	0.	0.0	624.	4.5
824	14927.	3.	0.0	1110.	7.4	2818.	18.9	0.	0.0	304.	2.0
825	10520.	29.	0.3	578.	5.5	861.	8.1	0.	0.0	575.	5.5
831	5196.	0.	0.0	218.	4.2	981.	18.9	0.	0.0	77.	1.5
832	4215.	9.	0.2	250.	5.9	681.	16.2	0.	0.0	122.	2.9
833	3688.	7.	0.2	257.	7.0	702.	19.0	0.	0.0	93.	2.5
834	6669.	0.	0.0	208.	3.1	503.	7.5	0.	0.0	662.	9.9
835	1792.	17.	0.9	203.	11.3	137.	7.6	0.	0.0	72.	4.0
836	3554.	14.	0.4	157.	4.4	242.	6.8	0.	0.0	432.	12.0
837	687.	2.	0.3	78.	11.4	37.	5.4	0.	0.0	119.	5.4
841	4303.	6.	0.1	448.	10.4	768.	17.9	0.	0.0	82.	3.9
842	2118.	0.	0.0	157.	7.4	379.	17.9	0.	0.0	82.	3.9
851	2451.	20.	0.8	175.	6.6	242.	9.1	0.	0.0	127.	4.8
852	1235.	9.	0.3	167.	13.5	268.	21.1	0.	0.0	51.	4.1
811A	1717.	5.	0.3	149.	8.7	207.	12.1	0.	0.0	48.	2.8
851	4869.	5.	0.1	333.	6.8	195.	4.0	0.	0.0	353.	7.2
865	3511.	2.	0.2	430.	12.2	500.	14.2	0.	0.0	93.	2.6
812A	966.	2.	0.2	131.	10.0	38.	3.9	0.	0.0	72.	7.5
813A	3851.	33.	0.9	254.	6.6	110.	2.9	0.	0.0	469.	12.2
821A	9562.	10.	0.1	351.	3.5	1761.	18.2	0.	0.0	100.	1.0
821A	13926.	7.	0.0	759.	5.4	1561.	11.2	0.	0.0	583.	4.2
822A	5731.	6.	0.1	462.	8.1	856.	15.0	0.	0.0	99.	1.7
823	6973.	0.	0.0	145.	2.1	798.	11.4	0.	0.0	68.	1.0
823A	6022.	0.	0.0	277.	4.6	176.	2.7	0.	0.0	802.	12.3
873	14877.	4.	0.0	518.	3.5	2063.	13.9	0.	0.0	103.	0.7
874	12494.	5.	0.0	815.	7.8	1842.	17.6	0.	0.0	105.	1.0
875	12275.	6.	0.0	354.	2.8	1810.	14.6	0.	0.0	425.	3.4
876	10831.	12.	0.1	678.	6.3	1633.	16.4	0.	0.0	261.	1.9
877	13707.	7.	0.1	1135.	8.2	977.	7.0	0.	0.0	1000.	7.2
881	2476.	305.	13.4	351.	15.4	109.	4.8	0.	0.0	188.	8.3
881A	1053.	31.	2.9	148.	14.1	211.	20.0	0.	0.0	10.	0.9
883	2741.	0.	0.4	333.	13.5	146.	6.5	0.	0.0	92.	4.1
891	7359.	20.	0.3	440.	6.0	648.	10.5	0.	0.0	786.	10.7
891	937.	0.	0.0	247.	29.3	88.	10.5	0.	0.0	2.	0.2

further developed to perform the data management required for mosaicking of digital images (Zobrist, Bryant and McLeod 1983). This has been done for Landsat MSS and RBV images and airborne scanner data sets. Since a large number of ground control points are required, tabular data mangement procedures are involved. IBIS therefore provides an easy interface between mathematical function processing and image processing, hence permitting the smooth transfer of co-efficients and observations from tabular to image domains. The resultant mosaic will provide a large cartographic data base for the geographic information system.

Example 4:

A photogrammetric urban information system

The author has also developed an urban information system based on input from aerial photography (Lo 1974; 1981). A photogrammetric plotting machine – the CP-1 plotter designed by the late E. H. Thompson of University College, London, UK is employed to provide input of data to the computer. The vector format is used and the recording unit is an individual building in the city. A stereomodel of the study area is set up in the photogrammetric plotting machine and the position of each building is determined by the polar pantograph of the CP-1 plotter. The polar coordinates can be converted to rectangular coordinates by the computer. For each building, the morphological characteristics, such as, roof form, height, area and type as well as its use have been extracted from the aerial photographs (Fig. 9.16). These data can be processed using statistical packages, graphical packages and trend surface analysis with the aid of the computer. Other nonphotographic data, notably census data, can be input into the system for correlation with the building characteristics. In this way, the morphological characteristics of the buildings are related to

Fig. 9.16
Flow diagram of the photogram-
metric urban information system
(*Source*: Lo 1981)

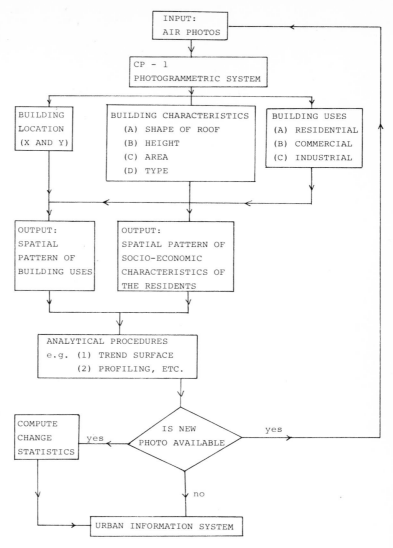

the socio-economic characteristics of the population from the census.
All the data stored can be displayed by the computer using SYMAP,
SYMVU and CALFORM packages as well as other specially written
programs for the Calcomp plotter. Further improvements of this
photogrammetric information system are carried out with attention
given to data structure and data organization (Yeung and Lo 1983).
Data are organized as data files in the system. A data file is a collection of related records which are made up of individual data items.
Table 9.1 shows an example of the data matrix for the urban area of
Kowloon, Hong Kong. For most operations, data relating to one or two
characteristics only are required. A better pattern of data arrangement is the sequential records (Table 9.2) and inverted file data structure (Table 9.3) because these forms are more suitable to data
updating. Another refinement of the system is to adopt a modular
system structure which divides tasks into pragmatic program units

Table 9.1
A sample land information data
matrix

	BSU#	Characteristics				
		Land Tenure	Landuse Type	Bldg Type	TPU #	. . . . .
	123	FH	I	MF	211	
	137	LH	I	MF	211	
	428	LH	C	OB	214	
	638	FH	C	HT	216	
Entity Identifiers	540	LH	R	AP	215	
	792	FH	R	AP	217	
	406	FH	C	OB	214	
	407	FH	C	OB	214	
	129	LH	R	TN	211	
	114	LH	C	HT	211	
	357	FH	C	OB	213	
	542	LH	R	OB	215	
	104	LH	G	GO	211	
	339	FH	R	TN	213	
	148	FH	C	OB	211	

Table 9.2
Sequence of related records
created from sample data matrix
in Table 9.1

Rec #	BSU #	Land Tenure	Landuse Type	Bldg Type	TPU #	. . . . . .
1	(123	FH	I	MF	211	. . . . . .)
2	(137	LH	I	MF	211	. . . . . .)
3	(428	LH	C	OB	214	. . . . . .)
4	(638	FH	C	HT	216	. . . . . .)
5	(540	LH	R	AP	215	. . . . . .)
6	(792	FH	R	AP	217	. . . . . .)
.						
.						
.						
.						
.						
14	(339	FH	R	TN	213	. . . . . .)
15	(148	FH	C	OB	211	. . . . . .)

known as 'modules'. This enables the user to keep quality control as
the data files are built-up and maintained over the period of system
use. The configuration of such a modular urban information system
is shown in Fig. 9.17. The photogrammetric approach has the major
advantage that it facilitates the extraction of accurate quantitative
and qualitative data from the aerial photography as input to the urban
information system. This approach encourages greater use of aerial
photography.

Table 9.3
Inverted file created from sample
data matrix in Table 9.1

Characteristics	Attribute values	Entity identifiers
Land tenure	FH	123, 638, 792, 406, 357, 339, 149
	LH	137, 428, 540, 129, 114, 542, 104
Land use type	I	123, 137
	C	428, 628, 406, 407, 114, 357, 149
	R	540, 792, 129, 542, 339
	G	104
Building type	MS	123, 237
	OB	428, 406, 407, 357, 542, 149
	HT	638, 114
	AP	540, 927
	TN	129, 339
	GO	104
TPU#	211	123, 137, 129, 114, 104, 149
	212	
	213	357, 339
	214	482, 406, 407
	.	
	.	
	.	

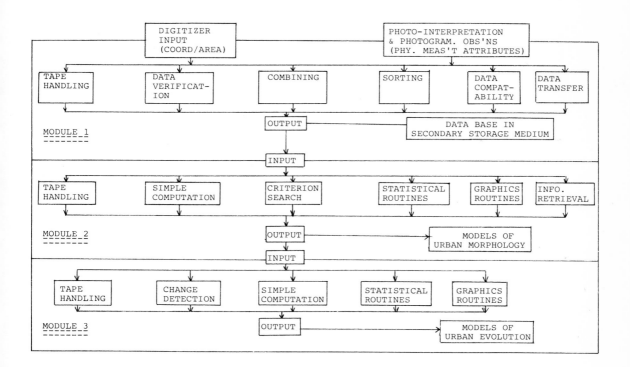

Fig. 9.17
Configuration of a modular land
information system (*Source*:
Yeung and Lo 1983)

Conclusions

In this chapter, the concepts and principles of geographic information
systems are discussed and some examples of systems are examined.
The limited number of examples of geographic information systems

discussed cannot do justice to the large number of geographic information systems of varying scales that have been designed and implemented. Most of these systems have been developed for resource management with remote sensing data input in mind (Mooneyhan 1982; McFarland 1982). The intimate relationship between remote sensing and information system suggests that this is the most effective way of integrating data from various sources required for planning decision making (Estes 1982). There is a great variety of environmental and resource data relating to the atmosphere, the lithosphere and the hydrosphere that one can extract from remotely sensed imagery. These data complement each other to provide a better insight into our terrestrial environment. A good example to illustrate this integrative approach is the monitoring of the desert locust habitat (Hielkema 1981). It was discovered that the desert locust had to depend on three main parameters: *precipitation* to provide suitable soil moisture conditions to complete egg development; *green vegetation* for food and shelter during the nymphal stages; a suitable *temperature* regime to develop and breed. Both the environmental and earth resources satellites offer us the data to estimate rainfall and temperature; to determine vegetation amount and conditions; and to map land use and/or land cover types in the desert area, all with the aid of the computer. This culminates in the mapping of the desert locust habitat for automated monitoring of the ecological parameters favourable to their breeding and development. This is an area of application which has a practical importance to mankind in the number of lives it may save. The application of geographic information system technology in this direction is most appropriate.

The rapid advancement in computer technology in recent years has spurred the development of geographic information systems, particularly in view of the availability of high-resolution data from the satellite platforms. However, it has been observed that the interface between geographic information systems and remote sensing systems is still weak (Marble and Peuquet 1983). To be useful, the geographic information system should be capable of updating its various spatial data elements continuously and be economical enough to operate. Remote sensing systems, on the other hand, should strive to incorporate more ground data to improve their classification accuracies. Further research in these directions is needed.

As a conclusion to this book, applied remote sensing should be aimed at benefiting mankind through the use of an interdisciplinary approach. Remote sensing as a space-age technology provides a convenient tool to scientists for a better monitoring of the earth – the human habitat.

References

Anderson, J. R., Hardy, E. E., Roach, J. T. and Witmer, R. E. (1976) *A Land Use and Land Cover Classification System for Use with Remote Sensor Data.* US Geological Survey Professional Paper 964: US Government Printing Office, Washington, DC.

Bryant, N. A. and Zobrist, A. L. (1976) IBIS: a geographic information system based on digital image processing and image raster data type, *Symposium Proceedings on Machine Processing of Remotely Sensed Data,* 29 June–1 July

1976; Laboratory for Applications of Remote Sensing: Purdue University, West Lafayette, Indiana USA.

Estes, J. E. (1982) Remote sensing and geographic information systems coming of age in the eighties: In Richason, Jr., B. J. (ed.), *Proceedings PECORA VII Symposium, Remote Sensing: An Input to Geographic Information Systems in the 1980s*, American Society of Photogrammetry, Falls Church, Virginia, pp. 23–40.

Hielkema, J. U. (1981) Desert locust habitat monitoring with satellite remote sensing, *ITC Journal* **No. 4**, 387–417.

Lo, C. P. (1974) A photogrammetric urban information system for town planning purposes, *Oriental Geographer* **18**: 89–104.

Lo, C. P. (1981) A photogrammetric urban information system using the CP 1 plotter, *Photogrammetric Record* **10:** 311–29

McFarland, W. D. (1982) Geographic data bases for natural resources: In Johannsen, C. J. and Sanders, J. L. (eds.) *Remote Sensing for Resource Management*. Soil Conservation Society of America: Ankeny, Iowa, pp. 41–50.

Marble, D. F. and Peuquet, D. J. (1983) Geographic information systems and remote sensing: In Colwell, R. N. (ed.) *Manual of Remote Sensing*. American Society of Photogrammetry: Falls Church, Virginia, pp. 923–58.

Meyer, W. L. (1982) Integration of remotely sensed data into geographic information: In Richason, Jr., B. F. (ed.) *Proceedings PECORA VII Symposium, Remote Sensing: An Input to Geographic Information Systems in the 1980s*. American Society of Photogrammetry: Falls Church, Virginia, pp. 3–14.

Mitchell, W. B., Guptill, S. C., Anderson, K. E., Fegeas, R. G. and Hallam, C. A. (1977) *GIRAS: A Geographic Information Retrieval and Analysis System for Handling Land Use and Land Cover Data*. Geological Survey Professional Paper 1059: US Government Printing Office, Washington, DC.

Mooneyhan, D. W. (1982) Organizing information for effective resource management: In Johannsen, C. J. and Sanders, J. L. (eds) *Remote Sensing for Resource Management*. Soil Conservation Society of America: Ankeny, Iowa, pp. 30–40.

Nagy, G. and Wagle, S. (1979) Geographic data processing, *Computing Surveys* **11**: 139–81.

Teicholz, E. (1980) Geographic information systems: the ODYSSEY project, *Journal of the Surveying and Mapping Division*; American Society of Civil Engineers, **106**: 119–35.

Tomlinson, R. F. (ed.) (1972) *Geographical Data Handling*. IGU Commission on Geographical Data Sensing and Processing: Ottawa, Canada.

Yeung, A. K. W. and Lo, C. P. (1983) The software environment for land information systems, *Proceedings of the Second South East Asian Survey Congress*. The Hong Kong Institute of Land Surveyors and the Royal Institution of Chartered Surveyors (Hong Kong branch): Hong Kong, pp. L12-1–L12-12.

Zobrist, A. L., Bryant, N. A. and McLeod, R. G. (1983) Technology for large digital mosaics of Landsat data, *Photogrammetric Engineering and Remote Sensing* **49**: 1325–35.

Index

DATE DUE